P9-BBQ-499

CLEAN COASTAL WATERS

Understanding and Reducing the Effects of Nutrient Pollution

Committee on the Causes and Management of Coastal Eutrophication
Ocean Studies Board
and
Water Science and Technology Board

Commission on Geosciences, Environment, and Resources

National Research Council

NATIONAL ACADEMY PRESS
Washington, D.C.

NATIONAL ACADEMY PRESS • 2101 Constitution Avenue, NW • Washington, DC 20418

NOTICE: The project that is the subject of this report was approved by the Governing Board of the National Research Council, whose members are drawn from the councils of the National Academy of Sciences, the National Academy of Engineering, and the Institute of Medicine. The members of the committee responsible for the report were chosen for their special competencies and with regard for appropriate balance.

This report and the committee were supported by a grant from the National Oceanic and Atmospheric Administration, the U.S. Environmental Protection Agency, the U.S. Geological Survey, and the Electric Power Research Institute. The views expressed herein are those of the authors and do not necessarily reflect the views of the sponsors.

Library of Congress Cataloging-in-Publication Data

Clean coastal waters : understanding and reducing the effects of
nutrient pollution / Ocean Studies Board and Water Science and
Technology Board, Commission on Geosciences, Environment, and Resources,
National Research Council.
p. cm.
Includes bibliographical references and index.
ISBN 0-309-06948-3 (casebound)
1. Nutrient pollution of water—United States. 2. Marine
eutrophication—United States. 3. Coastal zone management—Government
policy—United States. 4. Water quality management—Government
policy—United States. 5. Watershed management—Government
policy—United States. I. National Research Council (U.S.). Ocean
Studies Board. II. National Research Council (U.S.). Water Science and
Technology Board.
TD427.N87 C58 2000
363.739'4--dc21
00-009621

Clean Coastal Waters: Understanding and Reducing the Effects of Nutrient Pollution is available from the National Academy Press, 2101 Constitution Avenue, NW, Box 285, Washington, DC 20055 (1-800-624-6242 or 202-334-3313 in the Washington metropolitan area; Internet: http://www.nap.edu).

Printed in the United States of America.

THE NATIONAL ACADEMIES

National Academy of Sciences
National Academy of Engineering
Institute of Medicine
National Research Council

The **National Academy of Sciences** is a private, nonprofit, self-perpetuating society of distinguished scholars engaged in scientific and engineering research, dedicated to the furtherance of science and technology and to their use for the general welfare. Upon the authority of the charter granted to it by the Congress in 1863, the Academy has a mandate that requires it to advise the federal government on scientific and technical matters. Dr. Bruce M. Alberts is president of the National Academy of Sciences.

The **National Academy of Engineering** was established in 1964, under the charter of the National Academy of Sciences, as a parallel organization of outstanding engineers. It is autonomous in its administration and in the selection of its members, sharing with the National Academy of Sciences the responsibility for advising the federal government. The National Academy of Engineering also sponsors engineering programs aimed at meeting national needs, encourages education and research, and recognizes the superior achievements of engineers. Dr. William A. Wulf is president of the National Academy of Engineering.

The **Institute of Medicine** was established in 1970 by the National Academy of Sciences to secure the services of eminent members of appropriate professions in the examination of policy matters pertaining to the health of the public. The Institute acts under the responsibility given to the National Academy of Sciences by its congressional charter to be an adviser to the federal government and, upon its own initiative, to identify issues of medical care, research, and education. Dr. Kenneth I. Shine is president of the Institute of Medicine.

The **National Research Council** was organized by the National Academy of Sciences in 1916 to associate the broad community of science and technology with the Academy's purposes of furthering knowledge and advising the federal government. Functioning in accordance with general policies determined by the Academy, the Council has become the principal operating agency of both the National Academy of Sciences and the National Academy of Engineering in providing services to the government, the public, and the scientific and engineering communities. The Council is administered jointly by both Academies and the Institute of Medicine. Dr. Bruce M. Alberts and Dr. William A. Wulf are chairman and vice chairman, respectively, of the National Research Council.

COMMITTEE ON THE CAUSES AND MANAGEMENT OF COASTAL EUTROPHICATION

ROBERT W. HOWARTH *(Chair)*, Cornell University, Ithaca, New York
DONALD M. ANDERSON, Woods Hole Oceanographic Institution, Massachusetts
THOMAS M. CHURCH, University of Delaware, Newark
HOLLY GREENING, Tampa Bay Estuary Program, St. Petersburg, Florida
CHARLES S. HOPKINSON, Marine Biological Laboratory, Woods Hole, Massachusetts
WAYNE C. HUBER, Oregon State University, Corvallis
NANCY MARCUS, Florida State University, Tallahassee
ROBERT J. NAIMAN, University of Washington, Seattle
KATHLEEN SEGERSON, University of Connecticut, Storrs
ANDREW N. SHARPLEY, U.S. Department of Agriculture, University Park, Pennsylvania
WILLIAM J. WISEMAN, Louisiana State University, Baton Rouge

Staff

DAN WALKER, Study Director, OSB
CHRIS ELFRING, Senior Program Officer, WSTB
JODI BACHIM, Project Assistant, OSB

OCEAN STUDIES BOARD

WATER SCIENCE AND TECHNOLOGY BOARD

Preface

The National Research Council exists to provide independent scientific advice to the nation, and in particular to help federal agencies with guidance on how best to address significant or controversial problems and make wise use of science in their programs and activities. Sometimes, the topics of study are narrow and the advice is targeted at a specific program. But more often, and perhaps more importantly, a study will focus on a complex issue and the committee will need to synthesize significant (and at times contradictory) information, and then provide clear, practical conclusions and recommendations—recommendations that will make a real difference in solving the problem at hand.

The NRC's Committee on Causes and Management of Coastal Eutrophication conducted exactly this type of nationally important study. Accelerated eutrophication is a real threat to the nation's coastal waters: for instance, eutrophication-caused oxygen-poor waters on the inner continental shelf of the northern Gulf of Mexico can extend over an area as great as 20,000 km^2. It has major impacts, from economic losses associated with reduced fisheries to potential human health impacts, and is likely to increase in severity as nutrient loading from upstream sources increases as a result of continuing urbanization, deforestation, agriculture, and atmospheric deposition. Given that the population in U.S. coastal communities now exceeds 141 million (over half of the U.S. population) and that 17 of the 20 fastest growing counties are located along the coast, nutrient pollution is certainly a national priority requiring attention.

But it is the kind of diffuse, complex problem that prohibits easy answers. It can be addressed best by coordinated actions at many levels—local, state, regional, and national. And success depends on having a solid scientific understanding of the causes of the problem and the full range of possible management alternatives.

To provide advice to federal, state, and local government agencies charged with addressing the growing problems associated with nutrient pollution and eutrophication, the National Research Council (NRC) charged the Committee on the Causes and Management of Coastal Eutrophication to review current knowledge of watershed, estuarine, and coastal processes and their roles in eutrophication; assess past and ongoing efforts to monitor and assess water quality on a variety of scales; and address barriers to implementation of effective management practices and regulatory strategies for preventing and reducing nutrient enrichment and its effects. In essence, the committee was asked to recommend actions that could provide a basis for improving strategies for watershed management to reduce coastal eutrophication in the future. The committee was composed of 11 members with expertise in estuarine biology, aquatic and freshwater ecology, watershed management, environmental engineering, chemistry, agricultural science, economics, and other related fields. The study was funded with contributions from the National Oceanic and Atmospheric Administration, Environmental Protection Agency, U.S. Geological Survey, and Electric Power Research Institute.

To conduct this study, the committee met six times to gather information, talk at length with other experts in the fields, deliberate, and write its report. Special effort was made to get input from regional scientific experts and managers in eutrophication-related programs to gain a practical view of the problems faced. In an effort to better understand the challenges facing local managers, the committee conducted a series of detailed interviews with local, state, and federal managers and scientists responsible for addressing nutrient over-enrichment in 18 estuaries around the country.

Given the technical complexity of the problem and the myriad players who have a role in addressing it, considerable thought was given to who the potential audiences of the report may be and how best to convey the findings and recommendations to this diverse group. We identified four main audiences: 1) coastal and watershed managers—these individuals directly or indirectly influence coastal ecosystems, whether by formulating strategies to deal with local or regional problems or through the various permitting responsibilities they often have. Thus their decisions affect significant sectors of local and state economies, and these decisions cannot be put off until greater information or scientific understanding can be obtained. 2) Scientists—these individuals conduct research that may provide greater understanding and possible solutions to many of the problems facing resource managers. 3) Federal agencies—these organizations are often placed in support or oversight roles as the federal government attempts to enable or ensure that local and state entities are able to address environmental problems effectively and for the national good. And 4) Congress and the Executive Branch—these entities directly influence the legal and administrative powers given to federal agencies as well as the fiscal and human resources needed to implement recommendations.

These four audiences vary greatly in their level of technical acumen. Thus the report has been essentially divided into three parts. The first section, which includes Chapters 1 and 2, provides an introduction to the topic and a summary of the committee's findings and recommendations, and is intended for a non-technical audience. The second section, which includes Chapters 3, 4, 5, and 6, provides detailed technical information on the nature of nutrient over-enrichment and its sources, effects, and the relative susceptibility of different types of systems. The third section, which includes Chapter 7, 8, and 9, addresses abatement strategies.

This report is the result of the committee's extensive efforts, and I would like to thank the committee's members for their hard work, patience, and cooperation. I would also like to offer thanks to the large number of people—local managers, state agency personnel, federal agency personnel, and others—who provided information and insights to the committee. The committee could not have done its job without this assistance. In particular, the committee would like to thank Suzanne Bricker, David Brock, Jim Cloern, David Davis, Scott Dawson, David Flemer, Jonathan Garber, Robert Goldstein, Jack Kelly, Brian Lapointe, Peggy Lehman, Tom Malone, Karen McGlathery, Paul Orlando, Glenn Patterson, Donald Scavia, and John Sowles. For the committee, I would also like to thank the NRC staff who supported our efforts, Jodi Bachim, Chris Elfring, Kirstin Rohrer, Kate Schafer, Dan Walker, and Jennifer Wright.

In accordance with NRC report review policies, this report has been reviewed by individuals chosen for their diverse perspectives and technical expertise. This independent review provided candid and critical comments that assisted the authors and the NRC in making the published report as sound as possible and ensured that the report meets institutional standards for objectivity, evidence, and responsiveness to the study charge. The content of the review comments and draft manuscript remain confidential to protect the integrity of the deliberative process. We wish to thank the following individuals for their participation in the review of the report: Larry P. Atkinson (Old Dominion University), James Baker (Iowa State University), Donald F. Boesch (University of Maryland), John

Boland (Johns Hopkins University), Scott Dawson (California Regional Water Quality Control Board), Thomas J. Graff (Environmental Defense Fund), Alan Krupnick (Resources for the Future), Pamela A. Matson (Stanford University), Hans Paerl (University of North Carolina), Nancy Rabalais (Louisiana Universities Marine Consortium), and Larry Roesner (Colorado State University). While these people provided many constructive comments and suggestions, responsibility for the final content rests solely with the authoring committee and the NRC.

Bob Howarth,
Chair, Committee on the Causes and
Management of Coastal Eutrophication

Contents

Executive Summary

What common thread ties together such seemingly diverse coastal problems as red tides, fish kills, some marine mammal deaths, outbreaks of shellfish poisonings, loss of seagrass habitats, coral reef destruction, and the Gulf of Mexico's "dead zone"? Over the past 20 years, scientists, coastal managers, and public decision-makers have come to recognize that coastal ecosystems suffer a number of environmental problems that can, at times, be attributed to the introduction of excess nutrients from upstream watersheds. The problems are caused by a complex chain of events and vary from site to site, but the fundamental driving force is the accumulation of nitrogen and phosphorus in fresh water on its way to the sea. For instance, runoff from agricultural land, animal feeding operations, and urban areas plus discharge from wastewater treatment plants and atmospheric deposition of compounds released during fossil-fuel combustion all add nutrients to fresh water before it reaches the sea.

The introduction of excess nutrients into coastal systems, or nutrient enrichment, has a number of impacts. One of the most common effects is acceleration of a natural process known as eutrophication—that is, the increasing organic enrichment of an ecosystem.[1] Large inputs of nutri-

[1] Eutrophication, as used in this report and defined in Nixon 1995, is the process by which a body of water becomes enriched with organic material. This material is formed in the system by primary productivity (i.e., photosynthetic activity); and may be stimulated to harmful levels by the anthropogenic introduction of high concentrations of nutrients (i.e.,

ents (nutrient over-enrichment) can lead to excessive, and sometimes toxic, production of algal biomass (including harmful red and brown tides), loss of important habitat such as seagrass beds and corals, changes in marine biodiversity and distribution of species (with impacts on commercial fisheries), and depletion of dissolved oxygen (hypoxia and anoxia) and associated die-offs of marine life. Each of these impacts carries associated costs. A single harmful algal bloom, taking place in a sensitive area during the right season, might cost the region millions of dollars in lost tourism or lost seafood revenues.

Nutrient over-enrichment is a significant problem for the coastal regions of the United States. Because rivers transport the vast majority of nutrients reaching coastal waters, the concentration of land-borne nutrients tends to be high near the mouths of rivers. These areas of mixed fresh and marine water, referred to as estuaries, tend to be relatively slow moving and biologically rich water bodies that are particularly susceptible to the effects of nutrient over-enrichment. Of 139 coastal sites (138 estuaries plus a portion of the Gulf of Mexico) examined in the only comprehensive examination of the extent of eutrophic coastal conditions conducted to date (Bricker et al. 1999), 44 were identified as experiencing conditions symptomatic of high overall levels of nutrient over-enrichment (e.g., showing symptoms such as low dissolved oxygen, nuisance and toxic algal blooms, loss of submerged aquatic vegetation). Problem areas occur on all coasts, including those of California, Florida, Louisiana, Maryland, Massachusetts, New York, North Carolina, Texas, and Washington, but problems are particularly severe along the mid-Atlantic coast and the Gulf of Mexico. Unless actions are taken to reduce inputs, conditions are predicted to worsen over the next 20 years at many of these sites.

Estuaries and coastal zones are among the most productive ecosystems on earth. There is strong concern that the natural resources they represent are in danger from eutrophication and other problems caused by excess input of nutrients. The major nutrients that cause eutrophication and other adverse impacts are nitrogen and phosphorus. In coastal

---

nutrient over-enrichment) such as nitrogen and phosphorus. The term eutrophication is sometimes loosely used to describe any result attributable to anthropogenic nutrient loading to a system, but eutrophication per se is not necessarily caused by human action. It is, however, one of the processes that can be triggered by nutrient over-enrichment. The distinction in this report between nutrient over-enrichment and eutrophication is an important one, since nutrient over-enrichment can lead to a number of problems other than just eutrophication of coastal waters (such as coral reef decline), and the excessive primary production associated with eutrophication often leads to a secondary set of problems (such as hypoxia). Confusing cause and effect can impede mitigation, as remediation efforts may not bring about desired effects if those efforts are improperly targeted.

marine ecosystems, nitrogen is of paramount importance in both causing and controlling eutrophication. This is in contrast to lakes and other freshwater systems, where eutrophication is largely due to excess inputs of phosphorus.

The effect of human activity on the global cycling of nitrogen is immense, and the rate of change in the pattern of use is extremely rapid. The single largest change in the nitrogen cycle comes from increased reliance on synthetic inorganic fertilizer, which was invented during World War I and came into widespread use in the late 1950s. Inorganic fertilizers account for more than half of the human alteration of the nitrogen cycle, and approximately half of the inorganic nitrogen fertilizer ever used on the planet has been used in the last 15 years. Although production of fertilizer is the most significant way human activity mobilizes nitrogen globally, other human-controlled processes contribute to the problem by converting atmospheric nitrogen into biologically available forms of nitrogen, such as combustion of fossil fuels and production of nitrogen-fixing crops (crops such as soybeans and other legumes that can make use of nitrogen taken directly from the atmosphere) in agriculture. Overall, human fixation of nitrogen (including production of fertilizer, combustion of fossil fuel, and production of nitrogen-fixing agricultural crops) increased globally some 2- to 3-fold from 1960 to 1990, and continues to grow.

The problems caused by nutrient over-enrichment are significant and likely to increase as human use of inorganic fertilizers and fossil fuels (the two dominant sources of nutrients) continues to intensify. Much remains to be learned about the geographic extent and changing severity of impacts caused by nutrient over-enrichment, the relative susceptibility of different coastal ecosystems (both large and small), and the most effective nutrient control strategies. There is also a great need to better translate existing knowledge into effective policy and management strategies. This requires an understanding of complex oceanic, estuarine, and watershed processes. With this better understanding, more effective techniques may be developed for reducing and preventing nutrient pollution, eutrophication, and associated impacts.

Nutrient over-enrichment and its adverse impacts can cause extremely complex and variable problems. Often, impacts are caused by the accumulation of nutrients contributed by multiple local sources, and thus solutions will, by necessity, need to involve grassroots participation. The complexity of sources, fates, and effects of nutrients, coupled with the complex socioeconomic and political issues associated with the problem, will require coordinated local, state, regional and federal efforts involving an extremely varied group of stakeholders. Developing an effective strategy for reducing the impacts of nutrient over-enrichment requires an understanding of which nutrients are important, the sources and transport mechanisms for those nutrients, how human activities have altered their abundance, and effective mechanisms to reduce their inputs.

A NATIONWIDE STRATEGY TO ADDRESS NUTRIENT OVER-ENRICHMENT

A number of state, regional, and federal programs are in place that strive to protect and restore coastal waters and habitat in various ways. However, there is no comprehensive national strategy to address excess nutrient inputs to coastal waters. There are no easily implemented and reliable methods or sources of data for citizens, elected officials, and agency staff who live in or are responsible for managing a coastal area (or a watershed that may drain into it) to determine sources of nutrients and potential impacts to coastal waters. In addition, although many federal agencies are making significant efforts to deal with different aspects of nutrient over-enrichment in coastal settings, coordination among these agencies remains inadequate.

Because of the severity of nutrient-related problems and the importance of the coastal areas at risk, the nation needs to develop and implement a national strategy to combat nutrient over-enrichment in coastal areas, with the goal of seeing significant and measurable improvement over the next 20 years. Because both the causes and effects of nutrient over-enrichment are site-specific, development of this National Coastal Nutrient Management Strategy does not mean the national implementation of either uniform source-reduction goals or uniform management or policy approaches. Rather, it means the development of a national, coordinated effort to provide local decision-makers and those responsible for implementing management activities with the information they will need to determine appropriate source reduction goals and methods at the local level. Providing local decision-makers and managers with this information base will allow site-specific and, where necessary, regional or even federal implementation of policies designed to yield significant and measurable improvement in the environmental quality of impaired coastal systems.

Specifically, the committee believes that implementation of the recommendations contained in this report will provide local decision-makers and managers with an information base that could be used to determine what can and should be done to halt the degradation of many of the coastal waters identified in the National Oceanic and Atmospheric Administration (NOAA) National Estuarine Eutrophication Assessment as demonstrating symptoms of severe or worsening eutrophication. The committee believes implementation of the recommendations will dramatically enhance efforts of coastal and watershed managers and other individuals or groups attempting to mitigate the effects of nutrient over-enrichment in these and other estuaries. Improvements in all impaired coastal bodies could be achieved over the next 20 years, while preserving the environmental quality of now-healthy areas.

What are reasonable goals for improvement? In the committee's opinion, at a minimum, federal, state, and local authorities should work with academia and industry to[2]:

- reduce the number of coastal water bodies demonstrating severe impacts of nutrient over-enrichment by at least 10 percent by 2010;
- further reduce the number of coastal water bodies demonstrating severe impacts of nutrient over-enrichment by at least 25 percent by 2020; and
- ensure that no coastal areas now ranked as "healthy" (showing no or low/infrequent nutrient-related symptoms) develop symptoms related to nutrient over-enrichment over the next 20 years.

It was beyond the charge and resources of the committee to identify specific coastal areas for priority attention. All 44 of the areas identified by NOAA's National Estuarine Eutrophication Assessment as exhibiting severe symptoms certainly should be considered as areas where greater effort is needed. Additional study could help further target priorities, especially if it included careful consideration of economic issues and opportunities for stakeholder input. Such work could take significant time and effort, and decision-makers should not be tempted to defer action while waiting for "perfect" knowledge. The committee believes that nationwide implementation of the recommendations in this report, across the full range of systems from small to large and problems from the simple to the complex, will start the nation on a course to achieve the goals stated above. Additional focus on areas subsequently identified for priority attention will then add to cumulative improvement. Thus, the goals listed above are intended to reflect nationwide achievement. Targeting some subset of the impaired coastal areas (for instance, focusing on impaired water bodies associated with small watersheds or simpler ecosystems) in an effort to simply meet these numeric goals would be contrary to the national interest and the spirit of this report.

Working to reduce the effects of nutrient over-enrichment nationwide over the next two decades will be a challenge, but the committee

[2] Measured in relation to the benchmarks determined by NOAA's National Estuarine Eutrophication Assessment (Bricker et al. 1999).

believes that these general goals are realistic. The setting of such numeric goals is somewhat subjective, but the committee believes that such targets are important to encourage action. The goals were set after thorough discussion and are, in the committee's view, both achievable given current methods and challenging enough to facilitate real progress. Many of the principles espoused in this report have already been implemented on a smaller scale in Europe (e.g., Rhine and Elbe watersheds) and the United States (e.g., Tampa Bay and Chesapeake Bay) and have resulted in significant reduction in nutrient loads received from nonpoint sources (Behrendt et al. 1999, Belval and Sprague 1999, Johansson and Greening 2000). Achievement of these goals should not be seen as an end in itself. Rather, they are a first step toward reversing the effects of nutrient over-enrichment in the nation's coastal waters and preventing impairment of "healthy" coastal areas.

How would these goals be accomplished? The key to addressing coastal nutrient problems is understanding that nutrient inputs to coastal waters are affected directly and significantly by activities in the watersheds and airsheds that feed coastal streams and rivers, and building this recognition into planning as well as implementation of management solutions. Thus, people involved as scientists, technicians, and managers for local watershed and coastal programs will play a fundamental role in an effective national strategy to address the problems associated with nutrient over-enrichment. These individuals will be the front line of both policy-making and project implementation. Chapter 2 presents the major findings and recommendations of this report and emphasizes the important role local decision-makers and program managers will play in this national effort. The recommendations suggest ways to develop and implement an effective nutrient management strategy at the local and state level, and the important role federal agencies must play now and in the future.

By focusing on source reduction, actions can be targeted to most effectively reduce and reverse the problems caused by nutrient over-enrichment in coastal areas. Watershed-specific sources like urban stormwater runoff and inappropriate nutrient management at the farm level often can be addressed most effectively by local activities under local leadership, with activities typically quite site-specific. Chapter 2 provides a detailed decisionmaking framework to assist local officials and program managers, including discussions of useful information sources and research needs. However, while significant improvements can be achieved through local action, these managers alone cannot be expected to bring adequate resources and knowledge to bear on such a complex problem, nor are they always able to work at the scale of larger watersheds. Sometimes, broader participation is necessary to have significant impact.

Thus, a truly national strategy must challenge local, state, and federal agencies to work together, and to create partnerships with academia and the private sector. Federal leadership is essential to support and coordinate the research and development needed to provide new approaches and technologies that can be used by local and state agencies charged with reducing and reversing the impacts of nutrient over-enrichment. Perhaps even more importantly, federal leadership is critical for dealing with nutrient sources in large watersheds that span multiple states or jurisdictions or sources distant from the coast. In particular, federal leadership is needed to ensure adequate management of atmospheric forms of nitrogen.

The key federal agencies involved in increasing understanding of nutrient issues, providing technical assistance to state and local managers, and developing new ways to address these problems are the Environmental Protection Agency (EPA), NOAA, U.S. Geological Survey (USGS), U.S. Department of Agriculture (USDA), and National Science Foundation (NSF). NOAA and EPA are primarily responsible for research, policymaking, and management related to eutrophication, in part through the tools provided by the Clean Air, Clean Water, and Coastal Zone Management Acts. USGS has important scientific and data-collection responsibilities, and USDA has a long history of addressing pollution from agriculture. NSF funds research into ecological and biological processes that may be affected by nutrient over-enrichment. Together, these federal agencies have the potential to offer significant resources to help local, state, and regional decision-makers address nutrient pollution problems.

RECOMMENDED FEDERAL ACTIONS

Implementation of an effective National Coastal Nutrient Management Strategy will require coordinated effort, and federal agencies will play an important role. Specifically, the committee recommends that the appropriate federal agencies take the following actions:

- **Expand monitoring programs so efforts to reduce the impacts of nutrient over-enrichment in coastal settings are supported by coherent, consistent information.** The United States lacks a coherent, consistent strategy to monitor the effects of nutrient over-enrichment in coastal settings. One consequence is that the economic and ecological impacts are difficult to estimate with any accuracy. A national monitoring program would involve a partnership of local, state, and federal agencies, as well as academic and research institutions. Participants would agree to use consistent measures of biological, physical, and chemical properties, as well as consistent procedures, quality control, and data management techniques. In addition, representative coastal systems (i.e., index sites) should be selected as sites for long-term, intensive monitoring and study to better understand the causes and impacts of nutrient enrichment on the structure and function of coastal systems and possible mitigation strategies. Accurate estimates of nutrient inputs to estuaries are essential for management, and data on long-term trends are invaluable for determining how changing land use practices or other human activities can change the nutrient load to an adjacent waterbody. Thus inland monitoring, such as is now done by the USGS, should be adapted to include the specific objective of assessing nutrient inputs to coastal areas, especially estuaries, and monitoring how these change over time. USDA should develop monitoring programs to track the long-term effectiveness of various management approaches, especially for recommended Best Management Practices to achieve reduction of nitrogen and phosphorus from nonpoint sources.

- **Develop more effective ways to provide consistent and competent data, information, and technical assistance to coastal decision-makers and managers.** This might include a federally-managed national clearinghouse that links federal, state, and local programs and access to on-request assistance. A web-based descriptive database would be extremely valuable, especially if it includes direct links to the information and data described.
- **Exert federal leadership on issues that span multiple jurisdictions, involve several sectors of the economy, threaten federally managed resources, or require broad expertise or long-term effort beyond the resources of local and state agencies.** There are many important roles for federal leadership in addressing nutrient problems. For instance, the federal government should continue to move toward setting clear guidelines for nutrient loads, which are essential to successful nutrient management strategies. EPA efforts to develop nutrient criteria and standards on a regional and watershed basis should continue, and should incorporate complexities such as the interaction among physical, chemical, and biological factors; seasonality and timing of inputs; and the random nature of hydrologic forcing functions. These efforts should, however, focus on identifying sources and setting maximum loads, rather than on limiting the ambient concentration of a given nutrient in a receiving waterbody. The federal government also can design incentives to encourage innovative source reduction and control, especially related to reducing the impacts of agricultural practices. Federal leadership will play a key role in successfully dealing with

atmospheric deposition of nitrogen, since this issue clearly involves multiple jurisdictions.

- **Develop and implement a process to identify and correct overlaps and gaps in existing and proposed federal programs that deal with nutrient over-enrichment.** This effort should give particular attention to ensuring that programs meet the needs of local managers and improving coordination among the many agencies and organizations with relevant programs. It should plan how gaps will be addressed; for instance, it should identify ways to improve understanding of sources, fate, transport, and impacts of atmospheric deposition of nutrients. Implementation of the Clean Water Action Plan would go a long way toward improving the federal effort. Given the widespread impacts of nutrient over-enrichment, nutrient management should be an important consideration during reauthorization of the Clean Water Act, Clean Air Act, and Coastal Zone Management Act.
- **Develop a susceptibility classification scheme that allows managers to understand the susceptibility of a given estuary to nutrient over-enrichment.** Coastal waters vary considerably in their susceptibility to nutrient over-enrichment based on many factors such as depth, water residence time, flushing, dilution, stratification, and biology. Management could be more effective, and costs could be reduced, if a mechanism for determining susceptibility were available. Because of the tremendous variability in how different coastal water bodies may respond to a given nutrient load, systematic use of such a classification scheme is a prerequisite for taking lessons from one site to another (and to avoid repeating mistakes). Much work remains to be done in this area, and many classification schemes and susceptibility indicators are under development. However, those that emphasize the role that circulation, stratification, mixing, dilution, and turbidity play in predicting how a given waterbody will respond to a specific nutrient load hold the greatest promise. When index sites are selected, they should reflect the variability that coastal waterbodies exhibit in terms of susceptibility.
- **Improve models so they are more useful to coastal managers.** Monitoring is expensive, so managers increasingly rely on models for understanding nutrient effects and forecasting trends. However, because coastal waterbodies vary greatly in their response to a given nutrient load, models must be verifiable and realistically reflect the complex set of processes at work. Creating a single model, or small group of models, that successfully addresses all the variability among waterbodies is probably unreasonable.

However, assembling a suite of models, each tailored to deal with a different class of waterbody, may offer greater promise and provide coastal managers with more options. In many cases, what is needed is a better or more accurate understanding of how a small number of parameters affect the response of an estuary, rather than a more complex or robust model.

- **Conduct periodic, comprehensive assessments of coastal environmental quality.** Lack of detailed study of the scope and impacts of nutrient over-enrichment limits our capability to understand impacts, predict trends, or determine if management actions are having the intended results. The nation needs to conduct a periodic (i.e., every 10 years), comprehensive reassessment of the status of nutrient problems in coastal waters, similar in scope to the NOAA National Estuarine Eutrophication Assessment (Bricker et al. 1999).
- **Expand and target research to improve understanding of the causes and impacts of nutrient over-enrichment.** In particular, work is needed to study atmospheric deposition of nutrients, including sources, fate, transport, and impacts. Research is also needed to understand the relative roles of nitrogen and phosphorus in different freshwater and marine systems, and how these may change seasonally. Better understanding is needed of the role of specific nutrients and conditions in causing harmful algal blooms and the implications for all levels of the food web, from fish to humans. Finally, research is needed that increases our understanding of the effects of nutrient inputs on economically valuable resources (e.g., oysters, fish stocks, etc.) so that we are better prepared to do the analyses necessary to compare costs and benefits and set acceptable restoration goals.

In general, the committee believes the most appropriate approaches for combating nutrient over-enrichment and its impacts will involve a combination of voluntary and regulatory mechanisms. Flexibility is key, especially for local problems, if programs are to achieve goals at minimal cost. It will be important to use an adaptive management approach, so that lessons are learned as techniques are tried and adjustments are made in response to improved information. Other regions or localities can learn from success and failure in particular situations. Because nutrient over-enrichment is caused by "upstream" activities, it may prove necessary to form commissions or other multi-jurisdictional groups to involve diverse groups of stakeholders.

PART I

Introduction and Overview

1

Understanding Nutrient Over-enrichment: An Introduction

Over the last 20 years, there has been a growing awareness that coastal ecosystems have been experiencing a number of environmental problems that can be attributed to the introduction of excess nutrients. At first glance, many of the diverse problems may seem unrelated and their causes are often not readily apparent. However, there is growing evidence that events such as the deaths of unusually large numbers of sea lions and manatees, unusual patterns of coral reef destruction, widespread fish kills, outbreaks of certain shellfish poisonings, disappearance of seagrasses, and the occurrence of the so-called "dead zone" in the Gulf of Mexico actually have much more in common than originally thought. All of these events reflect both subtle and not-so-subtle changes in the relative and absolute abundance of certain organisms near the very base of the food web. The abundance of these organisms is related, sometimes directly and at other times indirectly, to nutrients flowing into the system from upstream watersheds.

All living things must take in nutrients, respire, synthesize new organic molecules, and eliminate waste. The base of the food web that supports the majority of life on the planet is founded on the ability of photosynthetic organisms to take in nutrients and use the energy of sunlight to produce new organic matter. As takes place in yards, gardens, farms, and forests around the world, photosynthetic organisms in marine environments take in carbon, oxygen, nitrogen, phosphorus, and other elements in varying amounts and use sunlight to produce the simple and complex organic molecules necessary for life.

Each species of photosynthetic marine life uses these elements in specific ratios and concentrations and makes use of specific portions of the light spectrum of the sun, and thus each thrives in slightly differing conditions. If sunlight, nutrients, or any environmental conditions are inadequate to support the growth of these organisms, such conditions are commonly referred to as "limiting." That is, the lack of sunlight or nutrients or other factor limits the growth of that organism. When the limiting factor is available in adequate amounts, either naturally or by human activity, photosynthetic organisms grow and multiply until some new limiting condition is encountered. Given adequate nutrients, some organisms may multiply until, through their sheer numbers, they shade themselves and cause light limitation.

One fundamental challenge for scientists and the managers responsible for implementing activities to prevent or reduce coastal nutrient over-enrichment is understanding this complex chain of events and impacts. They must develop an understanding of how natural and human modification of the environment influences the functioning of coastal ecosystems, especially how changes in overall quantity and relative abundance of basic life-sustaining parameters affect populations of aquatic organisms and species composition (the relative number and types of species making up a given ecosystem). This is of particular importance when changes in the total or relative abundance of organisms have adverse impacts on the environmental quality of biologically rich coastal waters.

NUTRIENT OVER-ENRICHMENT IN COASTAL WATERS

The coastal regions of the United States are economically vital areas, supporting a diverse range of industries and large population centers. The nation's coasts—in this report defined to include terrestrial areas located immediately landward of the coastline, the ocean-land interface (including beaches, estuaries, and nearshore marine areas along the coast), and the shallow coastal ocean just offshore of the ocean-land interface—are complex environments characterized by rich biological diversity and natural resources. As society has increasingly populated the coasts, vacationed at the beaches, dammed the rivers feeding the beaches and coasts, harvested fish, disposed of waste, and used these areas for transportation, the deterioration of the coastal environment has become a critical issue.

Although U.S. coastal counties account for only 17 percent of the U.S. landmass, population in these counties exceeds 141 million (U.S. Bureau of the Census 1998). This means that over half of the U.S. population lives in less than one fifth of its total area, and this pattern is expected to continue. For example, 17 of the 20 fastest growing counties are located along the coast. Nearly 14,000 new housing units are built in coastal counties every week (NOAA 1998). Beaches have become one of the largest vacation destinations in America, with 180 million people visiting the coast every year (Cunningham and Walker 1996). This increase in recreational use, together with the impact of larger year-round populations, demonstrates the high importance individuals place on the environmental quality of coastal areas. But people's love of the coast puts increasing stress on coastal ecosystems and makes management of coastal areas increasingly challenging. These areas face a variety of major environmental problems, including habitat modification, degraded water resources, toxic contamination, introduction of non-indigenous species, shoreline erosion, and increased vulnerability to storms and tsunamis.

One of the most significant problems, however, is nutrient over-enrichment. Introduction of excess nutrients to these coastal areas leads to a number of impacts. One of the most common is eutrophication—that is, the process of increasing organic enrichment of an ecosystem where the increased supply of organic matter causes changes to that system (Nixon 1995). In coastal ecosystems, eutrophication can lead to excessive, and sometimes toxic, production of algal biomass (including red and brown tides); loss of important nearshore habitat such as seagrass beds (caused by light reduction); changes in marine biodiversity and species distribution; increased sedimentation of organic particles; and depletion of dissolved oxygen (hypoxia and anoxia). Furthermore, these effects can cause adverse impacts farther up the food web. For example, red tides and hypoxia can cause fish kills. Similarly, red tides and blooms of other toxic algae can harm marine mammals: sea lion deaths in California and manatee deaths in Florida have been linked to harmful algal blooms (Reguera et al. 1998; Scholin et al. 2000).

Human activities can greatly accelerate eutrophication by increasing the natural movement of nutrients from inland watersheds to coastal water bodies. The sources of these nutrients include agricultural practices, wastewater treatment plants, urban runoff, and burning of fossil fuels. Human activities on land affect both the quantity and quality (including nutrient content) of freshwater delivered to coastal areas. Because these factors play such a role in eutrophication, any approach to understanding and reducing the impacts of nutrient over-enrichment must consider freshwater inflow and nutrient loading patterns.

The extent and impacts of nutrient over-enrichment in coastal ecosystems are far-reaching: eutrophication-related oxygen-poor waters extend over an area as large as 20,000 km^2 on the inner continental shelf of the northern Gulf of Mexico (Rabalais et al. 1999), with significant impacts on important fisheries (Council for Agricultural and Science Technology 1999). Other major ecosystems at risk include Chesapeake Bay, Long

Island Sound, the Neuse-Pamlico Estuary, and San Francisco Bay, as well as portions of the Baltic, North, and Black seas in Europe and estuaries and bays in Australia, Japan, and elsewhere around the globe.

PURPOSE OF THIS STUDY

The problems caused by nutrient over-enrichment are significant and likely to increase as human use of inorganic fertilizers and fossil fuels—the two dominant sources of nutrients—continues to intensify, at least on a global basis. Much remains to be learned about the geographic extent and severity of eutrophication, the relative susceptibility of different coastal ecosystems, and the most effective nutrient control strategies. There is also a great need to better translate scientific knowledge into effective policy and management strategies, which requires an understanding of the complex oceanic, estuarine, and watershed processes that contribute to eutrophication. With this better understanding, more effective techniques can be developed for reducing and preventing nutrient pollution, eutrophication, and associated impacts.

To provide advice to federal, state, and local government agencies charged with addressing the growing problems associated with nutrient over-enrichment, the National Research Council (NRC) created the Committee on the Causes and Management of Coastal Eutrophication (Appendix A). The committee was asked to review the current knowledge of watershed, estuarine, and coastal processes and their roles in eutrophication and to assess past and ongoing efforts to monitor and evaluate water quality on a variety of scales. Based on this review, the committee was then charged (Appendix A) with: (1) recommending ways to help coastal and watershed managers achieve meaningful reductions in the impacts of nutrient over-enrichment in the near-term, and (2) identifying areas where future efforts hold the promise of long-term reductions in nutrient over-enrichment and its effects.

Ongoing federal efforts to address this problem are extensive and complex. The National Oceanic and Atmospheric Administration (NOAA) and the Environmental Protection Agency (EPA) are primarily responsible for research, policymaking, and management related to eutrophication, in part through the tools provided by the Clean Air, Clean Water, and Coastal Zone Management Acts (Box 1-1). However, other significant activities are under way at other agencies. For example, the U.S. Geological Survey (USGS) has important scientific and data-collection responsibilities, particularly with regard to monitoring of the nation's streams and rivers. The U.S. Department of Agriculture (USDA) has a long history of addressing pollution from agriculture. The National Science Foundation (NSF) funds research into ecological and biological processes related to

BOX 1-1
Key Federal Legislation

Three major pieces of legislation—the Clean Water Act (Water Pollution Control Act, PL 92-500), Clean Air Act (Air Pollution Prevention and Control Act, PL 91-604), and Coastal Zone Management Act (PL 92-583)—contain elements that directly address the causes and effects of nutrient over-enrichment, and these are described below. Other relevant federal, state, and local programs are described in Appendix C.

Clean Water Act

The Water Pollution Control Act, or Clean Water Act, is the primary federal law addressing pollution in lakes, rivers, and coastal waters. The goal of the act is "to restore and maintain the chemical, physical, and biological integrity of the nation's waters." The federal government, through EPA, sets basic water quality criteria. States are responsible for development of standards from these criteria, and for implementation of these standards. Each state sets effluent limitations for pollutant sources and sets water quality standards for water bodies, but they are required to be at least as stringent as the criteria established by EPA.

The Clean Water Act has been up for reauthorization since the end of 1990. A primary recommendation of the National Research Council's Committee on Watershed Management (NRC 1999a) was to encourage reauthorization of the Act to "allow bottom-up development of watershed agencies that respond to local problems rather than having a rigid institutional structure imposed upon them from the federal level," a recommendation strongly supported by the Committee on the Causes and Management of Coastal Eutrophication.

Clean Air Act

The Air Pollution Prevention and Control Act of 1977, commonly called the Clean Air Act, was enacted to maintain and improve air quality in the United States. In recent years, it has become clear that air pollution has a significant impact on water quality through atmospheric deposition of various compounds including toxins and nutrients, and consequently this act has taken on even broader significance.

For some waters, including major east coast estuaries, implementation of the Clean Air Act and its amendments and revisions could be an important tool in a national effort to reduce nutrient over-enrichment. EPA estimated the potential benefits that would occur if stationary sources of atmospheric nitrogen oxides met proposed national ambient air quality standards for ozone and particulate matter (EPA 1998a). If the proposed standards were implemented, biologically useable nitrogen compounds deposited from the atmosphere in 12 estuaries studied would drop as much as 17 percent, depending on the location of the estuary and the stringency of the regulations. Estimated "avoided costs" (i.e., costs to implement stormwater or point source controls if atmospheric controls are not implemented)

continued

BOX 1-1 Continued

for these estuaries ranged from $152 million to $248 million, depending on the regulatory alternative (EPA 1998a). As evidence presented in Chapter 5 demonstrates, previously published estimates of atmospheric deposition of nitrogen may be too low; thus, these "avoided costs" may be even greater.

Coastal Zone Management Act

The Coastal Zone Management Act of 1972 establishes a partnership between federal and state governments for management of the coast. States develop and implement coastal zone management programs with enforceable policies designed to meet national objectives. The federal government provides funds to implement these programs and requires federal agencies to act consistently with federally approved state programs (Millhouser et al. 1998).

To obtain federal approval of their coastal zone management programs, states must define a coastal boundary, designate critical areas of concern based on a coastal resources inventory, and adopt enforceable policies covering their most important objectives. Over 99.7 percent, or 153,083 km, of U.S. shoreline is managed by federally approved state coastal zone management programs.

The goal of the Coastal Zone Management Act is to protect and conserve the resources of the coastal zone by providing incentives and funding to coastal states (including those around the Great Lakes) to develop and implement management plans for their coastal areas. Unlike the Clean Water Act, participation by states in coastal planning is not compulsory. While the preservation of the coastal zone was the goal of this legislation, the writers recognized that the role of zoning and managing land and nearshore coastal areas was traditionally one of state and local jurisdiction. The act, therefore, provides for and encourages local decisions by allowing federal funding as an incentive for states to participate, based on the specific nature of many of the planning issues.

nutrient over-enrichment. Together, these many federal efforts have the potential to offer significant resources to help citizens and regional, state, and local managers.

However, nutrient over-enrichment and eutrophication pose an extremely complex and variable problem that occurs at a number of scales. The complexity of sources, fates, and effects of nutrients coupled with associated socioeconomic and political issues mean that solutions will require coordinated local, state, regional, and national efforts and the involvement of an extremely varied range of stakeholders. Because of this complexity, the committee gave considerable thought to the potential audiences for this report and how best to convey the findings and recommendations. Four main audiences were identified:

1. Coastal and watershed managers—these individuals directly or indirectly influence how coastal areas and watersheds are managed, whether by formulating strategies to deal with local or regional problems or through various permitting responsibilities. Their decisions affect significant sectors of local and state economies.
2. Research scientists—these individuals (within academia, industry, and government), conduct research that strives to provide greater understanding and possible solutions to many of the problems faced by resource managers.
3. Federal agencies—these organizations are charged to implement federal policy and advance scientific understanding. Often, they act in support or oversight roles as the federal government works to ensure that local and state entities have the information needed to address environmental problems effectively. They also represent national priorities where they might conflict with local perspectives.
4. Congress and the White House—these entities set policy and delegate specific legal and administrative powers to federal agencies. They also control the fiscal and human resources required to implement programs.

These four audiences vary in their level of technical training and interest. Thus, the report has been divided into three sections. This first section, which includes Chapters 1 and 2, is an introduction to the central issues and an overview of how local actions, supported by state and federal agencies, can lead to nationwide reductions in the adverse effects of nutrient over-enrichment. This overview provides suggestions for how the nation can deal with nutrient over-enrichment effectively, including discussions of source reduction and control, policy design and goal setting, law and regulations, and coordination and communication issues. It is intended for a nontechnical audience. The remainder of the report, Sections II and III, provides a detailed treatment of the topics necessary to understand nutrient over-enrichment and plan actions to combat it. Section II, which includes Chapters 3, 4, 5, and 6, serves as a primer on nutrient enrichment and its impacts, including discussions of sources of nutrients and water body susceptibility. Section III, which includes Chapters 7, 8, and 9, then examines various abatement strategies. Chapter 7 looks in depth at the state of the science related to monitoring and modeling, which are critical for understanding the nature, extent, and impact of nutrient over-enrichment and developing mitigation strategies and goals. Chapters 8 and 9 then address the approaches available for setting and achieving effective water quality goals.

WHY IS NUTRIENT OVER-ENRICHMENT A PROBLEM?

Impacts of Nutrient Over-Enrichment

Coastal waters provide habitat for some of the most productive ecosystems on earth. These resources are in danger from eutrophication and other problems caused by excess inputs of nutrients, especially nitrogen and phosphorus. Because rivers transport the vast majority of nutrients reaching coastal waters, the concentration of land-borne nutrients tends to be high near the mouths of rivers. These areas of mixed fresh and marine water, referred to as estuaries, tend to be relatively slow moving and biologically rich water bodies.

Bricker et al. (1999) concluded that "Nearly all estuarine waters now exhibit some symptoms of eutrophication, though the scale, intensity, and impact may vary widely, the level of nutrient inputs required to produce the symptoms also varies." This conclusion reflects, in part, the reality that coastal water bodies vary in susceptibility to nutrient loading (Box 1-2). As discussed in Chapter 6, many factors contribute to this variability. However, one of the most important appears to be the ability of the body to exchange water with the open ocean (which results in a reduced residence time for the nutrients to be taken up by local biota). Thus estuaries with low exchange rates with the ocean seem to be particularly vulnerable. Because watershed management offers real possibilities for reducing the nutrient runoff carried in rivers, estuaries can be the greatest benefactors of improved watershed management approaches. Although other marine environments are discussed, estuaries are the major focus of this report.

Nutrient over-enrichment can cause a range of economic and non-economic impacts, including eutrophication and associated anoxia and hypoxia, loss of seagrass beds and corals, loss of fishery resources, changes in ecological structure, loss of biotic diversity, and impairment of aesthetic enjoyment, each of which has associated costs. Impacts resulting from nutrient over-enrichment during a single summer can cost millions of dollars in lost revenue from tourism or harm to the seafood industry. Although the costs for a single high profile problem can be substantial, they may not appear to warrant significant changes in human behavior. However, when the market and non-market costs of multiple events (Box 1-3) are compiled over time and when local costs are aggregated over regions, it becomes clear why many coastal areas are willing to consider expensive options for reducing nutrient loading. For instance, a wastewater treatment plant capable of removing nutrients from wastewater can cost several million dollars and yet will only address a small part of the total nutrient load delivered from a watershed to an estuary.

BOX 1-2
Understanding the Extent of the Problem: NOAA's National Estuarine Eutrophication Assessment

Although eutrophication is recognized as a growing problem in many of the nation's estuaries and coastal areas, ranging from Long Island Sound to the Chesapeake Bay to the Mississippi delta region, the nation's capability to respond has been limited by lack of knowledge about the extent, severity, and characteristics of eutrophication. To fill this void, NOAA designed the National Estuarine Eutrophication Assessment to gather consistent data nationwide and provide the basis for a national strategy for research, monitoring, and management of this pervasive problem.

NOAA began collecting and synthesizing data and information about nutrient related water quality parameters in 1992. More than 300 federal, state, and academic scientists and environmental managers provided information for 138 estuaries and the Mississippi River Plume through a written survey and a series of workshops. At a national synthesis workshop in 1999, data and information for several water quality parameters were integrated to arrive at an overall assessment of the eutrophic condition of each estuary. Evaluations were made regarding the influence of natural and human related factors in the development of these conditions, estuarine use impairments, and how conditions might change in the next 20 years. Participants made recommendations for management, research, and monitoring to address problems and to prevent worsening conditions.

In 1999, NOAA published its synthesis report, The National Estuarine Eutrophication Assessment (Bricker et al. 1999), which provides a comprehensive summary of the assessment results. This report concludes that symptoms of eutrophication are present in many of the nation's estuaries, with high expression of symptoms in roughly one-third of the estuaries studied (44 of the 139 sites studied; Figure 1-1). Furthermore, the report concludes that left unabated, two out of every three of the estuaries studied will have impaired use by 2020. Problems occur in estuaries along all coasts, but are most prevalent in estuaries along the Gulf of Mexico and mid-Atlantic coasts where human influences are substantial and exchange with the open ocean tends to be slow.

FIGURE 1-1 Forty-four estuaries along all the nation's coasts were showing high expressions of nutrient over-enrichment by the NOAA National Estuarine Eutrophication Assessment. The Middle Atlantic and Gulf of Mexico have the highest percentages of estuaries affected by nutrient over-enrichment. The Pacific and South Atlantic regions contain the highest percentage of estuaries that lack sufficient information to assess eutrophication conditions confidently. An additional 36 estuaries (not shown) show moderate effects of nutrient over-enrichment. Conditions are not necessarily related in whole to anthropogenic loading; to various degrees natural causes and other human disturbances may also play a role. For instance, some estuaries in Maine are typified by natural occurrences of toxic algae, which drift in from the open ocean. However, once in the estuary, these blooms may be sustained by human nutrient inputs (modified from Bricker et al. 1999).

Figure 1-1 on next page

BOX 1-2 Continued

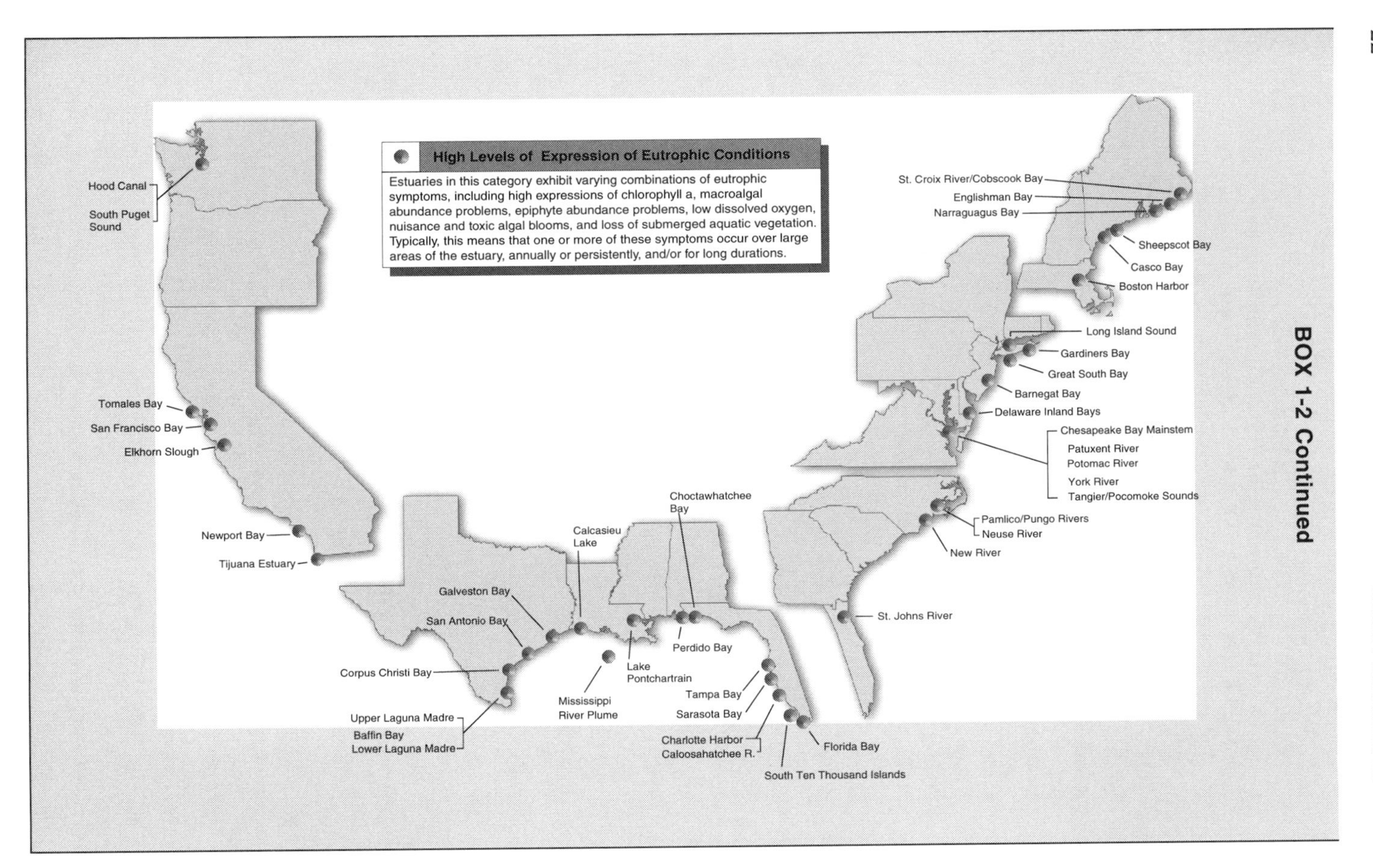

BOX 1-3
Measuring the Value of Water Quality Improvements

The biological and ecological effects of nutrient over-enrichment reduce the value of the nation's coastal waters to the country. That value derives both from the direct use of the resource for pursuits like recreation or commercial fishing ("use value") and from the value that individuals place on the existence of a healthy coastal and marine environment, even if they do not directly use the resource ("non-use value"). This value is reduced when eutrophication degrades water quality. The fact that individuals value a coastal environment with high water quality implies that, even if they do not (and perhaps should not) *have* to pay for water quality improvements, there is some amount they would be *willing* to pay for those improvements. This "willingness-to-pay" is a measure of what the improvement is worth to them. Economists have devised various methodologies for estimating "willingness to pay," and empirical studies show that individuals often place a high value on water quality improvements. For example, a study by Boyle et al. (1998) of lakefront property owners in Maine found that a one-meter improvement in water clarity (a measure of the eutrophic state of the lake) would increase the average property prices for lakefront property on selected lakes by $3,545 to $5,604. The average willingness to pay for this improvement was $3,765. Similarly, in a study of water quality in the Albemarle and Pamlico Sounds in North Carolina, Huang et al. (1997) found that on average, a single household would be willing to pay between $82 and $188 annually (depending on the assumptions and methods used) to restore water quality to 1981 levels. Of this, 55 percent is estimated to be derived from non-use value. While these estimates capture only a component of total value (e.g., that associated with lakefront use), they suggest that individuals place a high value on water quality improvements.

Given the growing magnitude of the problem and the significance of the resources at risk, nutrient over-enrichment represents the greatest pollution threat faced by the coastal marine environment. The impacts of nutrient over-enrichment are discussed in detail in Chapter 4; however, a brief overview of the most significant and common impacts is included here.

Eutrophication

As noted earlier, eutrophication is the process of increasing organic enrichment of an ecosystem where the increased rate of supply of organic matter causes changes to that system (Nixon 1995; see Chapter 4). In moderation, increasing organic matter can sometimes be beneficial, such as when an increased rate of primary production leads to greater fishery production and, ultimately, increased harvests (Nixon 1988; Hansson and

Rudstam 1990; Rosenberg et al. 1990). However, far more often the impacts of high levels of nutrients are negative. Eutrophication can be one of many responses to the introduction of excessive amounts of nutrients. The general process known as eutrophication occurs in both freshwater lakes and in coastal marine ecosystems, with some similarities and some differences in what causes the problem and what impacts result. Marine systems that are most susceptible are those that have limited exchange with the adjacent ocean (e.g., fjords, estuaries, lagoons, and inland seas), but eutrophication can also occur on the continental shelf if nutrient inputs are sufficiently high (see Chapter 6).

Because increased organic content often results from nutrient over-enrichment, the term eutrophication is sometimes loosely used to describe a whole host of environmental problems that can result from nutrient over-enrichment of a system (Rosenberg 1985; Hinga et al. 1995). The distinction in this report between nutrient over-enrichment (cause) and eutrophication (effect) is an important one, as increased primary productivity is only one of many possible responses a coastal ecosystem may have to nutrient over-enrichment. For example, changes in relative abundance of certain nutrients may trigger adverse changes in the relative abundance of some species without triggering an overall increase in net primary productivity. Furthermore, the excessive primary production associated with eutrophication often leads to a secondary set of problems such as dissolved oxygen deficiency, or hypoxia (Box 1-4). Confusing cause and effect can impede mitigation efforts because proposed changes may not bring about desired effects.

What can be considered high in terms of nutrients will vary among systems and in relation to particular uses. The sensitivity of estuaries and coastal systems to accelerated nutrient inputs varies; currently, there is no widely accepted framework for classifying coastal ecosystems by their sensitivity. Mixing, stratification, flushing, dilution, depth, and other physical factors play a role in how sensitive a site is, as do biological factors such as the community structure.

Loss of Seagrasses and the Habitat They Form

Most coastal waters are shallow enough that benthic plant communities contribute significantly to primary production as long as sufficient light penetrates the water column to the seafloor. In good conditions, dense populations of seagrasses and perennial macroalgae can grow and attain rates of net primary production that are as high as the most productive terrestrial ecosystems (Charpy-Roubaud and Sournia 1990).

Benthic organisms such as seagrasses provide important habitat for many species of finfish and shellfish and help stabilize sediment on the

BOX 1-4
The Gulf of Mexico "Dead Zone"

Each spring in the Gulf of Mexico, the oxygen levels near the bottom become too low to allow most fish and crustaceans to live in a vast region stretching from the Mississippi River westward along the Louisiana and Texas coasts, creating what has come to be called the "Dead Zone." The cause is complex, but clearly related to nutrient over-enrichment. In essence, the process occurs because nutrients carried in the waters of the Mississippi River lead to rapid growth of phytoplankton in the Gulf, which in turn use up the available oxygen and lead to a condition known as hypoxia. The problem is exacerbated because when the Mississippi's freshwater enters the sea, it floats over the denser, saltier water, resulting in a two-layered or "stratified" system. This stratification intensifies in the summer, as surface waters warm and the winds that normally mix the water subside, thus preventing the diffusion of oxygen from the surface waters to the lower layer. The low or non-existent oxygen levels drive away fish, shrimp, and crabs, and most bottom-dwellers such as snails, clams, starfish, and worms eventually die.

Research indicates that the Dead Zone is caused by a combination of natural and human influences (Rabalais et al. 1991). For instance, the summer stratification is a natural condition. The key driver, however, is excess nutrients. Over the past four decades the amount of nitrogen delivered to the Mississippi River, and carried to the Gulf of Mexico, has tripled and phosphorus loads have doubled (Turner and Rabalais 1991; Justic et al. 1995; Goolsby et al. 1999).

Nitrogen loads in the Mississippi Basin come from many sources: industrial discharge, urban runoff, atmospheric deposition, fertilizer runoff, animal wastes, and decomposition of leguminous crops. Over half the nitrogen can be attributed to agriculture, primarily runoff of nitrate from fertilizers (Howarth et al. 1996; Turner and Rabalais 1991). The key source areas are southern Minnesota, Iowa, Illinois, Indiana, and Ohio; streams draining Iowa and Illinois alone contribute on average about 35 percent of the total nitrogen ultimately discharged by the Mississippi River to the Gulf of Mexico (Goolsby et al. 1999).

estuarine floor. Thus, extensive stands of seagrass not only indicate a healthy ecosystem but play an important role in preserving the environmental quality of estuaries and other coastal settings. These perennial macrophytes are less dependent on water column nutrient levels than phytoplankton and ephemeral macroalgae, and light availability is usually the most important factor controlling their growth (Sand-Jensen and Borum 1991; Dennison et al. 1993; Duarte 1995).

In temperate systems, perennial seagrasses largely obtain their nutrient requirements by using stored nitrogen pools, internal recycling, and nutrient sources in the sediment (Pedersen and Borum 1996). As a result, excess nutrient enrichment rarely stimulates these populations. Instead, nutrient inputs cause a shift to less desirable phytoplankton or bloom-forming benthic macroalgae. Even in tropical waters, where seagrasses may be more limited by nutrient (phosphorus) availability (Short et al. 1990), fast-growing phytoplankton and macroalgae have more rapid nutrient uptake potential and can replace seagrasses as the dominant primary producers in enriched systems (Duarte 1995; Hein et al. 1995). These fast growing, "nuisance" macroalgae are typically filamentous or sheet-like forms (e.g., *Ulva, Cladophora, Chaetomorpha*) that can accumulate in extensive thick mats over the seagrass or sediment surface. Massive and persistent macroalgal blooms ultimately displace seagrasses and perennial macroalgae through shading effects (Valiela et al. 1997). In addition to causing the loss of important habitat, these nuisance macroalgae are usually unsightly (Chapter 4).

Harmful Algal Blooms

Harmful algal blooms (HAB) include, but are not restricted to, those events referred to as red or brown tides, and are characterized by the proliferation and occasional dominance of particular species of toxic or harmful algae. As with most phytoplankton blooms, this proliferation results from a combination of physical, chemical, and biological mechanisms and interactions that are, for the most part, poorly understood.

Among the thousands of species of microscopic algae at the base of the marine food web are a few dozen that produce potent toxins or that cause harm to humans and marine mammals, fisheries resources, or coastal ecosystems. These species make their presence known in a variety of ways, ranging from massive blooms of cells that discolor the water (giving rise to the term red or brown tide) to dilute, inconspicuous concentrations of cells noticed only because of the harm caused by their potent toxins. The impacts of these phenomena include mass mortalities of wild and farmed fish and shellfish; human illness or even death from contaminated shellfish or fish; alterations of marine trophic structure through adverse effects on larvae and other life history stages of commercial fisheries species; and death of marine mammals, seabirds, and other animals. HABs and related phenomena such as *Pfiesteria* outbreaks have attracted intense public and political attention (Box 1-5).

One major category of HAB impact occurs when toxic phytoplankton are filtered from the water as food by shellfish, which then accumulate the algal toxins to levels harmful or lethal to humans or other consumers (Shumway 1990). These poisoning syndromes have been given the names paralytic, diarrhetic, neurotoxic, and amnesic shellfish poisoning (PSP, DSP, NSP, and ASP). Whales, porpoises, seabirds, and other animals can be victims as well, receiving toxins through the food web via contami-

BOX 1-5
Pfiesteria piscicida: Implications for Nutrient Over-Enrichment

Prior to 1990, problems attributable to nutrient over-enrichment rarely made national news, but a once little-known species called *Pfiesteria piscicida*[1] gained wide public attention in the 1990s and inadvertently served to increase public understanding of these types of problems. Interest began in May 1991, when a fish kill in the Ablemarle-Pamlico estuarine system in North Carolina was attributed to *Pfiesteria piscicida* (Burkholder 1997). But wide attention began in earnest in August 1997, when hundreds of dead and dying fish were found in a tributary to Chesapeake Bay, the Pocomoke River near Shelltown, Maryland, prompting state and local officials to close a portion of the river. Subsequent fish kills and observations of *Pfiesteria*-like organisms led to successive closing of segments of the Manokin and Chicamacomico rivers in Maryland. Soon, Maryland's Department of Health and Mental Hygiene presented preliminary evidence that adverse human health effects could result from exposure to the toxins released by *Pfiesteria piscicida* or *Pfiesteria*-like organisms (Grattan et al. 1998).

With the publicity, and despite the fact that the fish most commonly affected by *Pfiesteria piscicida* are Atlantic menhaden (a fish used primarily as an ingredient in animal feed), the local seafood industry suffered as restaurants and stores stopped selling Chesapeake Bay seafood (Weinraub 1997). In September 1997, the State of Maryland appointed a Citizens *Pfiesteria* Action Commission, which convened a forum of scientists to provide advice. The final report is referred to as the Cambridge Consensus (Maryland Department of Natural Resources 2000).

The scientists discussed questions that had been raised in the scientific community concerning the relationships between *Pfiesteria*-like dinoflagellates (which included *P. piscicida*) and nutrients. After thorough analysis, they concluded there was a likely connection between nutrients, toxic outbreaks of *Pfiesteria*-like dinoflagellates, and fish kills. Also, they determined that it is improbable that toxic contaminants (such as pesticides and trace metals) are primarily responsible for outbreaks of *Pfiesteria*-like dinoflagellates. The scientists noted that while most evidence comes from North Carolina and environmental conditions vary, their findings apply to the mid-Atlantic region in general. Specifically, they found:

- In laboratory cultures, growth of non-toxic stages of *Pfiesteria piscicida* can be stimulated by addition of inorganic and organic nutrients.

continued

---

[1] Neither a true plant nor animal, *Pfiesteria* is a dinoflagellate within the Kingdom Protista. It has a complex life cycle, which makes identification of *Pfiesteria* species by nonexperts extremely difficult. It spends much of its life span as a nontoxic predatory organism feeding on bacteria and algae, or as encysted cells existing in a dormant state in muddy substrates. However, when large schools of oily fish (e.g., Atlantic menhaden) swim into an area and linger to feed, their excreta may trigger encysted cells to emerge and secrete potent toxins. These toxins make the fish lethargic, so that they tend to remain in the area where they are susceptible to direct attack by the *Pfiesteria* cells. This, either alone or as a result of concurrent attacks by bacteria or fungi, may lead to open sores on the fish.

BOX 1-5 Continued

- Nutrient enrichment stimulates the growth of algae and other microbes on which *Pfiesteria*-like dinoflagellates can feed and grow.
- At this point, it cannot be concluded that either phosphorus versus nitrogen or inorganic versus organic nutrients are relatively more important in directly or indirectly stimulating the growth of *Pfiesteria*-like dinoflagellates.
- High nutrient concentrations are not necessarily required for *Pfiesteria*-like dinoflagellates to transform into toxic stages.

Based on their review, the scientists explained that excessive nutrient loading helps create an environment rich in the microbial prey and organic matter that both *Pfiesteria* and menhaden use as a food supply. By stimulating an increase in *Pfiesteria* concentrations, nutrient inputs increase the likelihood of a toxic outbreak when adequate numbers of fish are present. However, the presence of excess nutrients appears to be only one of many factors involved in *Pfiesteria* outbreaks. Stream hydraulics, water temperature, and salinity also seem to play important roles.

The work done to understand *Pfiesteria piscicida* and *Pfiesteria*-like dinoflagellates has implications for managers concerned with reducing nutrients in marine systems. In the long term, decreases in nutrient loading will reduce eutrophication, thereby improving water quality, and in this context will likely lower the risk of toxic outbreaks of *Pfiesteria* and harmful algal blooms. However, even drastic decreases in nutrient loading will not completely eliminate the risks of toxic outbreaks of these organisms, which are indigenous species adapted to use toxins to attack fish when presented with the opportunity. While the outbreaks of *Pfiesteria piscicida* and *Pfiesteria*-like organisms do represent grounds for concern, the number and distribution have been relatively small compared with other impacts of nutrient over-enrichment. Thus, the *Pfiesteria* outbreaks to date may be most significant for the attention they have drawn to the larger threat posed by excess nutrient loading of freshwater and coastal systems nationwide.

nated zooplankton or fish (Geraci et al. 1990). Another type of HAB impact occurs when marine animals are killed by algal species that release toxins and other compounds into the water, or that kill without toxins by physically damaging gills or by creating low oxygen conditions as bloom biomass decays. Farmed fish mortalities from HABs have increased considerably in recent years, and are now a major concern to fish farmers (and their insurance companies).

HABs have become more frequent and longer in duration in recent decades (Figure 1-2). Although not all HABs are caused by nutrient loading, many are at least in part associated with the general change in ecological structure that accompanies eutrophication. The causal mecha-

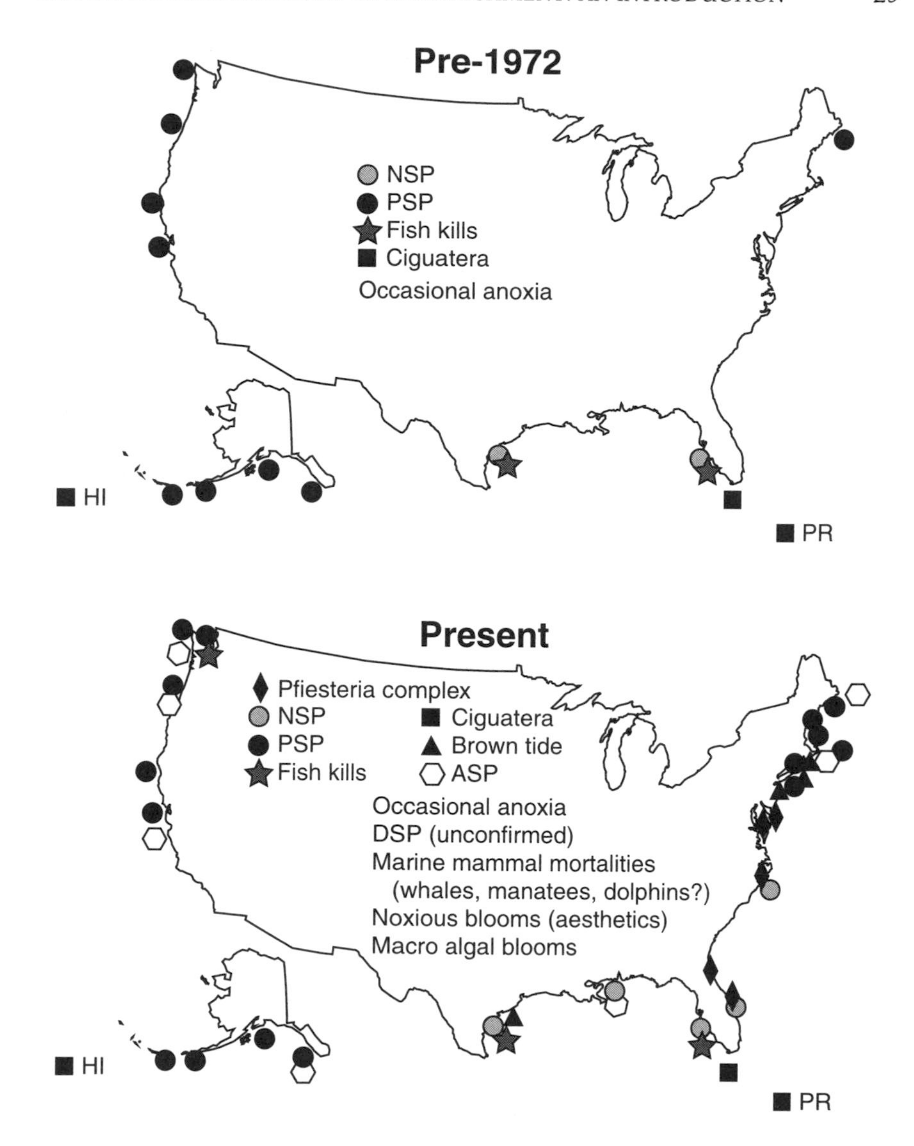

FIGURE 1-2 Expansion of harmful algal bloom (HAB) problems in the United States. These maps depict the HAB outbreaks known before and after 1972. This is not meant to be an exhaustive compilation of all events, but rather an indication of major or recurrent HAB episodes. In addition to the toxic impacts shown, harmful microalgal and macroalgal species have caused whale and other marine mammal mortalities, occasional anoxia, habitat destruction, and a general decline in coastal aesthetics in many coastal areas during the last 20 years. Neurotoxic shellfish poisoning = NSP, paralytic shellfish poisoning = PSP, and amnesic shellfish poisoning = ASP (Anderson 1995).

nisms for such blooms remain poorly known, and some blooms have always occurred and are entirely natural. However, other blooms are tied to nutrient availability, thus leading to more frequent and longer lasting blooms as human-induced nutrient over-enrichment becomes more common in coastal waters. Although reducing the overall availability of nutrients will reduce the likelihood of certain HABs, more research is needed to better understand the role of specific nutrients in the occurrence of various blooms and gain a complete understanding of anthropogenic influences and mitigation options.

Coral Reef Decline

Coral reefs are among the most productive and diverse ecosystems in the world. They grow as a thin veneer of living coral tissue on the outside of the hermatypic (reef-forming) coral skeleton. The world's major coral reef ecosystems are distributed in nutrient-poor surface waters in the tropics and subtropics. Coral reefs are a paradox because their high gross productivity and biodiversity occur in waters with very low concentrations of dissolved and particulate nutrients. The abundant sunlight characteristic of the earth's equatorial zones, supported by tight nutrient recycling within the coral-zooxanthellae[2] symbiosis (Muscatine and Porter 1977), allows coral reefs to attain high rates of productivity. Thus, the worldwide decline in coral reefs is particularly disturbing. In the 1970s, offshore reefs in the Florida Keys were composed primarily of coral, and some contained more than 70 percent coral cover (Dustan 1977). But now, the "best" reefs have only about 18 percent coral cover. More resilient turf and macroalgae now dominate these reefs, accounting for 48 to 84 percent cover (Chiappone and Sullivan 1997).

Early references to coral reef ecosystems thriving in areas of upwelling or other nutrient sources were incorrect (Hubbard, D. 1997). It is now recognized that high nutrient levels are detrimental to reef health (Kinsey and Davies 1979). That view is supported by observations of phase shifts away from corals and coralline algae toward dominance by algal turf or macroalgae in coastal areas experiencing eutrophication from expanding human activities (Lapointe 1997).

Myriad other direct and indirect effects of coastal nutrient enrichment are known to affect coral reefs. One direct impact associated with elevated nutrients is decreased calcification, which results in dramatic decreases in the growth of reefs as a whole (Kinsey and Davies 1979; Marubini and Davies 1996). Indirect effects of nutrient over-enrichment

---

[2] Zooxanthellae are green and brown algae-like photosynthetic cells that live symbiotically in many coelenterates, especially corals.

include increased phytoplankton biomass (Caperon et al. 1971) that alters the quality and quantity of particulate matter and the optical properties of the water column in a predictable fashion, with subsequent effects on reefs (Yentsch and Phinney 1989). Research has shown that outbreaks of the "Crown-of-Thorns" starfish in the South Pacific, which prey on living coral tissue, are related to the effects of nutrient-rich runoff on starfish larval development (Birkeland 1982). Because sea urchins and other marine herbivores are limited by dietary nitrogen (Mattson 1980), increased nitrogen availability, in particular, increases populations of these organisms. Because some organisms that increase in abundance in response to abnormally high nutrient levels, such as sponges and sea urchins, can damage reef formations (Glynn 1997), nutrient over-enrichment of coastal waters can ultimately lead to the destruction of both the reef framework and also adjacent shorelines due to increased erosion.

Controlling the Right Nutrients

The major nutrients that cause eutrophication and other adverse impacts associated with nutrient over-enrichment are nitrogen and phosphorus. Nitrogen is of paramount importance in both causing and controlling eutrophication in coastal marine ecosystems (Box 1-6). Other elements—particularly silicon and iron—may also be of importance in regulating HAB occurrences in coastal waters and in determining some of the consequences of eutrophication, but their importance with respect to nutrient over-enrichment in coastal waters is secondary to nitrogen.

BOX 1-6
Why Focus on Nitrogen?

The key to controlling eutrophication in freshwater systems is managing phosphorus inputs. Conversely, the key to controlling eutrophication in marine systems is managing nitrogen inputs. This conclusion follows significant debate, and even now some policymakers and the press continue to question the relative role of nitrogen versus phosphorus in coastal eutrophication. But marine scientists recognized the prominent role nitrogen plays in coastal eutrophication decades ago, and the report *Managing Wastewater in Coastal Urban Areas* (NRC 1993a) clearly concludes that the marine scientific community has reached consensus about the primary importance of nitrogen as the prinicpal cause of nutrient over-enrichment in coastal systems.

continued

BOX 1-6 Continued

Why is the scientific community so clear that the key to controlling eutrophication in coastal systems is managing nitrogen inputs? First, the experimental evidence is much clearer than in the past. Most early studies of nutrient limitation and eutrophication in coastal waters either relied on fairly short-term and small-scale enrichment experiments to infer limitation by nitrogen, or made inferences from pure-culture studies. When applied to the problem of lake-eutrophication in the 1960s and early 1970s, these approaches often led to the erroneous conclusion that nitrogen or carbon rather than phosphorus was limiting in lakes. Later, whole-lake experiments clearly showed that phosphorus and not nitrogen or carbon was the nutrient most regulating eutrophication in lakes (Schindler 1977). Consequently, the scientific community that studied eutrophication in lakes and the water-quality management community that dealt with freshwater systems became skeptical about any results obtained from similar small-scale experiments (NRC 1993a).

However, the information produced by more recent estuarine studies is much more reliable. Since 1990, the results of three large-scale enrichment "experiments" have been published from studies carried out in Narragansett Bay, in a portion of the Baltic Sea, and in Laholm Bay in Sweden. Each study was similar in scope and methodology to the pivotal lake experiments of the late 1970s, and all three showed nitrogen limitation in the systems studied (Granéli et al. 1990; Rosenberg et al. 1990; Oviatt et al. 1995; Elmgren and Larsson 1997). Overall, available data from these major studies and from bioassay and nutrient ratio data from many estuaries all give credence to the generalization that nitrogen availability is the primary regulator of eutrophication in most coastal systems.

But why should nitrogen usually control eutrophication in coastal marine systems while phosphorus controls eutrophication in lakes? Primary production by phytoplankton is generally thought to be a function of the relative availability of nitrogen and phosphorus in the water. For instance, phytoplankton require approximately 16 moles of nitrogen for every mole of phosphorus they assimilate (the Redfield ratio of nitrogen:phosphorus = 16:1). If the ratio of available nitrogen to available phosphorus is less than 16:1, primary production is nitrogen-limited. If the ratio is higher, production is phosphorus-limited.

Lakes receive nutrient inputs from upstream terrestrial ecosystems and from the atmosphere, while estuaries and coastal marine systems receive nutrients from these sources as well as from neighboring oceanic water masses. For estuaries such as those along the northeastern coast of the United States, the ocean-water inputs of nutrients tend to have a nitrogen:phosphorus ratio well below the Redfield ratio due to denitrification on the continental shelves (Nixon et al. 1995, 1996). Thus, given similar nutrient inputs from land, estuaries will tend to be more nitrogen-limited than will lakes.

The biogeochemical processes operating in freshwater lakes and coastal estuaries, as well as their watersheds, are complex, and thus whether biological activity at any one location at any given time is nitrogen- or phosphorus-limited is dependent on a number of complex factors. However, despite the complexity of the processes involved, nitrogen-limiting conditions are much more common in estuaries than in lakes, and effective management of these areas and their associated watersheds requires much greater focus on nitrogen.

Developing an effective strategy for reducing the impacts of nutrient over-enrichment requires an understanding of which nutrients are important, the sources and transport mechanisms for those nutrients, and how human activities have altered the abundance of each (Figure 1-3). Seen broadly, the earth is a closed system and the total amount of nitrogen or phosphorus in all forms is essentially fixed. These elements are constantly recycled, changing forms as they combine with different elements to form a variety of chemical compounds. These varied compounds are stable at different temperatures and pressures, and thus each may be more or less soluble in water, or more or less volatile than its predecessor, and may be used by organisms in different ways. Each element has a specific biogeochemical cycle—it is distributed or moves in a way dictated by its unique chemistry within the air, freshwater lakes and streams, the ocean, and land.

Nitrogen and phosphorus have different chemical properties, and thus each reacts differently to form a different set of compounds, many of which again behave differently. When human activity drastically alters the distribution or relative abundance of the various compounds containing these element forms, they alter the overall biogeochemical cycle of

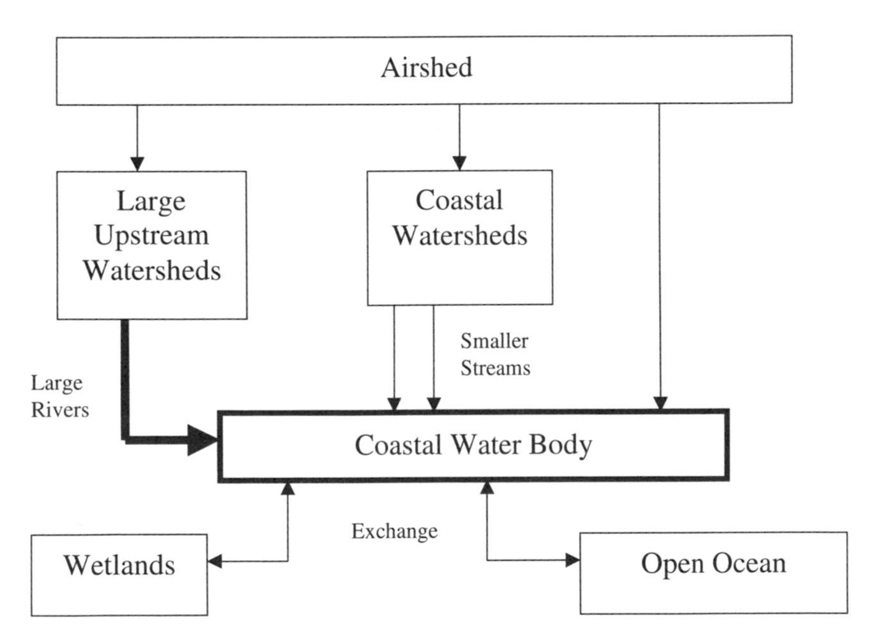

FIGURE 1-3 Schematic showing general sources of nutrients and main routes of transport to coastal waters.

these elements. Thus, one measure of how dramatically human activity is altering the environment is to examine the degree to which this activity is altering the normal geochemical cycle of a number of key elements including nitrogen, phosphorus, and also carbon.

For phosphorus, global fluxes (reflecting the net change from one segment of the geochemical cycle to another) are dominated by the essentially one-way flow of phosphorus carried in eroded soils and wastewater from the land to the oceans, where it is ultimately buried in oceanic sediments (Hedley and Sharpley 1998). The size of this flux is large—estimated at 22 Tg P yr^{-1} (teragrams of phosphorus per year) (Howarth et al. 1995). It is estimated that the flow of phosphorus prior to increased human agricultural and industrial activity was around 8 Tg P yr^{-1} (Howarth et al. 1995). Thus, current human activities cause an extra 14 Tg P to flow into the ocean sediment sink each year, or approximately the same as the amount of phosphorus fertilizer (16 Tg P yr^{-1}) applied to agricultural land each year.

The effect of humans on the global cycling of nitrogen is immense, and the rate of change in the pattern of use is extremely rapid (Galloway et al. 1995). The single largest change globally in the nitrogen cycle comes from increased reliance on synthetic inorganic fertilizer, which was invented during World War I and came into widespread use in the late 1950s (Box 1-7).

Inorganic fertilizers account for more than half of the human alteration of the nitrogen cycle (Vitousek et al. 1997). Approximately half of the inorganic nitrogen fertilizer ever used on the planet has been used in the last 15 years. The rate of use increased steadily until the late 1980s, when the collapse of the former Soviet Union led to great disruptions in agriculture and drops in fertilizer use in Russia and much of eastern Europe. This caused a slight decline in global nitrogen fertilizer use for a few years (Matson et al. 1997). By 1995, however, the global use of inorganic nitrogen fertilizer was again growing rapidly, with much of the growth driven by use in China. As of 1996, use was approximately 83 Tg N yr^{-1}.

The increased use of commercial fertilizer over the last 50 years has contributed to a dramatic increase in per acre crop yields. But it has also brought problems (e.g., adverse changes in soil properties and offsite environmental problems caused by runoff). Problems are exacerbated because fertilizers are frequently over applied. Crop absorption of applied nitrogen can be extremely variable, depending on the crop, plant growth, and the method and timing of fertilizer application (NRC 1989). Although some improvements in fertilizer efficiency may have occurred in the last 10 years, the importance of choosing appropriate fertilizer application methods, amounts, and timing remains.

BOX 1-7
Atmospheric Nitrogen as a Source for Inorganic Nitrogen Fertilizer

Unlike phosporus- or potassium-based fertilizers, whose abundance is limited by the extraction of source materials by mining, nitrogen-based fertilizers are largely derived from the direct chemical conversion of inert elemental nitrogen, N_2, in the atmosphere to biologically useable forms of nitrogen (typically compounds of nitrogen plus oxygen or hydrogen). Elemental nitrogen is the most abundant gas in the earth's atmosphere, thus there is an essentially inexhaustible supply of inorganic, nitrogen-based fertilizer. The process used, originally developed to address Germany's needs for nitrate to produce munitions during World War I, remains the most economical method for the commercial fixation of nitrogen, and with modifications is one of the basic processes of the chemical industry. Over the last decade, ammonia derived from natural gas has emerged as another important source of inorganic fertilizer.

When nitrogen-based fertilizer is applied to a field, it can move through a variety of flow paths to downstream aquatic ecosystems. Some fertilizer leaches directly to groundwater and surface waters, varying from 3 to 80 percent of the fertilizer applied, depending on soil characteristics, climate, and crop type (Howarth et al. 1996). On average for North America, about 20 percent leaches directly to surface waters (NRC 1993a). Some fertilizer is volatilized directly to the atmosphere; in the United States, this averages 2 percent of the fertilizer applied, but the value is higher in tropical countries and also in countries that use more ammonium-based fertilizers, such as China (Bouwman et al. 1997). Much of the nitrogen from fertilizer is incorporated into crops and is removed from the field in the crops when they are harvested. An NRC report (NRC 1993a) suggests that on average, 65 percent of the nitrogen applied to croplands in the United States is harvested, although other estimates are somewhat lower (Howarth et al. 1996). Given these paths and rates, about 13 percent of the nitrogen applied builds up in soils or is denitrified back to elemental nitrogen (a gas) and released to the atmosphere.

To fully understand nitrogen transport, it is important to trace the eventual fate of the nitrogen harvested in crops. Some nitrogen is consumed directly by humans eating vegetable crops—in North America this constitutes perhaps 10 percent of the amount of nitrogen originally applied to the fields (Bouwman and Booij 1998). Perhaps 10 percent of the nitrogen originally applied to fields is lost during food processing and ends up in landfills or released to surface waters from food-processing plants. The largest part of the nitrogen is fed to animals in feed crops, estimated to be about 45 percent (Bouwman and Booij 1998).

Of the nitrogen consumed by animals, much is volatilized from animal wastes to the atmosphere as ammonia. In North America, this volatilization is roughly one-third of the nitrogen fed to animals (Bouwman et al. 1997), or 15 percent of the amount of nitrogen originally placed on the fields. This ammonia is deposited back onto the landscape, often near the source of volatilization, although some of it travels long distances through the atmosphere (Holland et al. 1999). Some of the nitrogen in animals is consumed by humans, an amount roughly equivalent to 10

continued

BOX 1-7 Continued

percent of the amount of nitrogen fed to the animals, or 4 percent of the nitrogen originally applied to fields. The rest of the nitrogen—over 25 percent of the amount of nitrogen originally applied to the fields—is in animal waste that is accumulating somewhere in the environment. Much of this may be leached to surface waters.

Of the nitrogen consumed by humans, either through vegetable crops or meat, some is released through wastewater treatment plants and from septic tanks. In North America, this is an amount equivalent to approximately 5 percent of the amount of nitrogen originally applied to fields (Howarth et al. 1996). The rest is placed as food wastes in landfills or is denitrified to nitrogen in wastewater treatment plants and septic tanks.

In conclusion, fertilizer leaching from fields is only a portion of the nitrogen that potentially reaches estuaries and coastal waters. Probably of equal or greater importance in many regions of North America is the nitrogen tied up in ammonia, which is volatilized to the atmosphere or released to surface waters from animals' wastes and landfills. Since food is often shipped over long distances in the United States, the concentration and subsequent environmental effect of nitrogen overenrichment can occur well away from the original fertilized cropland.

Although production of fertilizer is the most significant way human activity mobilizes nitrogen globally, other human-controlled processes, such as combustion of fossil fuels and production of nitrogen-fixing crops in agriculture, convert atmospheric nitrogen into biologically available forms of nitrogen. Overall, human fixation of nitrogen (including production of fertilizer, combustion of fossil fuel, and production of nitrogen-fixing agricultural crops) increased globally some 2- to 3-fold from 1960 to 1990, and continues to grow (Galloway et al. 1995). By the mid 1990s, human activities made new nitrogen available at a rate of some 140 Tg yr^{-1} (Vitousek et al. 1997), or a rate roughly equivalent to the natural rate of biological nitrogen fixation on all of the land surfaces of the world (Vitousek et al. 1997; Cleveland et al. 1999). Thus, the rate at which humans have altered nitrogen availability globally far exceeds the rate at which humans have altered the global carbon cycle.

2

Combating Nutrient Over-enrichment: Findings and Recommendations

This committee was charged to recommend ways to help coastal and watershed managers achieve meaningful reductions in the impacts of nutrient over-enrichment in the near-term. The committee was charged further to identify areas where scientific uncertainty and imperfect knowledge limit the nation's ability to achieve long-term reductions in nutrient over-enrichment and its effects. The dichotomy in its charge required the committee to explore both current scientific understanding and resource management practice. This chapter, which summarizes the committee's major conclusions, also reflects this dichotomy, with many of the key findings and recommendations organized to emphasize their relationship to common coastal or watershed management practice. If the nation is to address coastal nutrient over-enrichment successfully, efforts by local, state, and federal agencies must be coordinated nationwide. Thus, by keying the major components of a national nutrient management strategy to a common decision process followed by the managers working on-the-ground at local, state, and regional levels, the committee hopes to emphasize that, with a few important exceptions such as where problems span multiple jurisdictions, involve multiple sectors of the economy, threaten federally held resources, or fall under federal regulations like the Clean Air Act, nationwide improvement can best be achieved through coordinated local and regional actions.

DEVELOPING A NATIONWIDE STRATEGY TO ADDRESS NUTRIENT OVER-ENRICHMENT

Recent efforts to determine the extent of nutrient over-enrichment, such as the National Oceanic and Atmospheric Administration's (NOAA) National Estuarine Eutrophication Assessment (Bricker et al. 1999), have been hampered by a lack of systematic monitoring and reporting and inadequate coverage of the nation's coasts. Nonetheless, the available data clearly demonstrate that problems associated with nutrient over-enrichment occur nationwide and that impacts will worsen if current trends continue. As discussed earlier, NOAA's examination of 139 coastal water bodies found that nearly one-third (44) are experiencing severe or worsening impacts caused by nutrient over-enrichment (Figure 1-1). The nutrient sources are diverse, often rooted in major changes in technology and human behavior over the last 50 years—ranging from significant changes in how agriculture is conducted to expanded use of fossil fuels. Solutions to the problems caused by nutrient over-enrichment are multi-faceted and vary from region to region.

Although there are large national programs that aim to "protect and restore coastal waters and habitat," there is no nationwide strategy designed specifically to address excess nutrient inputs to coastal waters. At present, there is little accessible information or easily implemented and reliable methods for a decisionmaker or program manager in a coastal area to determine the sources of excess nutrients or the potential impacts of those nutrients to a specific coastal waterbody.[1] Although many federal agencies are making significant independent efforts to help local jurisdictions deal with the effects of nutrient over-enrichment in coastal settings, the degree of coordination among these agencies and efforts remains inadequate.

The severity of nutrient-related problems and the importance of the coastal areas at risk demand the development and implementation of a National Nutrient Management Strategy. The National Nutrient Management Strategy should coordinate local, state, regional, and national efforts to combat nutrient over-enrichment in coastal areas, with the goal of seeing significant and measurable improvement in the environmental quality of impaired coastal ecosystems.

---

[1] This report places significant emphasis on the role of local decisionmakers to formulate and implement local actions. Because this authority and responsibility to formulate and implement policy is vested in different entities in all the various jurisdictions involved, it is impractical for the report to identify specific actors for recommended actions at the local level. Thus, wherever the term "local decisionmaker or manager" is used in this report, the committee is referring to the appropriate entity responsible for formulating and implementing policy in any given jurisdiction.

The effects of nutrient over-enrichment are site-specific, and the sources of nutrients vary greatly among regions and among particular sites. Consequently, development of a national strategy must allow for variation among sites and regions in the implementation of source-reduction goals and in the management and policy approaches used. However, some federal oversight is essential for issues such as the movement of nutrients across state boundaries. Many important coastal systems, including the "Dead Zone" in the Gulf of Mexico, Chesapeake Bay, and Long Island Sound, receive nutrient inputs from many states, and often from far away. Further, national policies are necessary to deal with nutrient sources from agriculture and from the combustion of fossil fuels, to ensure that pollution sources are not simply shifted from one region to another.

The National Nutrient Management Strategy must also facilitate the development of a national, coordinated effort to provide local decisionmakers and managers with the information they will need to determine appropriate source reduction goals and methods at the local level. Providing local decisionmakers and managers with this information base will allow site-specific and, where necessary, regional or even federal implementation of policies designed to yield significant and measurable improvement in the environmental quality of impaired coastal systems.

If the national strategy is to achieve a "measurable improvement" in the quality of impaired coastal systems, some systemic measure of change nationwide must be instituted. Even though the recent National Estuarine Eutrophication Assessment (Bricker et al. 1999) was hindered by inconsistent and inadequate data sets in many areas, and there was some subjectivity in the assessment process, it still represents the best measure of the extent to which nutrient over-enrichment has impaired coastal environmental quality. Thus the committee suggests that a similar assessment, repeated at roughly ten-year intervals, would be a useful mechanism to determine whether "measurable improvement" has, in fact, occurred. Beyond this, the national strategy should encourage more uniform approaches to monitoring of coastal systems across the country so that future assessments can be made with greater efficiency and accuracy.

Another goal of the national strategy should be "significant improvement" in coastal water quality, but what constitutes "significant improvement"? Local or state managers often face the dilemma of having to frame specific and achievable performance metrics or goals. As discussed in Chapter 8, establishing such goals requires input and commitment from a large number of stakeholder groups. Employing the kind of stakeholder process needed to set goals for a nation as large and complex as the United States is well beyond the scope of this study, which was designed to propose solutions to the intellectual and logistical barriers associated with nutrient over-enrichment. However, based on a review of current scientific understanding and nutrient management practice, discussed in detail in Chapters 3-9, the committee concluded that meaningful reductions in nutrient loads to coastal waters are achievable.

Specifically, the committee believes that implementation of the recommendations contained in this report would provide local decisionmakers and managers with an information base that could be used to determine what can and should be done to halt the degradation of many of the coastal waters identified in the NOAA National Estuarine Eutrophication Assessment as demonstrating symptoms of severe or worsening eutrophication. The committee believes implementation of the recommendations would dramatically enhance efforts of coastal and watershed managers and other individuals or groups attempting to mitigate the effects of nutrient over-enrichment in these and other estuaries. Improvements in all impaired coastal bodies could be achieved over the next 20 years, while preserving the environmental quality of now-healthy areas.

What are reasonable goals for improvement? In the committee's opinion, at a minimum federal, state, and local authorities should work with academia and industry to[2]:

- reduce the number of coastal water bodies demonstrating severe impacts of nutrient over-enrichment by at least 10 percent by 2010;
- further reduce the number of coastal water bodies demonstrating severe impacts of nutrient over-enrichment by at least 25 percent by 2020; and
- ensure that no coastal areas now ranked as "healthy" (showing no or low/infrequent nutrient-related symptoms) develop symptoms related to nutrient over-enrichment over the next 20 years.

It was beyond the charge and resources of the committee to identify specific coastal areas for priority attention. All 44 of the areas identified by NOAA's National Estuarine Eutrophication Assessment as exhibiting severe symptoms certainly should be considered as areas where greater effort is needed. Additional study could help further target priorities, especially if it included careful consideration of economic issues and opportunities for stakeholder input. Such work could take significant time and effort, and decisionmakers should not be tempted to defer action while waiting for "perfect" knowledge. The committee believes that

[2] These goals are all in relation to the benchmarks determined by NOAA's National Estuarine Eutrophication Assessment (Bricker et al. 1999, see Box 1-2). That report found 44 of 139 sites suffering high eutrophic conditions; 38 of the 139 sites studied showed low or no nutrient-related symptoms.

nationwide implementation of the recommendations in this report, across the full range of systems from small to large and problems from the simple to the complex, will start the nation on a course to achieve the goals stated above. Additional focus on areas subsequently identified for priority attention will then add to cumulative improvement. Thus, the goals listed above are intended to reflect nationwide achievement. Targeting some subset of the impaired coastal areas in an effort to simply meet these numeric goals (for instance, focusing on impaired water bodies associated with small watersheds or simpler ecosystems) would be contrary to the national interest and the spirit of this report.

Working to reduce the effects of nutrient over-enrichment nationwide over the next two decades will be a challenge, but the committee believes these general goals are realistic. The setting of such numeric goals is somewhat subjective, but the committee believes that such targets are important to encourage action. The goals were set after thorough discussion and are, in the committee's view, both achievable given current methods and challenging enough to facilitate real progress. Many of the principles espoused in this report have already been implemented on a smaller scale in Europe (e.g., Rhine and Elbe watersheds) and the United States (e.g., Tampa Bay and Chesapeake Bay) and have resulted in significant reduction in nutrient loads received from nonpoint sources (Behrendt et al. 1999; Belval and Sprague 1999; Johansson and Greening 2000). However, achievement of these goals should not be seen as an end in itself. Rather, they are a first step toward reversing the effects of nutrient over-enrichment in the nation's coastal waters and preventing impairment of "healthy" coastal areas.

How would these goals be accomplished? The key to addressing coastal nutrient problems is understanding that nutrient inputs to coastal waters are affected directly and significantly by activities in the watersheds and airsheds that feed the nation's streams and rivers, and building this recognition into planning as well as implementation of management solutions. Thus, an effective National Nutrient Management Strategy must recognize the fundamental role that local watershed and coastal managers play. These individuals will be the front line of both policy-making and project implementation.

The committee believes that by focusing on source reduction, actions can be targeted to most effectively reduce and reverse the problems caused by nutrient over-enrichment in coastal areas. Watershed-specific sources like urban stormwater runoff and inappropriate nutrient management at the farm level often can be addressed most effectively by local activities under local leadership, with activities typically site-specific. However, while significant improvements can be achieved through local action, local managers alone cannot be expected to bring adequate resources and knowledge to bear on such a complex problem, nor are they always able to work at the scale of larger watersheds. Sometimes, broader participation is necessary to bring about significant improvement.

Thus, what is required is a National Nutrient Management Strategy that emphasizes the need for local, state, and federal agencies to work together, and to create partnerships with academia and the private sector. First, federal leadership is essential to support and coordinate the research and development needed to provide new approaches and technologies that can be used by local and state agencies charged with reducing and reversing the impacts of nutrient over-enrichment. Perhaps even more importantly, federal leadership will be needed to deal with nutrient sources in large watersheds that span multiple states or jurisdictions. For example, the burning of fossil fuel by both mobile and stationary sources far from the coast can account for a significant component of the overall contribution of nitrogen from nonpoint sources in some watersheds. These watersheds, in turn, deliver that nitrogen to the sea, often hundreds of miles and many states away from the original source of emissions. Similarly, livestock feed is now shipped great distances to large, concentrated animal feeding operations. The cattle, hogs, or chickens in turn produce huge amounts of nitrogen- and phosphorus-rich wastes, which are ultimately released into a watershed that, again, may be several hundred miles and many states away from the original source of the nutrients.

Implementation of a National Nutrient Management Strategy to improve the understanding and management of nutrient over-enrichment and eutrophication requires action at two levels, local and federal. To facilitate these actions, the committee proposed two interrelated sets of recommendations. First are recommendations for a process to use at the local level now. Second are recommendations that address the development and implementation of federal activities to provide the long-term information, data, and analyses needed to address nutrient over-enrichment in coastal waters and support effective nutrient management strategies at the national, regional, and local levels.

A RECOMMENDED APPROACH FOR LOCAL MANAGERS

Figure 2-1 shows a decision-making framework that outlines the elements necessary in a process to help local, state, and regional managers make decisions about what steps and methods are appropriate to manage nutrients effectively in their area, recognizing their particular problems. This process is based on a number of recommended strategies, plus the experiences of local managers (EPA 1989; NRC 1990, 1999a; Schueler 1996; CENR 1998; Bricker et al. 1999).

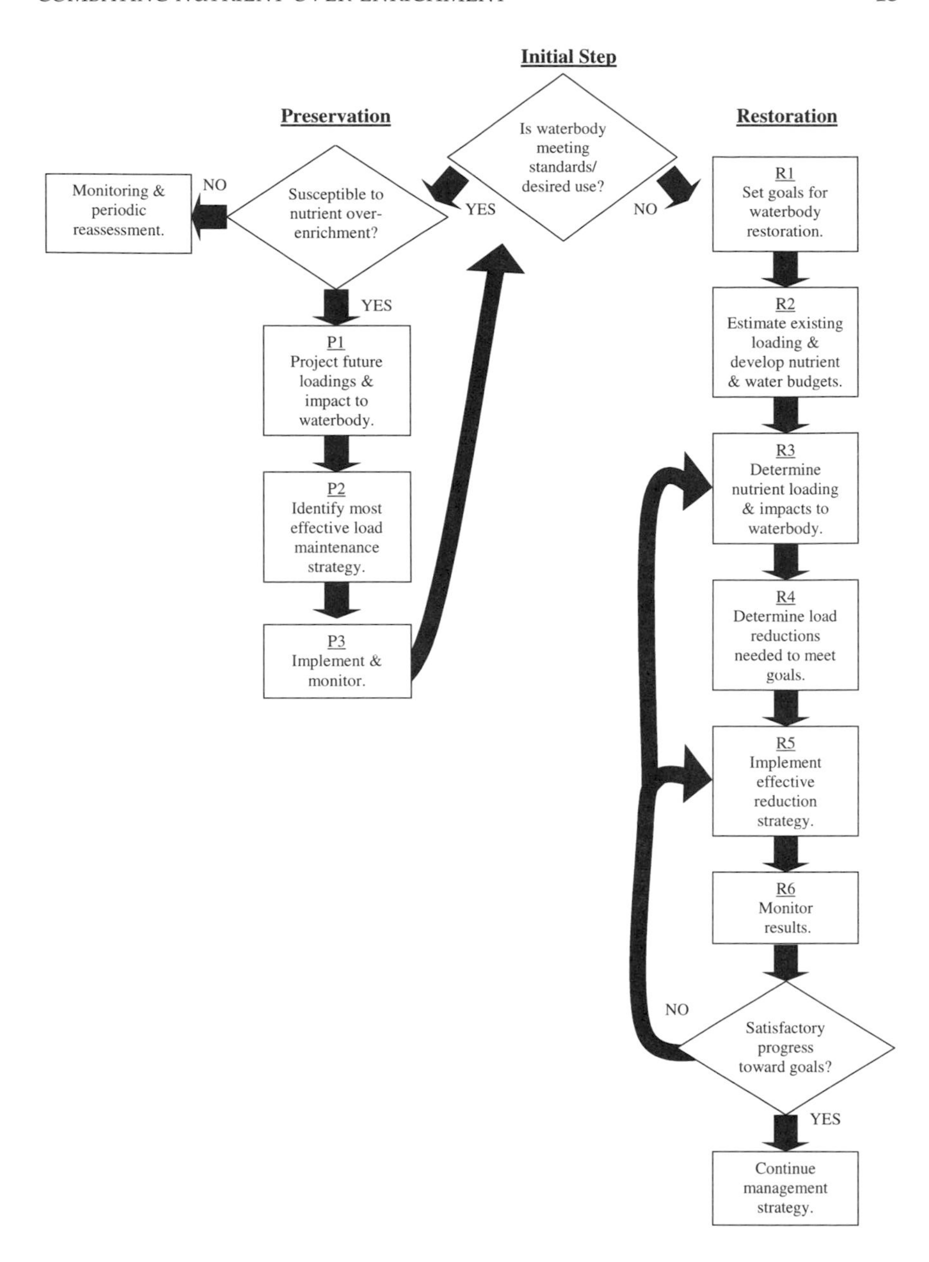

FIGURE 2-1 Key decision points for developing and implementing a site-specific nutrient management strategy.

Based on extensive discussions with managers and personnel from relevant federal support programs, review of published guidance documents, and detailed phone interviews with coastal managers from over 25 estuaries nationwide, the following text expands on the decision-making process to discuss what tools, analytical techniques, and data and information resources are available to help implement each step. The steps (keyed to Figure 2-1) are presented as a series of questions or decisions that the responsible manager would ask or make. The text also identifies what resources are available to support local managers. Thus, coastal or watershed managers dealing with nutrient management and its associated problems for the first time should find this a useful primer (references to chapters in this document where the reader can find more detailed information on the issue are included). The text also attempts to identify areas where greater resources are needed (in the vast majority of instances, federal or state leadership will be required). These areas are summarized, and specific federal actions are proposed to address them, in the last section.

Initial Step

Determine Status of Coastal Water Body: Is it Meeting Standards, Criteria, and/or Desired Uses?

The initial step of the decision-making framework is to determine whether the specific coastal water body in question is currently meeting standards, criteria, and/or desired uses. Characterization of the extent and severity of eutrophic symptoms (Chapter 4) may also include determining whether eutrophic conditions are natural or anthropogenic, and whether symptoms are seasonal or are exhibited throughout the year (Chapter 6). Existing standards and criteria vary from water body to water body; and many coastal areas currently do not have regulatory or non-regulatory guidelines for eutrophication or nutrient loading (Chapter 8).

Existing tools and information to assist with characterization of eutrophic symptoms include NOAA's National Estuarine Eutrophication Assessment and an initial susceptibility index as developed by NOAA (see also Chapters 4 and 6).

Needed resources and research to support the characterization step include:

- development of standards or guidelines for nutrient criteria or loads, including total maximum daily loads (TMDLs) that a waterbody can assimilate without exceeding criteria or affecting desired uses (Chapter 8);
- greater information on how long coastal water bodies can expect symptoms to persist following nutrient loading reductions (Chapters 4 and 6); and
- greater information on the contribution of specific sources to nutrient loading (Chapter 7).

If the water body is not meeting standards or goals and the causes appear to be anthropogenic, the strategy moves to the "restoration" series of steps, labeled R1 through R7. If the water body is meeting standards or goals, then the decision-maker moves to the maintenance or "preservation" series of steps, labeled P1 through P4.

Restoration Steps

R1. Set Goals for Waterbody Restoration

If restoration is deemed appropriate, the first step is the setting of measurable goals for restoration. Goals can be regulatory (e.g., dissolved oxygen concentrations) or those adopted by stakeholders, such as seagrass acreage or water clarity (Chapter 8). If goals are stakeholder-determined, commitment by the stakeholders to participate in the process is essential for successful implementation.

Existing tools and information include guidance as summarized in a previous NRC report (NRC 1999a), guidance prepared for the National Estuary Programs (EPA 1989) and other guidance efforts (Schueler 1996; ASCE and WEF 1998).

Needed resources and research include:

- historical information on the state of the water body in question (so that remediation goals can be more easily tied to previous conditions);
- a compilation of experience from existing programs (to capitalize on the success of other efforts); and
- ambient environmental data (to establish quantitative goals).

R2. Estimate Existing Loading and Develop Nutrient and Water Budgets

The next step is to develop nutrient and water budgets, including estimating nutrient loads from all sources to the coastal water body. This can be accomplished in several ways. Whenever adequate data from the contributing watershed are available, the recommended method is to use real measurements of discharge (flow) and nutrient concentrations to estimate the load contributed by surface water and groundwater (Chapter 7). However, in many instances adequate data are not currently available to calculate these estimates or to establish the actual sources (e.g., runoff from croplands, animal feeding operations, atmospheric sources), and a modeling approach must be used to estimate the surface water and groundwater load. These approaches range from simple spreadsheets to complex mechanistic techniques (Chapters 5 and 7).

Existing tools and information include flow and nutrient concentration data collected by the U.S. Geological Survey (USGS) and other state, local and federal entities, atmospheric deposition collected by the National Atmospheric Deposition Program and NOAA (very limited data from coastal areas), and various modeling approaches as reviewed in Chapter 7.

Needed resources and research include:

- ambient environmental data (to more clearly identify sources and loads and to support or validate more robust modeling efforts as needed);
- measurements of atmospheric deposition (wet and dry) to coastal surface waters (to better identify the relative role atmospheric sources play in contributing nutrients; Chapter 5);
- evaluation of processes and retention of atmospheric deposition on various land-use covers (Chapter 5);
- expanded USGS flow and nutrient concentration data collection in coastal watersheds (Chapter 7);
- assessment of modeling approaches to address eutrophication (Chapter 7); and
- evaluation of transferability of findings between existing studies, including scaling (Chapter 6).

R3. Determine Relationship between Loadings and Impact to Waterbody

The next step is to determine the responses of a water body to variations in nutrient loading. Responses may include changes in ambient nutrient concentrations, algal biomass and turbidity levels, and other measurable changes (Chapter 4). Models currently used to calculate responses to variable nutrient loads range from simple regression approaches to linked watershed:water body hydrodynamic models, as reviewed in Chapter 7. Estimating the relationships between nutrient loads and responses in the water body (i.e., developing dose/response curves) and the relative susceptibility to changes in nutrient inputs for various classes of coastal waters will provide managers with critical tools to help answer "what if" questions concerning nutrient reductions or increases (Chapter 6). This will be a difficult task, requiring research.

Existing tools and information include various modeling approaches as reviewed in Chapter 7 and the initial susceptibility index developed by NOAA (Bricker et al. 1999) and discussed in Chapter 6.

Needed resources and research include:

- the development of techniques to better predict the response of a given class of water body to a specific load (to allow for greater predictability and reduce reliance on trial and error approaches; Chapter 6);
- further development and implementation of the classification scheme (to help assess susceptibility to nutrient over-enrichment; Chapter 6); and
- assessment of the use and effectiveness of various models (Chapter 7).

R4. Determine Nutrient Load Reductions Needed to Meet Goals

The next step is to calculate the difference between existing loads (estimated in Step R2) and loads that would result in meeting water body goals (estimated in Step R3). An important consideration is that many coastal waters will not show immediate response to reductions or changes in external nutrient loadings, and managers should expect a lag time of months to years after nutrient input changes have been initiated (Chapter 9).

R5. Identify and Implement Most Effective Load Reduction Strategy and Projects

Stakeholder participation is critical in identifying the most effective strategies and projects (Chapter 8). Each watershed and coastal water body will have unique sets of potential strategies (regulatory and non-regulatory) and projects to help meet goals. Strategy development should include consideration of the effectiveness of management practices (Chapter 9), economic assessments and incentives (Chapter 8), and an evaluation of the most cost-effective ways to meet goals (Chapter 8). Implementation will require strong and long-term commitment by the participants.

Existing tools and information include those sources listed under R1.

Needed resources and research, in addition to many of the recommendations listed in the previous steps, include:

- compilation and assessment of methods used to identify relative contributions from each type of nutrient source (to support more effective management strategies; Chapters 5, 8, and 9);
- compilation of economic studies that examine the relative costs of various approaches in a variety of settings, organized so that coastal decisionmakers can more readily identify relevant results and approaches;
- identification, compilation, and making accessible a list of potential management options for each type of source, including costs and effectiveness of existing best management practices for urban, agricultural, and residential areas (to help achieve source reductions; Chapter 9) and recognizing the roles of existing regulations;
- continued development of improved best management practices (Chapter 9);
- evaluation of effective management structures for implementation (Chapter 8); and
- continued identification of potential barriers (local, state, and federal levels) and development of ways to address these barriers.

R6. Monitor Results

Monitoring results of the implementation of a defined management approach can include several elements: ambient monitoring of the water body, monitoring of loadings, and monitoring of specific projects to determine effectiveness (Chapters 7 and 9). Measuring progress (or lack of progress) towards reaching goals provides a crucial "feedback loop" for participants and managers (Chapter 8).

Existing tools and information include ambient monitoring programs where available.

Needed resources and research include:

- definition and implementation of a national ambient monitoring program, including source monitoring (Chapter 7).

R7. Progress Being Made toward Reaching Goals

Managers must ultimately be aware of whether the actions taken result in progress toward stated goals. Progress can be assessed by a variety of means, from casual observations by citizens and stakeholders, to simple visual examination of monitoring trends, to use of more sophisticated quantifiable methods of assessing monitoring data. Reporting results to stakeholders and the public on a timely basis is an important element of this step.

Existing tools and information include an example of a three-phase reporting system (as defined by the Chesapeake Bay Program) to report to scientists, managers, and the general public (NRC 1990).

Needed resources and research include development of quantifiable methods for assessing progress towards goals.

If progress is not found to be satisfactory, the manager should return to Step R3 (if new data may provide a revised relationship) or Step R5.

Maintenance or Preservation Steps

If, as is often the case, the initial evaluation step finds that the water body is currently meeting standards, criteria and/or stakeholder goals, the manager should work his or her way through a series of steps designed to ensure maintenance or preservation of the quality of the water body. These steps are:

P1. Evaluate Potential for Future Nutrient Over-Enrichment

Because the water body is currently meeting standards and goals, steps taken in this sequence should be simple and cost-effective, and designed to evaluate future potential for eutrophication. More detailed susceptibility evaluations can be used if the simple methods indicate possible problems in the future (Chapter 6). Simple evaluations (i.e., "red flags") that indicate that the water body may be susceptible include:

- large watershed size,
- the watershed is experiencing rapid land use change,
- long residence time in receiving water body,

- a seagrass-dominated system,
- often contains high levels of dissolved organic material (DOM), and/or
- low turbidity.

The initial susceptibility index developed by NOAA (Bricker et al. 1999) and used to assess current and future status of the nation's estuaries in NOAA's National Estuarine Eutrophication Assessment, in conjuction with other approaches discussed in Chapter 6, may be a useful starting point for determining the risk faced by a given water body and should be considered during this preliminary evaluation.

Existing tools and information include NOAA's initial susceptibility index used in NOAA's National Estuarine Eutrophication Assessment (Bricker et al. 1999).

Needed resources and research focus on further development of a systematic estuarine classification scheme and dose/response curves (Chapter 6). If the water body is considered potentially susceptible to nutrient over-enrichment in the future, continue to P2.

P2. Project Future Nutrient Loadings and Impacts to the Water Body

Tools similar to those described in Steps R2 and R3 to estimate existing loads can be used for this step, but in this case the effort should be more modest (e.g., measurements taken at regular, but less frequent intervals).

P3. Identify Most Effective Load Maintenance Strategy

This step will include the same basic considerations outlined in Step R5, with the objective of maintaining loadings below those that would result in detrimental effects. Determination of total maximum or "acceptable" loads (TMDLs or other determinations) for nutrients will be an important tool in this step.

P4. Implement Maintenance Strategy and Monitor Water Body

The monitoring recommendations outlined in Step R6 are also relevant for this step, although in this series they should be implemented at a modest or targeted level.

RECOMMENDED FEDERAL ACTIONS

To combat coastal nutrient over-enrichment in an effective, coordinated way, federal agencies, in concert with the White House and with the support of Congress, should develop and implement a National Nutrient Management Strategy. This national strategy should contain elements to address source identification, impacts, management approaches, and other local needs identified above. Furthermore, because many of the problems faced locally reflect regional processes beyond the purview of local jurisdictions, the strategy should include mechanisms to coordinate efforts at local, regional, and national levels. To minimize potential competition between agencies for limited funds, and to reduce unnecessary and costly duplication of effort, the mechanisms chosen to implement the National Nutrient Management Strategy will be critical and should build on existing efforts, like the Clean Water Action Plan (Box 2-1). These mechanisms could include convening multi-entity steering/oversight panels (possibly modeled after the Executive Office's Committee on Environment and Natural Resources [CENR]; Box 2-2), which include representatives from relevant federal programs but also have strong state, regional, and local program participation, including citizen and industry groups. Regional or state oversight committees may be necessary to obtain adequate local and regional input. Federal support, both financial and technical, will be critical for the successful implementation of the national strategy. The proposed National Nutrient Management Strategy should strive to increase coordination and efficiency of ongoing efforts, promote technical exchange, strengthen monitoring, modeling, and research efforts, and support, to the degree possible, local management efforts.

Identify and Address Program Gaps and Overlaps

One of the first federal actions taken under the National Nutrient Management Strategy should be to develop and implement a process to assess overlaps and gaps in existing and proposed federal programs for all aspects of nutrient over-enrichment, with particular attention to the needs of local managers. This assessment should identify specific roles and responsibilities carried out by various federal agencies while recognizing the important roles that state and local governments, industry, and nongovernment groups play. Assessment results should be used to redirect resources from redundant efforts to areas where additional work is needed. An evaluation of the combined experiences of existing nutrient management programs and local programs, particularly their successful (and less successful) methods for determining sources of nutrients and

BOX 2-1
Clean Water Action Plan: 1999 Progress

The Clean Water Action Plan is a multi-agency effort initiated in 1998 in commemoration of the 25th anniversary of the Clean Water Act (USDA and EPA 1998a). The Plan contains 111 key actions, many of which are relevant to nutrient over-enrichment in coastal waters, including:

- steps to improve water quality using wetland restoration and preservation actions. The goal is to achieve a net increase of 100,000 acres of wetlands each year, beginning in 2005. As of 1998, roughly 212,000 acres had been enrolled in the Wetlands Reserve Program, a voluntary program that offers financial support to landowners for wetlands restoration.
- several actions to improve assessment and response to coastal HABs. The National Harmful Algal Bloom Research and Monitoring Strategy (U.S. Department of Interior et al. 1997) outlines a long-term strategy for federally supported research and monitoring on problems of HABs. An emergency response plan was distributed in August 1998 and will continue to be refined and expanded.
- calls for the development of a Coastal Nonpoint Pollution Control Program (as required by the Coastal Zone Management Act Reauthorization amendments of 1990 that require each coastal state to develop and implement programs by June 1998, with full approval by December 1999). All 29 participating state and territorial programs had been conditionally approved, but as of late 1999, none had been fully approved.
- actions designed to improve coordination and information sharing among federal agencies and with state and local entities. As of 1999, a draft outline of the Coastal Research and Monitoring Strategy had been developed by an interagency workgroup. This strategy seeks to coordinate existing programs; however, it is not clear how and when implementation will take place or whether a new initiative will be needed to implement recommendations from the workgroup.
- calls for the Environmental Protection Agency (EPA) to develop nutrient criteria for U.S. surface waters that will define numeric criteria for nutrients that are tailored to reflect the types of water bodies (lakes, streams, estuaries) and ecoregions of the country. EPA has drafted a multi-year strategy to help develop and implement nutrient criteria and standards, and state implementation of these criteria is scheduled for 2003. Scientific peer review of the procedural document is ongoing, with completion expected in 2001.
- recognizes the need to assess and reduce atmospheric deposition of nitrogen, and calls for:
  - better quantification of the risks of atmospheric deposition of nitrogen and other airborne pollutants to water bodies;
  - evaluation of the linkage of air emissions to water quality impacts; and
  - employment of both the Clean Water Act and Clean Air Act authorities to reduce air deposition of nitrogen compounds and other pollutants that adversely affect water quality.

continued

BOX 2-1 Continued

- plans for addressing concentrated animal feeding operations as included in the unified EPA/U.S. Department of Agriculture (USDA) National Animal Feeding Operations Strategy. At the end of 1999, USDA and EPA had developed the Strategy, with the primary goal of implementing comprehensive nutrient management plans at all animal feeding operations by 2008. The strategy includes technical assistance and back-up regulatory approaches.
- proposes several incentives to reduce polluted runoff, including encouragement of growth management (or "smart growth") and use of financial incentives to encourage farmers and ranchers to voluntarily remove sensitive lands from agricultural use. At the end of 1999, approximately $976 million in federal funds had been committed to six states that had signed up to participate in the Conservation Reserve Enhancement Program.
- outlines efforts to increase access to the many different sources of data and information generated by federal programs through a new internet-based Water Information Network. During 1999, the first version of the Water Information Network was released to the public and refinements are continuing.

potential impacts, could provide much needed information to both local managers and national policy makers. Specific actions to increase coordination at all levels would include:

- **Increase Attention Given to Atmospheric Deposition of Nutrients**—Due to the geographic extent of airsheds (often many times larger than the watersheds that managers use as boundaries), federal programs, such as EPA's Great Waters program, are encouraged to increase their efforts to quantify atmospheric deposition of nutrients to the nation's coastal waters. Local programs should be encouraged to participate in a national monitoring program (such as the National Atmospheric Deposition Program) through offers of technical and funding assistance for development of monitoring sites, sample collection and analyses, and data analyses and interpretation. (The existing NADP database could be considered as the core for data management of atmospheric deposition.)
- **Consider Need for Nutrient Management During Reauthorization of the Clean Water, Clean Air, and Coastal Zone Management Acts**—Obviously, the movement and concentration of nutrients among the biosphere, atmosphere, and freshwater and marine

Box 2-2
CENR: Is It a Model for Implementing a National Nutrient Management Strategy?

The Committee on Environment and Natural Resources (CENR) was established by President Clinton to foster a multi-agency, interdisciplinary approach to environment and natural resources research and development and to coordinate federal efforts totaling approximately $5 billion in 1999. CENR was created in recognition that the traditional single-agency, single-discipline way of solving environmental and natural resource problems was no longer adequate. CENR addresses science policy and research and development efforts that cut across agency boundaries and provides a formal mechanism for interagency coordination on domestic and international issues. Twelve federal agencies participate on CENR committees (NSTC 1998).

The CENR, a committee of the National Science and Technology Council (NSTC), provides advice regarding the effectiveness and productivity of federal research and development efforts in the area of the environment and natural resources. The NSTC is a standing cabinet-level body chaired by the President and composed of the Vice-President, the Assistant to the President for Science and Technology, the cabinet secretaries and agency heads with responsibilities for significant science and technology programs, and other White House officials. The organizational structure of CENR includes full-committee and subcommittee vice-chairs (science and policy co-chairs) who bring high-level policy perspectives.

CENR consists of seven subcommittees representing areas of important policy that transcend the interest of any single agency: global change; biodiversity and ecosystem dynamics; resource use and management; water resources and coastal and marine environments; air quality; toxic substances and hazardous and solid waste; and natural disaster reduction. Coastal nutrient over-enrichment, including hypoxia and coordinated monitoring, is one responsibility within these subcommittees. CENR is recognized for its success in reducing coordination and interagency barriers, and could be used as a model for the proposed national strategy to combat nutrient problems. If used as a model, the structure would need to be adapted to include significant local and state participation.

systems is rarely affected by the jurisdictional boundaries of cities, counties, states, or nations. Thus, the sources and effects of nutrient over-enrichment will rarely be confined to a single political jurisdiction. Constructing effective regional or national policies or regulations to deal with the problems associated with nutrient over-enrichment will involve many of the issues addressed by these three acts. Thus, the implications of nutrient over-enrichment should be an important consideration when Congress addresses the re-authorization of these important components of national environmental policy.

- **Complete and Implement the Clean Water Action Plan**—To a very large degree, the actions called for throughout this report are represented or discussed in the Clean Water Action Plan currently under consideration. After revising the plan to address issues raised by this report and elsewhere, as appropriate, the plan should be implemented as quickly as is practical.

Accessible Data, Information, and Expertise

Local programs and agencies are using a variety of assessment and management tools to study and reduce the effects of nutrient over-enrichment in coastal settings, ranging from linked hydrodynamic water quality models to a purely technology-based approach. Although the complexity ranges among programs, managers believe that their processes are providing or are expected to provide adequate information to initiate management strategies; all, however, identified needs for additional assessment tools. In general, local managers cannot wait for answers from the scientific community. They must use what they perceive to be the best available or most appropriate information to compile nutrient budgets by source type, to estimate potential impacts from management practices, and to develop and implement management strategies. However, some managers may not be aware of better sources of information, indicating a communication problem between the scientific and management communities. Managers recognize that the data sources and tools that they currently use need improvement.

Better mechanisms for communicating information could lead to rapid improvement in management plans. As one of its initial actions, those implementing the National Nutrient Management Strategy should create mechanisms to provide consistent and competent technical assistance from federal agencies to local decisionmakers and agency staff. This might include development of a national clearinghouse and access to on-request assistance and review. The following should be considered high priority and parallel actions:

- **Develop a national information clearinghouse**—Understanding how the effects of nutrient over-enrichment are manifest or vary from estuary to estuary is an important step for coastal managers who must deal with nutrient over-enrichment. A web-based clearinghouse for information on the effects of nutrient over-enrichment should be established with links to federal, state, and local assistance programs, on-request technical assistance, contacts, and the metadatabase recommended below. In addition, relevant economic studies should be integrated and organized so that decisionmakers can readily identify available information on the economic viability of different management approaches from similar coastal settings or for similar nutrient sources. Many federal and state programs already have extensive websites or databases that should be directly linked to the clearinghouse.

- **Develop a metadatabase of distributed information and data**—Determining the validity or applicability of information requires an understanding of the data and techniques used to collect it. The metadata[3] supporting the information clearinghouse should be easily accessible for all users. All partners in the coordination effort should be encouraged to link their existing sites to the metadata site, although partners should continue to be responsible for updating and controlling the quality of their own databases. As various databases are developed to meet emerging local, state, regional, or national needs for information relevant to nutrient over-enrichment, they should be linked to both the metadata website and included in the clearinghouse discussed above.

Expand Federal Leadership

Federal leadership is critical to address issues that span multiple jurisdictions, involve several sectors of the economy, threaten federally held resources, or fall under existing federal regulations such as the Clean Air Act. This leadership should be manifest in several ways including specific actions to help establish credible goals and mechanisms, including:

- **Set Clear Guidelines for Nutrient Loads**—The development of critical nutrient loads (above which nutrient over-enrichment and eutrophication symptoms may be expected) are essential to successful nutrient management strategies (Chapter 8). EPA's efforts to develop nutrient criteria and TMDLs should incorporate interaction among physical, chemical, and biological factors, seasonal

[3] Metadata refer to information about the origin and provenance of information contained in a database. With adequate metadata, a user can determine the validity or applicability of information before using it. Such information is essential if data from a wide number of sources are to be shared effectively among a large number of users. By creating a metadatabase to complement the national information clearinghouse discussed in the preceding paragraph, the utility of both would be greatly enhanced.

and timing imports, and the nature of hydrologic forcing functions (Chapter 8). These efforts should, however, focus on identifying sources and setting maximum loads, rather than on limiting the ambient concentration of a given nutrient in a receiving water body (Chapter 5).

- **Reduce Impact of Agriculture Practices**—Technological and organizational advances in agriculture have made North America one of the leading producers of foodstuffs worldwide. However, information is still often lacking on the impacts of agriculture on various ecosystems. National and regional strategies are needed to help address the introduction of excess nutrients from various agricultural practices. Farmers' decisions are often influenced by regional or even global economics. At these scales, farmers have little or no control over these economic pressures and the resulting changes in nutrient flows and distribution. Therefore, new ways of using incentives to help farmers implement innovative source reduction and control will be needed (Chapter 9).
- **Evaluate Existing Efforts to Determine Elements of Success**—Conduct an evaluation of the combined experiences of the local, state, and federal programs, highlighting successful (and less successful) methods for determining sources of nutrients, potential impacts, and management strategies, to provide needed information to both local managers and national policy makers. Identify, compile, and make accessible a list of potential management options for each type of source, included costs and effectiveness of existing BMPs for urban, agricultural and residential areas (Chapter 9); continue research and development of new best management practices (Chapter 9).

Expand Monitoring Capabilities

The United States lacks a coherent and consistent strategy to monitor the effects of nutrient over-enrichment in coastal settings on a regular and consistent basis. The NOAA assessment effort (Bricker et al. 1999) is an admirable one, but it is limited by the inconsistency of data collection among estuaries. One consequence is that the full economic and ecological impact of nutrient over-enrichment is not currently demonstrable.

Implementation of a nationally consistent monitoring program will be a critical component of the proposed National Nutrient Management Strategy because monitoring brings better characterization of the spatial extent and temporal trends of nutrient over-enrichment in estuaries and coastal waters. Such a program must be commensurate with the scale of the issue. The best approach is probably to use a partnership of efforts by local, state, and federal agencies, as well as academic and research institutions where appropriate. Consistent procedures, criteria, quality control, and data management and reporting are essential. Monitoring should include biological, physical, and chemical properties on time and space scales relevant to capture the necessary variability and linkages between variables. Often, this will mean that biological and chemical measurements will need to be made at finer scales than is presently the norm, while additional collection of long-term data will be needed to detect subtle change. Monitoring programs should be regularly evaluated by independent panels to determine their effectiveness. Further, monitoring programs should be adaptive, incorporating new technology and scientific understanding while preserving the long data time series necessary to detect trends. Selection of monitoring sites should be made with consideration of classification schemes that illustrate estuarine susceptibility to nutrient over-enrichment. It is likely that incentives will need to be developed to encourage widespread state and local implementation of consistent quality control and metadata standards (Chapter 7).

Representative coastal systems (e.g., index sites) should be selected to serve as sites for long-term, intensive research programs to better understand the effects of nutrient enrichment on estuarine structure and function, and to track how changes in management affect coastal systems. Index sites must be representative of the range of estuarine "types" included in the classification and should be selected in parallel with development of an overall classification scheme for susceptibility to nutrient over-enrichment, as described in Chapter 6. Index sites should be chosen to show varying degrees of human impacts. Research at index sites can help explain temporal patterns of change revealed through monitoring and assessment programs, and is essential to develop better predictive models for management.

Estimates of nutrient inputs to estuaries are essential for management, and data on long-term trends on nutrient inputs are invaluable for determining sources of nutrients. Throughout the United States, the USGS is the best and primary source of data on nutrient inputs to estuaries from upstream rivers. The data it collects are invaluable, and continuation of this monitoring is essential. However, the USGS monitoring networks were not designed to assess inputs to coastal regions, and should be expanded to include this role.

Other recommendations that support the development of a nationally consistent monitoring program include:

- Implement a national monitoring framework including the adoption of a three-tiered national monitoring program similar to that recommended in the CENR Draft Coastal Research and Monitoring Strategy (Chapters 6 and 7).

- Establish representative or index sites (as defined by the susceptibility classification) where long-term, intensive research programs are conducted to better understand the mechanisms controlling eutrophication processes and the effects of nutrient enrichment on estuarine structure and function for the various classifications (Chapter 7).
- Include monitoring of the effectiveness of nutrient management projects and strategies, including BMPs (Chapter 9).
- USGS monitoring should be expanded with the specific objective of assessing nutrient inputs to estuaries and monitoring how these change over time. Further, monitoring data collected by the state and local agencies should be used more fully. Often, these data are collected for other purposes (such as assuring drinking water quality), yet they could provide useful information on nutrient inputs to estuaries if adequate quality control were maintained (Chapter 7).
- Develop and implement regional or national monitoring and management strategies for atmospheric deposition. Expand deposition monitoring to better represent urban and coastal areas. Improve dry deposition monitoring and model efforts (Chapter 5).

Conduct Periodic Comprehensive Assessments of Coastal Environmental Quality

One key deficiency in the nation's approach to coastal water quality deficiencies is the lack of periodic, comprehensive analysis like the recent NOAA National Estuarine Eutrophication Assessment. In the future, such efforts will be particularly important because they would provide information about how systems have changed, which is critical for understanding whether policy and management choices have been effective in causing improvements. Thus, the nation needs to conduct a periodic (every 10 years) reassessment of the status of eutrophication in the nation's coastal waters (similar in scope to NOAA's 1999 National Estuarine Eutrophication Assessment).

Develop a Susceptibility Classification Scheme

The National Nutrient Management Strategy should encourage further development and use of a classification scheme to determine a given estuary's susceptibility to nutrient over-enrichment (Chapter 6). An important goal of coastal zone managers is to accommodate human actions while minimizing the impact on coastal ecosystems. Successful management requires considerable information at a variety of levels, including an understanding of systems in their natural, pristine condition as well as how natural systems respond to human activities (Karr and Chu 1997). In coastal waters, the situation is particularly acute because the different types of estuaries, embayments, and shelf systems differ in their response to nutrient enrichment. Thus, a given nutrient input results in different response trajectories in various types of systems.

A widely accepted estuarine classification scheme is a prerequisite for a systematic approach to extending lessons learned and management options from one estuary or affected coastal water body to others. Such a classification scheme should allow categorization of relatively poorly known systems on the basis of a minimum suite of measurements. Quantitative classifications that provide insights into the relative importance of the different factors controlling estuarine dynamics have the most potential for predictive analysis. A high priority should be the development of a national framework of "index sites," within which there would be an integration and coordination of environmental monitoring and research with the goal of developing a predictive understanding of the response of coastal systems to both nutrient enrichment and nutrient reduction.

Improve Models to Support Coastal Managers

Largely due to the high cost associated with monitoring, it is impractical for managers to collect and assimilate enough observational data to fully understand the effects of changes in land use patterns. Furthermore, even if adequate observational data could be obtained, understanding the processes involved sufficiently to predict future conditions is difficult. Consequently reliance on complex and simple models to forecast environmental conditions has become widespread. This approach couples the benefits of reducing the need for prohibitively large monitoring systems with rapid exchange of the information generated by increasingly sophisticated computer technology. More effort needs to be made to convey modeling methods and results widely (Chapters 6 and 7). This will also require additional efforts to develop a database of typical input values (organized by watershed and receiving waterbody type) for use in existing models. Such efforts will be instrumental for developing quantifiable methods for assessing progress towards goals, an essential element of effective resource management (Chapter 8).

Expand and Target Research

The potential impacts of atmospheric deposition of nutrients on coastal waters and its contribution to the effects of nutrient over-enrichment is just beginning to be estimated and fully recognized by local, state, and federal agencies and managers. As noted earlier, due to the large geographic extent of airsheds, local and state programs cannot adequately address this issue alone. In addition to the expanded monitoring effort called for earlier to address the atmospheric deposition of nutrients, additional efforts should also be directed by national programs toward quantifying sources, fate, transport, and impacts (including economic) of atmospheric deposition of nutrients on watersheds.

In addition, federal programs that fund basic research (such as EPA, the National Science Foundation [NSF], and NOAA) should provide competitive grants for academic support for research into the role atmospheric deposition plays in nutrient over-enrichment. The implications for reducing the effects of nutrient over-enrichment through implementation of the Clean Air Act should be a major component of the national dialog concerning coastal environmental quality.

Additional research is needed to address the relative role that nitrogen and phosphorus nutrients play in specific freshwater and marine systems, and how those roles vary seasonally (Chapter 3). Greater research effort is needed in order to better understand the role of specific nutrients in the occurrence of various harmful algal blooms, and how toxic algae of all types can endanger fish and birds, as well as humans and other organisms at higher levels of the food web (Chapter 4). Finally, research is needed that builds understanding of the effects of nutrient inputs on economically valuable resources (e.g., oysters, fish stocks, etc.) so we are better prepared to do the analyses necessary to compare costs and benefits and set acceptable restoration goals.

Support Local Management Initiatives

The strategies used by local programs and agencies for minimizing the effects of nutrient over-enrichment range from entirely educational and non-regulatory to primarily regulatory. In many instances, the most appropriate approach is a combination of voluntary and regulatory approaches that grant flexibility and are designed to achieve goals at minimum costs.

However, few local programs to date report that their strategies for reducing the effects of nutrient over-enrichment actually seems to be providing observable improvements. Others report that it is either too early to tell, or that results are mixed. Reasons for lack of observable effects vary, from inadequate modeling or data, to lag times in seeing results, to ineffectual management actions. In general, many of these programs could benefit from independent, objective analysis of the effectiveness of their planning, assessment, and management processes.

The local, state, and federal elements of the proposed National Nutrient Management Strategy could provide information and assistance with the development and implementation of effective management at all levels, and a means for objective independent review. As noted many times in this report, effective management is site-specific and unique for every estuary and coastal water body, with no universal "right answers." An adaptive management approach, using accessible and emerging tools, knowledge of successful techniques, coupled with and supported by a strong monitoring program, appears to provide the highest probability of long-term success.

Part II

Understanding the Problem

3

Which Nutrients Matter?

KEY POINTS IN CHAPTER 3

This chapter reviews the nutrients that exert the greatest control on eutrophication:

- Phosphorus is the nutrient usually controlling freshwater lake eutrophication.
- In contrast, eutrophication in most coastal marine ecosystems is primarily controlled by nitrogen.
- Even though nitrogen usually controls eutrophication of coastal systems, it is important to manage both phosphorus and nitrogen inputs since phosphorus is important in some of systems and since managing only nitrogen without also managing phosphorus inputs can lead to a situation where phosphorus becomes the nutrient controlling eutrophication.
- A variety of ecological and biogeochemical mechanisms lead to these differences between freshwater and coastal marine ecosystems, including the relative inputs of nutrients from adjoining systems, the preferential storage or recycling of nitrogen versus phosphorus within the ecosystem, and the extent to which nitrogen fixation can alleviate nitrogen shortages.
- Eutrophication of coastal systems is often accompanied by decreased silica availability and increased iron availability, both of which may promote the formation of harmful algal blooms.

The major nutrients that cause eutrophication and other adverse impacts associated with nutrient over-enrichment are nitrogen and phosphorus. In this chapter, we discuss why nitrogen is of paramount importance in both causing and controlling eutrophication in coastal marine ecosystems. This is in contrast to lakes, where eutrophication is largely due to excess inputs of phosphorus. Also discussed in this chapter are other elements—particularly silicon and iron—that may be important in regulating harmful algal blooms in coastal waters and in determining some of the consequences of eutrophication.

NITROGEN AND PHOSPHORUS IN ESTUARIES AND LAKES

After extensive study in the early 1970s, a consensus developed that phosphorus was the nutrient most responsible for nutrient over-enrichment in freshwater lakes (Edmondson 1970; Vollenweider 1976; Schindler 1977), and since then better control of phosphorus loadings to lakes has gone a long way toward mitigating freshwater eutrophication (Carpenter et al. 1998). In contrast, research indicates that in numerous estuaries and coastal marine ecosystems (at least in the temperate zone) nitrogen generally is more limiting to primary production by phytoplankton, and nitrogen inputs are more likely to accelerate eutrophication (Howarth 1988; Vitousek and Howarth 1991; Nixon 1995; Paerl 1997). Note that the concept of "nutrient limitation" is often poorly defined and used rather loosely; the committee follows the definition of control of the potential rate of primary production, allowing for potential changes in the composition of the ecosystem (Howarth 1988). Thus, a nutrient is limiting if its addition to the system increases the rate of net primary production.

There are exceptions to the generality that nitrogen is limiting in coastal ecosystems. For instance, certain temperate estuaries, such as the Apalachicola on the Gulf coast of Florida and several estuaries on the coast of the Netherlands in the North Sea appear to be phosphorus limited (Myers and Iverson 1981; Postma 1985; Brockman et al. 1990). In the case of the North Sea estuaries, phosphorus limitation is probably the result of extremely high nitrogen inputs combined with fairly stringent control of phosphorus inputs (Howarth et al. 1995, 1996). In the case of the Apalachicola, phosphorus limitation results from a relatively high ratio of nitrogen to phosphorus in nutrient inputs, although in this case the high ratio may reflect the relatively small amount of human disturbance in the watershed and the relatively low nutrient inputs overall (Howarth 1988; Billen et al. 1991).

For nearshore tropical marine systems, it is commonly believed that phosphorus is more limiting of primary production (Howarth et al. 1995). This is probably true for many tropical lagoons with carbonate sands that are relatively unaffected by human activity (Smith and Atkinson 1984; Short et al. 1990). However, such lagoons may move toward nitrogen limitation as they become eutrophic (McGlathery et al. 1994; Jensen et al. 1998). Also, even oligotrophic tropical seas may be nitrogen limited away from shore; for example, much of the Caribbean Sea away from the immediate shorelines appears to be nitrogen limited (Corredor et al. 1999).

Nutrient limitation of primary production switches seasonally between nitrogen and phosphorus in some major estuaries, such as Chesapeake Bay (Malone et al. 1996) and in portions of the Gulf of Mexico, including the "dead zone" (Rabalais et al. 1999). Even in these systems, nitrogen is probably the nutrient responsible for the major impacts of eutrophication. The production of most of the biomass that sinks into bottom waters and leads to low-oxygen events is more likely to be controlled by nitrogen than by phosphorus; when primary production is phosphorus limited in these systems, relatively little of the production tends to sink out of the water column (Gilbert et al. 1995; Malone et al. 1996; Rabalais et al. 1999).

Acceptance of the need to better control nitrogen inputs to coastal marine waters has been slower than acceptance of phosphorus control to manage eutrophication in freshwater systems. Many marine scientists recognized the nitrogen problem decades ago, yet the need for nitrogen control was hotly debated throughout the 1980s (NRC 1993a). For some locations, such as the Baltic Sea, the debate continues (Hellström 1996; Elmgren and Larsson 1997; Hecky 1998; Howarth and Marino 1998). Nonetheless, efforts to manage coastal eutrophication by controlling nitrogen inputs lag far behind the widespread success in managing lake eutrophication by controlling phosphorus inputs (NRC 1993a).

EVIDENCE FOR NITROGEN LIMITATION IN COASTAL MARINE ECOSYSTEMS

Most of those who in the 1980s disagreed with the assertion that nitrogen is the key to regulating marine eutrophication in coastal marine systems instead argued that phosphorus is the critical nutrient, as in lakes. In general, they either challenged the type of evidence used by marine scientists to infer nitrogen limitation of primary production (Smith 1984; Hecky 1998; Hecky and Kilham 1988), or doubted that there was a reason to believe that eutrophication in coastal marine systems is different in any fundamental way from eutrophication in lakes. Both arguments, however, have now largely been refuted.

Most early studies of nutrient limitation and eutrophication in coastal waters relied on fairly short-term and small-scale enrichment experiments to infer limitation by nitrogen (Ryther and Dunstan 1971; Vince and Valiela 1973) or made inferences from pure-culture studies (Smayda 1974). When applied to the problem of lake eutrophication in the 1960s and early 1970s, these approaches often led to the erroneous conclusion that nitrogen or carbon, rather than phosphorus, was limiting in lakes. Later, whole-lake experiments clearly showed that phosphorus and not nitrogen or carbon was the key nutrient regulating eutrophication in lakes (Schindler 1977). Consequently, the scientific community that studied lake eutrophication and the lake water quality management community developed an appropriate skepticism for bioassay experiments (NRC 1993a). Other types of evidence also have been used to infer nitrogen limitation in coastal ecosystems, including relatively low ratios of dissolved inorganic nitrogen to phosphorus (Boynton et al. 1982). These approaches also can be criticized, since concentrations of dissolved inorganic nutrients do not always accurately reflect their biological availabilities (Howarth 1988; Howarth and Marino 1990).

Ecosystem-scale experiments were the galvanizing force that led to the clear conclusion that eutrophication in lakes is best managed through controlling phosphorus inputs (Schindler 1977; NRC 1993a). A decade ago, there were no comparable experiments testing the relative importance of nitrogen and phosphorus as regulators of eutrophication in coastal marine ecosystems (Howarth 1988). However, since 1990 the results of three larger-scale enrichment experiments in estuaries have been published, all clearly showing nitrogen limitation in the systems (Howarth and Marino 1998).

One of these experiments, a mesocosm experiment conducted at the Marine Ecosystem Research Laboratory (MERL) on the shores of Narragansett Bay in Rhode Island, was specifically designed to see if coastal systems respond to nutrient additions in the same manner as lakes (Oviatt et al. 1995). Large mesocosms containing water and sediment from Narragansett Bay were maintained for a period of four months; many previous studies in MERL mesocosms has demonstrated that these systems accurately mimic much of the ecological functioning of Narragansett Bay. In this experiment, mesocosms received no nutrient enrichment or were enriched with nitrogen, phosphorus, or both. The level of nitrogen and phosphorus enrichment paralleled those used in a whole-lake eutrophication experiment at the Experimental Lakes Area in Canada, an experiment where phosphorus inputs clearly led to eutrophication and where nitrogen had no effect on rates of primary production (Schindler 1977). In sharp contrast, the addition of nitrogen (either alone or with phosphorus) but not of phosphorus alone to MERL coastal mesocosms caused large increases in both rates of primary production (Figure 3-1A) and the abundance of phytoplankton (Figure 3-1B; Oviatt et al. 1995).

Another whole-ecosystem estuarine study followed the impacts of experimental alteration of nutrient releases from a sewage treatment plant into Himmerfjarden, an estuary south of Stockholm, Sweden, on the Baltic Sea. The response of the estuary to nutrient inputs from sewage (the

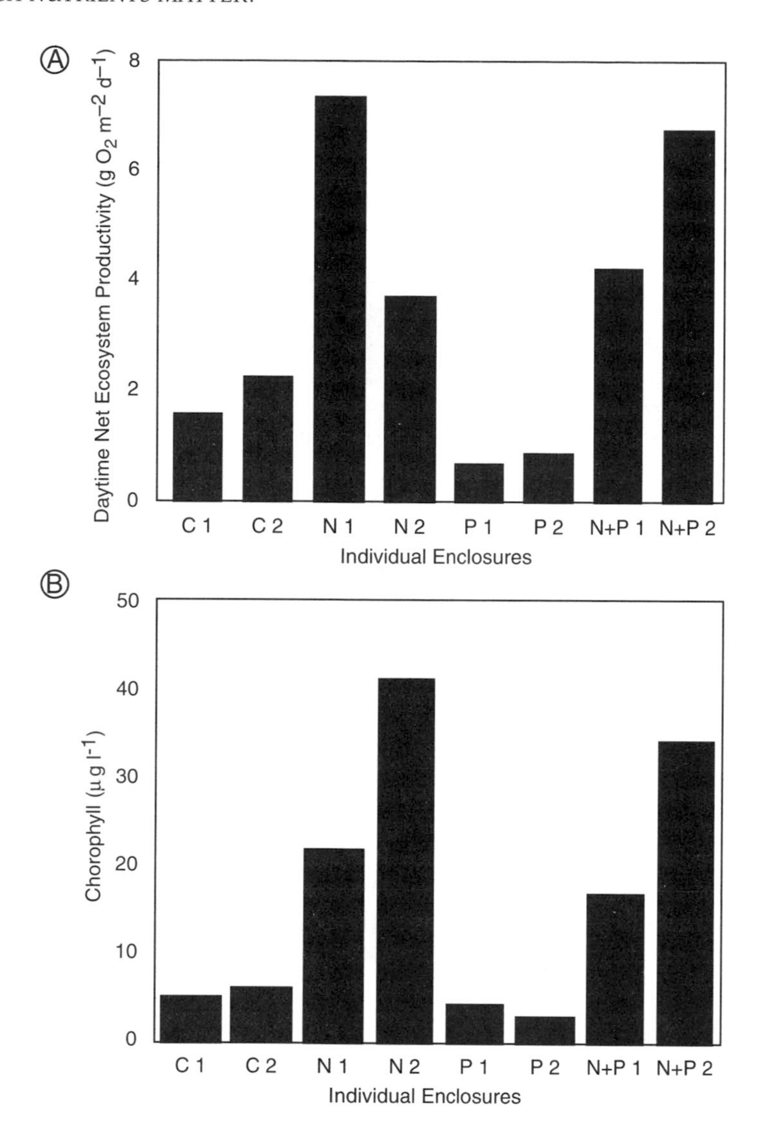

FIGURE 3-1 Response of estuarine mesocosms in Narragansett, Rhode Island, to experimental nutrient additions, clearing showing nutrient limitation by nitrogen but not phosphorus. (A) Mean daytime rate of net ecosystem production. (B) Mean chlorophyll concentration (an indicator of phytoplankton biomass). On the x-axis, C are control systems that received no nutrient additions, N are systems that were enriched with nitrogen, P are systems that were enriched with phosphorus, and N and P are systems that received both nitrogen and phosphorus. The numbers show replicate systems (two replicates for each treatment). The experiment ran for two months in the summer (modified from Oviatt et al. 1995).

primary input to this system) was studied from 1976 to 1993 (Elmgren and Larsson 1997). For the first 12 years, nitrogen loads gradually increased while phosphorus loads gradually decreased. For a one-year period beginning in the fall of 1983, phosphorus additions were greatly increased (by stopping the phosphorus removal during sewage treatment). Subsequently, phosphorus removal was again used, but nitrogen inputs were increased by 40 percent in 1985 as a result of an increase in population served by this particular sewage treatment plant. Finally, nitrogen removal technology was gradually introduced to the sewage treatment plant between 1988 and 1993, gradually reducing the nitrogen load to the value originally seen in 1976.

Throughout the 17 years of observation, the concentration of total nitrogen tended to reflect the nitrogen input from the sewage treatment plant (Elmgren and Larsson 1997), and both abundances of phytoplankton (Figure 3-2A) and water clarity (Figure 3-2B) were clearly related to the total nitrogen concentration. Total phosphorus concentrations varied independently of total nitrogen over time in Himmerfjarden, and total phosphorus was a poor predictor of phytoplankton abundances. This is strong evidence that nitrogen was the element most controlling eutrophication in this estuary. During the year that phosphorus loadings were experimentally increased, there was no effect on primary production; however, there was an unusually large bloom the following spring, probably due both to some residual high levels of phosphorus and to an unusually high input of nitrogen from spring floods (Elmgren and Larsson 1997).

A third whole-ecosystem study explored long-term changes in Laholm Bay, an estuary on the southwestern coast of Sweden (Figure 3-3). Early signs of eutrophication appeared there in the 1950s and 1960s and steadily increased over time (Rosenberg et al. 1990). The earliest reported signs of eutrophication were changes in the community composition of macroalgae species, and over time filamentous algae typical of eutrophic conditions have become more prevalent. Harmful algal blooms have become much more common, particularly in the 1980s (Rosenberg et al. 1988, 1990). During the early stages of eutrophication in Laholm Bay, inputs of both phosphorus and nitrogen to the estuary were increasing. However, from the late 1960s through the 1980s, phosphorus inputs decreased by a factor of almost two, while nitrogen inputs continued to increase (more than doubling) (Rosenberg et al. 1990). During this same period, plankton blooms continued, clearly indicating that nitrogen controlled the Laholm Bay eutrophication.

These three ecosystem-scale experiments show only that nitrogen controlled eutrophication in Narragansett Bay, Himmerfjarden, and Laholm Bay. Importantly, however, the finding in each of these three systems is

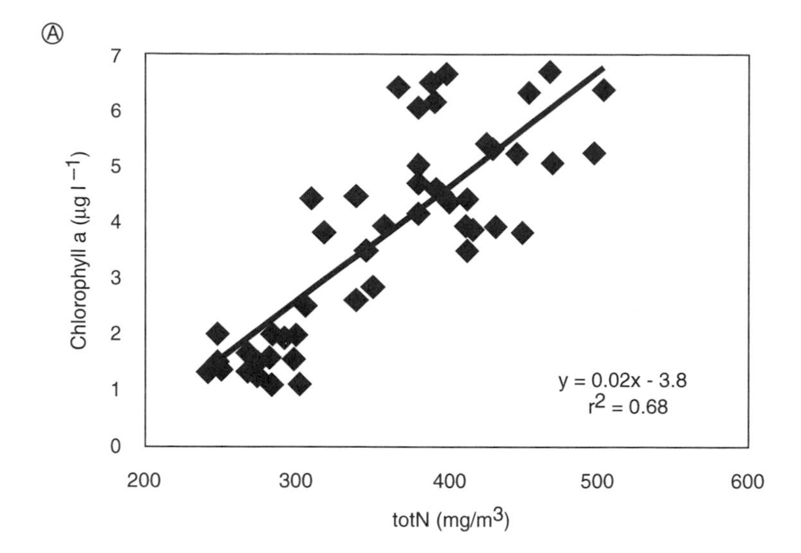

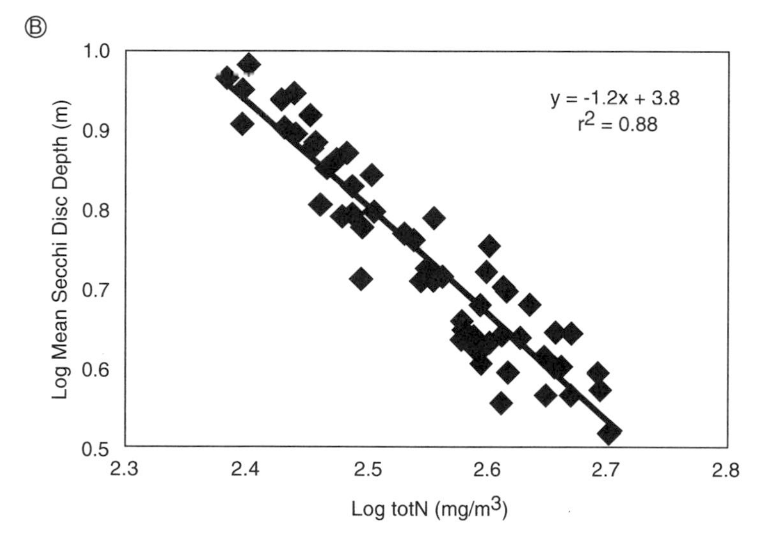

FIGURE 3-2 Long-term observations in the Himmerfjarden estuary south of Stockholm, Sweden, over a period of years in which nitrogen and phosphorus additions were experimentally altered through changes in sewage treatment. (A) The relationship between the mean concentration of total nitrogen and the chlorophyll a in the surface water layer (modified from Elmgren and Larsson 1997). (B) The relationship between the mean concentration of total nitrogen in the surface water layer and the water clarity (secchi disc depth) (modified from Elmgren and Larsson 1997). Note that the major period of eutrophication in this estuary coincided with a period in which nitrogen inputs were increasing yet phosphorus inputs were decreasing.

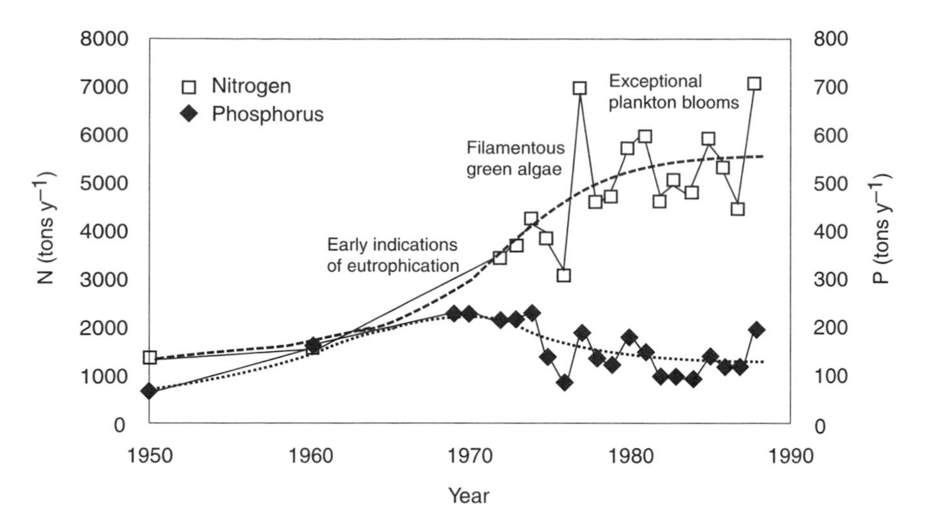

FIGURE 3-3 Transport of nutrients to Laholm Bay, Sweden. Periods of significant changes in the marine biota are also indicated (modified from Rosenberg et al. 1990).

consistent with conclusions drawn from short-term bioassay studies and from ratios of dissolved inorganic nitrogen:phosphorus in these ecosystems (Granéli et al. 1990; Oviatt et al. 1995; Elmgren and Larsson 1997; Howarth and Marino 1998). These three ecosystem experiments therefore add credence to the application of bioassay data and inorganic nutrient data in assessing whether nitrogen or phosphorus is more limiting in estuaries. The large preponderance of bioassay data in estuaries and coastal marine systems indicates nitrogen limitation (Howarth 1988), as does the generally low inorganic nitrogen:phosphorus ratio found in most estuaries at the time of peak primary production (Figure 3-4; Boynton et al. 1982). Thus, taken together the three whole-ecosystem scale and bioassay data from many sites lead to the conclusion that nitrogen availability is the primary regulator of eutrophication in most coastal systems.

MECHANISMS THAT LEAD TO NITROGEN LIMITATION IN COASTAL MARINE ECOSYSTEMS

What ecological or biogeochemical mechanisms can lead to nitrogen control of eutrophication in most coastal marine systems and to phosphorus control in so many freshwater lakes? This question was reviewed by Howarth (1988) and Vitousek and Howarth (1991). Here we summarize

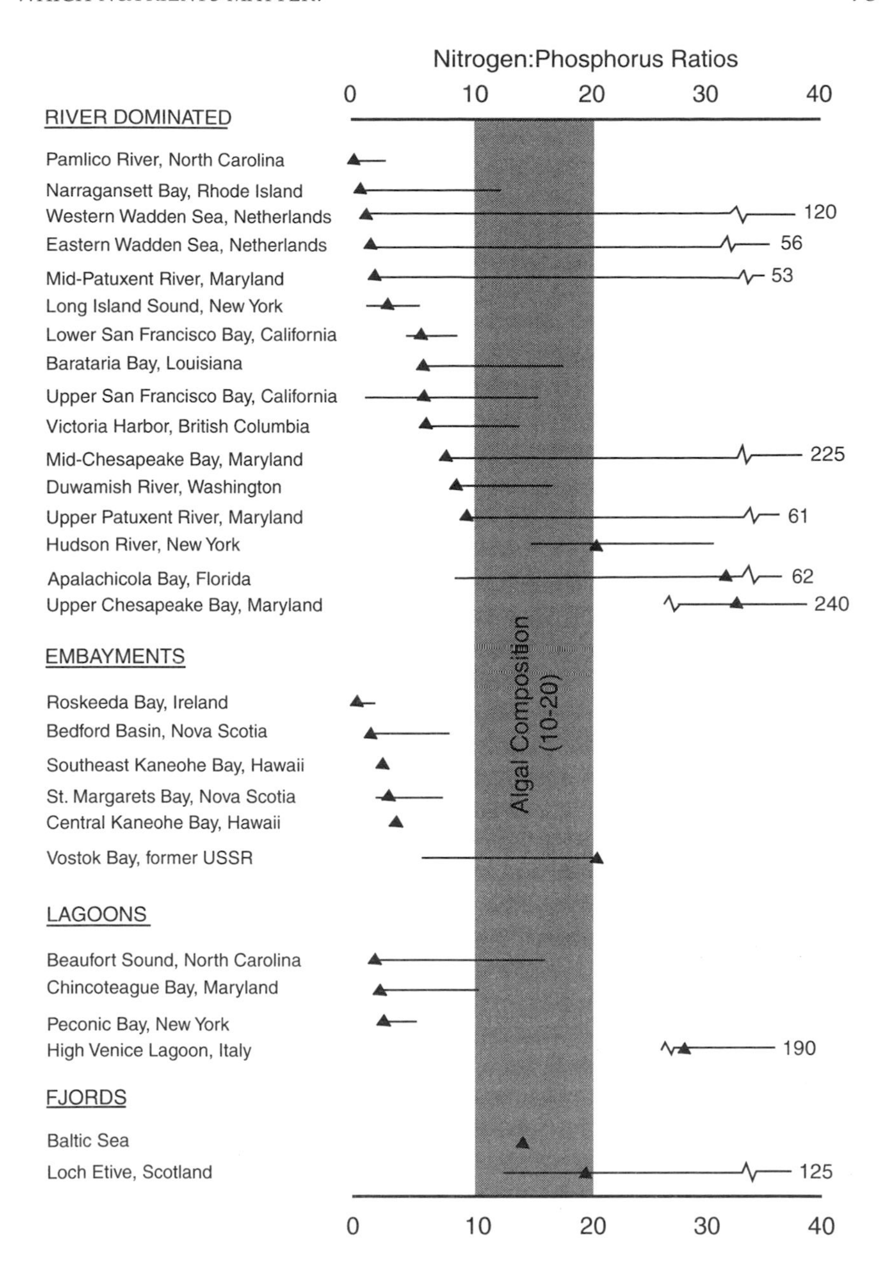

FIGURE 3-4 Summary of nitrogen:phosphorus ratios in 28 sample estuarine ecosystems. Horizontal bars indicate the annual ranges in nitrogen:phosphorus ratios; solid triangles represent the ratio at the time of maximum productivity. Vertical bands represent the typical range of algal composition ratios (modified from Boynton et al. 1982).

and update those reviews. Whether primary production by phytoplankton is nitrogen or phosphorus limited is a function of the relative availabilities of nitrogen and phosphorus in the water. Phytoplankton require approximately 16 moles of nitrogen for every mole of phosphorus they assimilate (the Redfield ratio of nitrogen:phosphorus=16:1) (Redfield 1958). If the ratio of available nitrogen to available phosphorus is less than 16:1, primary production will tend to be nitrogen limited. If the ratio is higher, production will tend to be phosphorus limited.

The relative availabilities of nitrogen and phosphorus to the phytoplankton is determined by three factors (Figure 3-5):

- the ratio of nitrogen to phosphorus in inputs to the ecosystem;
- preferential storage, recycling, or loss of one of these nutrients in the ecosystem; and
- the amount of biological nitrogen fixation.

For each of these factors, there are reasons why nitrogen limitation tends to be more prevalent in coastal marine ecosystems than in lakes. For instance, lakes receive nutrient inputs from upstream terrestrial ecosystems and from the atmosphere, while estuaries and coastal marine systems receive nutrients from these sources as well as from neighboring oceanic water masses. For estuaries such as those along the northeastern coast of the United States, the ocean-water inputs of nutrients tend to have a nitrogen:phosphorus ratio well below the Redfield ratio due to denitrification on the continental shelves (Nixon et al. 1995, 1996). Thus, given similar nutrient inputs from land, estuaries are more likely to be more nitrogen limited than are lakes.

Another factor to consider is that the ratio of nitrogen to phosphorus in nutrient inputs from land will tend to reflect the extent of human activity in the landscape. As the landscape changes from one dominated by forests to one dominated by agriculture and then industry, total nutrient fluxes from land increase for both nitrogen and phosphorus, but the change is often greater for phosphorus and so the nitrogen:phosphorus ratio tends to fall (Billen et al. 1991; Howarth et al. 1996). This, too, influences why nitrogen limitation is of primary importance in estuaries (NRC 1993a). The occurrence of phosphorus limitation in the Apalachicola estuary, for instance, may be the result of the relative low level of human activity in most of the watershed. This suggests that there is a tendency for estuaries to become more nitrogen limited as they become more affected by humans and as nutrient inputs increase overall (Howarth et al. 1995).

The biogeochemical processes active in an aquatic ecosystem affect the availability of nutrients to phytoplankton in that particular system.

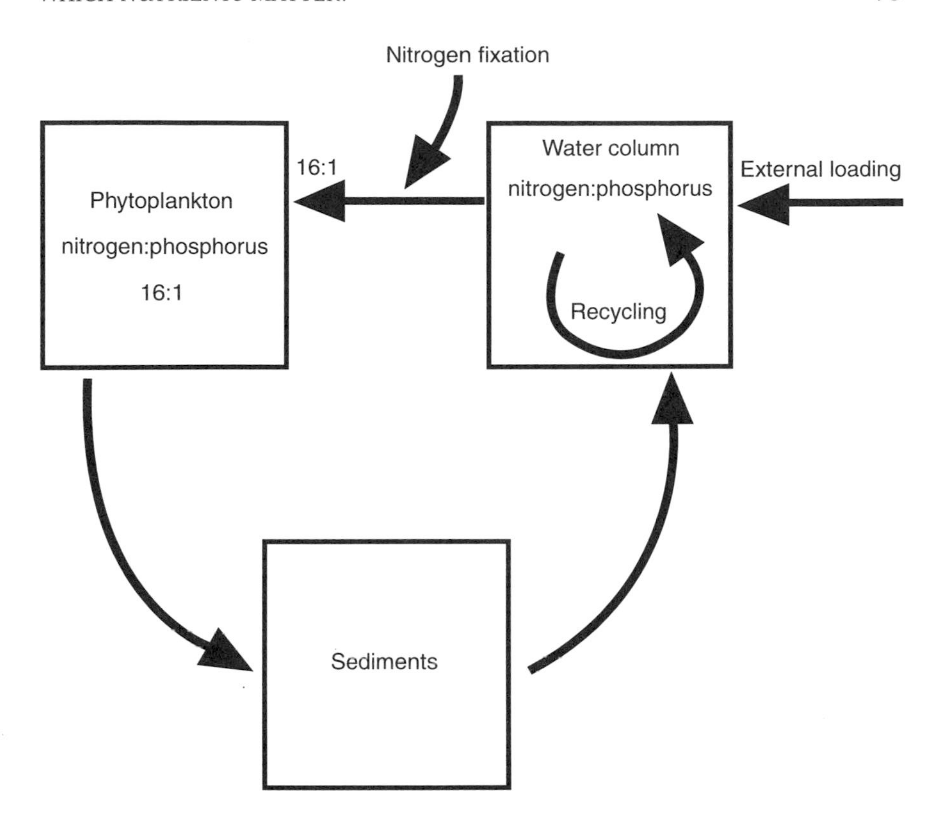

FIGURE 3-5 Factors that determine whether nitrogen or phosphorus is more limiting in aquatic ecosystems, where one of these macronutrients is limiting to net primary production. Phytoplankton use nitrogen and phosphorus in the approximate molar ratio of 16:1. The ratio of available nitrogen in the water column is affected by: 1) the ratio of nitrogen:phosphorus in external inputs to the ecosystem; 2) the relative rates of recycling of nitrogen and phosphorus in the water column, with organic phosphorus usually cycling faster than organic nitrogen; 3) differential sedimentation of nitrogen in more oligotrophic systems; 4) preferential return of nitrogen or phosphorus from sediments to the water column due to processes such as denitrification and phosphorus adsorption and precipitation; and 5) nitrogen fixation (modified from Howarth 1988; Howarth et al. 1995).

Of these processes, the sediment processes of denitrification and phosphate adsorption are the dominant forces that affect the relative importance of nitrogen or phosphorus limitation on an annual or greater time scale. Other processes, such as preferential storage of phosphorus in zooplankton (Sterner et al. 1992), act only over a relatively short period.

Denitrification is often a major sink for nitrogen in aquatic ecosystems, and it tends to drive systems toward nitrogen limitation unless counterbalanced by other processes such as phosphorus adsorption and storage (Howarth 1988; Seitzinger 1988; Nixon et al. 1996). The overall magnitude of denitrification tends to be greater in estuaries than in freshwater ecosystems, but this may simply be a result of greater nitrogen fluxes through estuaries (Seitzinger 1988). When expressed as a percentage of the nitrogen input to the system lost through denitrification, there appears to be relatively little difference between estuaries and freshwater ecosystems (Nixon et al. 1996). That is, available evidence indicates that denitrification tends to drive both coastal marine and freshwater ecosystems toward nitrogen limitation, with no greater tendency in estuaries. In fact, the tendency toward nitrogen limitation—based on this process alone—might be greater in lakes, since lakes generally have a longer water residence time, and the percent nitrogen loss through denitrification is greater in ecosystems having a longer water residence time (Howarth et al. 1996; Nixon et al. 1996).

A sediment process counteracting the influence of denitrification on nutrient limitation is phosphorus adsorption. Sediments potentially can absorb and store large quantities of phosphorus, making the phosphorus unavailable to phytoplankton and tending to drive the system toward phosphorus limitation. This process is variable among ecosystems (Howarth et al. 1995). At one extreme, little or no phosphorus is adsorbed by the sediments of Narragansett Bay, and virtually all of the phosphate produced during decomposition in the sediments is released back to the water column (Nixon et al. 1980). This, in combination with nitrogen lost through denitrification, is a major reason that Narragansett Bay is nitrogen limited (Figure 3-6; Nixon et al. 1980; Howarth 1988). Caraco et al. (1989, 1990) suggested that lake sediments have a greater tendency to adsorb and store phosphorus than do estuarine sediments; if this were true, this differential process would make phosphorus limitation more likely in lakes than in estuaries. However, the generality of a difference in phosphorus retention between lakes and coastal marine sediments has yet to be established. It is also important to note that eutrophication may lead to less denitrification since the coupled processes of nitrification and denitrification are disrupted in anoxic waters.

Among estuaries, the ability of sediments to adsorb phosphorus is variable (Howarth et al. 1995), with little or no adsorption occurring in systems such as Narragansett Bay and almost complete adsorption of inorganic phosphate in some other systems, such as those along the coast of the Netherlands (van Raaphost et al. 1988). Chesapeake Bay sediments show an intermediate behavior, with some of the inorganic phosphorus released during sediment decomposition being adsorbed and some

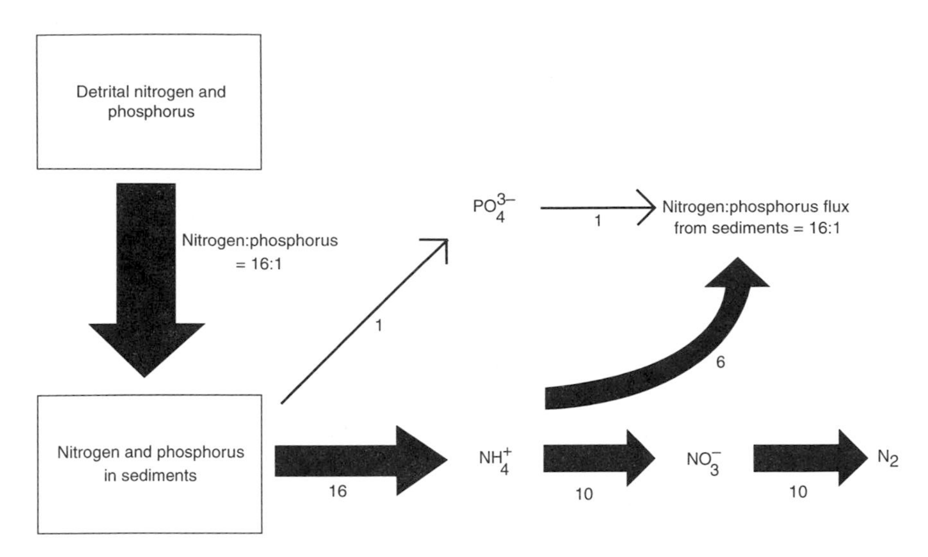

FIGURE 3-6 Schematic diagram showing nutrient regeneration from the sediments of Narragansett Bay, Rhode Island. Nitrogen and phosphorus enter the sediments as particulate matter in approximately the Redfield ratio of 16:1. Phosphorus mineralized during decomposition is released back to the water column, whereas much of the mineralized nitrogen is lost through the combined processes of nitrification and denitrification (modified from Nixon et al. 1980; Howarth 1988).

released to the overlying water (Boynton and Kemp 1985). The reasons for this difference in behavior among systems are not well understood. However, there is some indication that the ability of coastal marine sediments—both in tropical and in temperate systems—to adsorb and store phosphorus decreases as an ecosystem becomes more eutrophic, at least until they become extremely hypereutrophic, as in the case of some of the estuaries in the Netherlands (Howarth et al. 1995). For temperate systems, the lessened ability to sorb phosphate as a system becomes more eutrophic results from decreased amounts of oxidized iron and more iron sulfides in the sediments; for tropical carbonate systems, the rate of sorption of phosphate decreases as the phosphorus content of the sediment increases. These changes result in an increase in phosphorus availability in eutrophic systems, intensifying nitrogen limitation and encouraging the growth of algae and other organisms (including some heterotrophic organisms, such as the heterotrophic life stages of *Pfiesteria*) with high phosphorus requirements.

The process of nitrogen fixation clearly has different effects on nutrient limitation in freshwater lakes and coastal marine ecosystems. If a lake of moderate productivity is driven toward nitrogen limitation, blooms of heterocystic, nitrogen-fixing cyanobacteria ("blue-green algae") occur, and these tend to fix enough nitrogen to alleviate the nitrogen shortage (Schindler 1977; Howarth et al. 1988a). Primary productivity of the lake remains limited by phosphorus (Schindler 1977). This was demonstrated experimentally in whole-lake experiments at the Experimental Lakes Area in northwestern Ontario, where a lake was fertilized with a constant amount of phosphorus over several years. For the first several years, the lake also received relatively high levels of nitrogen fertilizer, so that the ratio of nitrogen:phosphorus of the fertilization treatment was above the Redfield ratio of 16:1 (by moles). Under these conditions, no nitrogen fixation occurred in the lake. The regime was then altered so that the lake received the same amount of phosphorus, but the nitrogen input was decreased so that the nitrogen:phosphorus ratio of the inputs was below the Redfield ratio. Nitrogen-fixing organisms quickly appeared and made up the nitrogen deficit (Schindler 1977; Flett et al. 1980). This response is a major reason that nitrogen limitation is so prevalent in mesotrophic and eutrophic lakes (Schindler 1977; Howarth 1988).

Estuaries and eutrophic coastal waters provide a striking contrast to this behavior. With only a few exceptions anywhere in the world, nitrogen fixation by planktonic, heterocystic cyanobacteria is immeasurably low in mesotrophic and eutrophic coastal marine systems, even when they are quite nitrogen limited (Horne 1977; Doremus 1982; Fogg 1987; Howarth et al. 1988b; Paerl 1990; Howarth and Marino 1990, 1998). This major difference in the behavior between lakes and estuaries allows nitrogen limitation to continue in estuaries (Howarth 1988; Vitousek and Howarth 1991).

Much research has been directed at the question of why nitrogen fixation by planktonic organisms differs between lakes and coastal marine ecosystems, with much of this focused on single-factor controls, such as short residence times, turbulence, limitation by iron, limitation by molybdenum, or limitation by phosphorus (Howarth and Cole 1985; Paerl 1985; Howarth et al. 1988a; Howarth et al. 1999; Paerl and Zehr 2000). A growing consensus has developed, however, that nitrogen fixation in marine systems—estuaries, coastal seas, as well as oceanic waters—probably is regulated by complex interactions of chemical, biotic, and physical factors (Howarth et al. 1999; Paerl and Zehr 2000). With regard to estuaries and coastal seas, recent evidence indicates that a combination of slow growth rates caused by low availabilities of trace metals required for nitrogen fixation (iron and/or molybdenum) and grazing by zooplankton and benthic animals combine to exclude nitrogen-fixing heterocystic cyanobacteria (Figure 3-7; Howarth et al. 1999).

Nitrogen fixation by planktonic cyanobacteria does occur in a few coastal marine ecosystems, notably the Baltic Sea and the Peel-Harvey inlet in Australia (Howarth and Marino 1998). In the Baltic, rates of nitrogen fixation are not sufficient to fully alleviate nitrogen limitation (Granéli et al. 1990; Elmgren and Larsson 1997). The reason that nitrogen fixation occurs in the Baltic but not in most other estuaries and seas remains disputed (Hellström 1998; Howarth and Marino 1998), but a model based on the interplay of trace metal availability and grazing as controls on nitrogen fixation correctly predicts that nitrogen fixation would occur in the Baltic but not in most estuaries (Howarth et al. 1999); this model result is driven by the greater availability of trace metals at the low salinity of the Baltic compared to most estuaries. The reason why nitrogen fixation occurs in the Peel-Harvey (Lindahl and Wallstrom 1985; Huber 1986), and also a similar estuary in Tasmania (Jones et al. 1994), remains unknown. One hypothesis is that this is a result of extreme eutrophication, which has driven these systems anoxic, increasing trace metal availability and lowering grazing by animals (Howarth and Marino 1998; Howarth et al. 1999). Nitrogen fixation in both estuaries has only begun in the recent past, and only as they became extremely eutrophic.

Oceanographic scientists have long believed that phosphorus is the long-term regulator of primary production in the oceans as a whole (Redfield 1958; Broecker 1974; Howarth et al. 1995). In this view, nitrogen limitation can occur in oceanic surface waters, but this is a transient effect that is made up for by nitrogen fixation over geological time scales. Recently, Tyrrell (1999) formalized this concept with a simple, 6-variable model of nutrient cycling and primary production in the world's oceans. The basic concept is appealing, if as yet unproven, in that it explains the strong correlation of dissolved nitrogen and phosphorus compounds over depth profiles in the oceans.

Based on this conceptual view of the interaction of nitrogen and phosphorus over geological time scales, Tyrrell (1999) concluded that coastal eutrophication is largely a phosphorus problem, and that "removal of nitrates in the river supply should lead to increased nitrogen fixation, no significant effects on final nitrate concentrations, and no significant effect on eutrophication." The Committee on Causes and Management of Coastal Eutrophication disagrees strongly. While nitrogen fixation in oceanic waters may alleviate nitrogen deficits over tens of thousands of years, nitrogen fixation simply does not occur in most estuaries and coastal seas and does not alleviate nitrogen shortages. Therefore, decreasing nitrogen inputs to estuaries will not in general lead to increased nitrogen fixation. The Tyrrell model operates on geological time scales for oceans as the

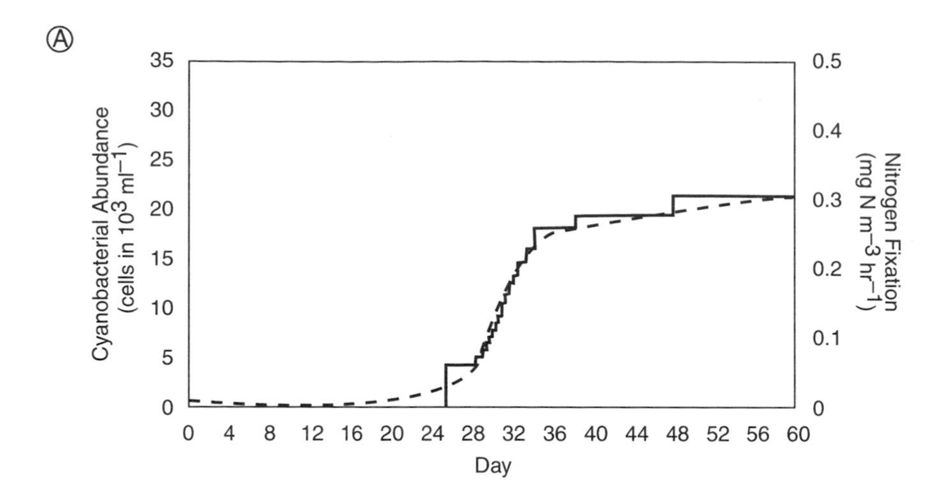

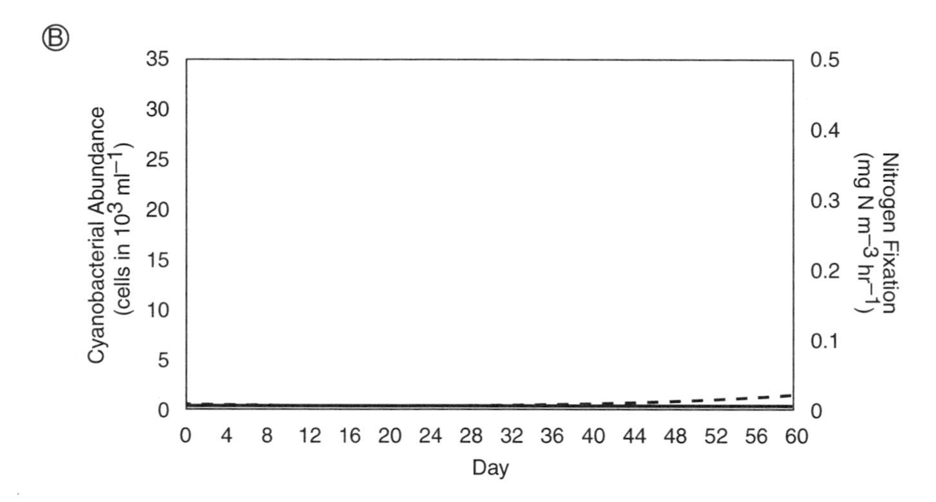

FIGURE 3-7 Results from a simple simulation model designed to show the importance of top-down and bottom-up factors as regulators of nitrogen fixation in lakes and estuaries. (A) Shown are the number of cyanobacterial cells (dashed line) and rates of nitrogen fixation (solid line) over time under typical conditions for a freshwater lake and (B) under identical conditions except for a lower availability of molybdenum as one would find in an estuary. Molybdenum is required for nitrogen fixation, and its lower abundance leads to slower growth rates by cyanobacteria. In combination with typical rates of grazing by zooplankton, this slower rate of growth in estuaries excludes nitrogen-fixing organisms that flourish in lakes (modified from Howarth et al. 1995).

whole, a time scale not applicable to estuaries, coastal seas, and continental shelves where water residence time varies from less than one day to at most a few years. Nitrogen fixation does occur in the Baltic Sea, yet even there the water residence time is on the scale of a few decades, thousands-fold shorter than the time scale of response by nitrogen fixation in Tyrrell's model. While debate continues as to whether or not nitrogen fixation completely alleviates nitrogen shortages in the Baltic, much evidence shows that it does not and that much of the Baltic Sea remains nitrogen limited (Granéli et al. 1990; Elmgren and Larsson 1997; Hellström 1998; Howarth and Marino 1998; Savchuck and Wulff 1999).

THE IMPORTANCE OF SILICA AND IRON IN COASTAL SYSTEMS

Although nitrogen is the element primarily controlling eutrophication in estuaries and coastal seas, and phosphorus is the element primarily controlling eutrophication in lakes, other elements can have a major influence on the community structure of aquatic ecosystems and can influence the nature of the response to nutrients. A key element in this regard is silica, an element required by diatoms. The availability of silica in a waterbody has little or no influence on the overall rate of primary production, but when silica is abundant, diatoms are one of the major components of the phytoplankton. When silica is in low supply, other classes of algae dominate the phytoplankton composition.

Inputs of biologically available silica to aquatic systems come largely from weathering of soils and sediments. The major human influence on silica delivery to coastal marine systems is to decrease it, as eutrophication in upstream ecosystems tends to trap silica before it reaches the coast (Schelske 1988; Conley et al. 1993; Rabalais et al. 1996; Turner et al. 1998). Thus, the concentration of silicate in Mississippi River water entering the Gulf of Mexico decreased by 50 percent from the 1950s to the 1980s (Figure 3-8), a time during which nitrogen and phosphorus fluxes and concentrations increased (Goolsby et al. 1999). Eutrophication in a system can further decrease silica availability as it is incorporated into diatoms and stored in bottom sediments, as demonstrated in the Baltic Sea (Wulff et al. 1990). As discussed in Chapter 4, a decrease in silica availability, particularly if accompanied by increases in nitrogen, may encourage the formation of some blooms of harmful algae as competition with diatoms is decreased (NRC 1993a; Rabalais et al. 1996). As noted by Smayda (1989), for all cases where long-term data sets are available on silica availability in coastal waters, a decrease in silica availability relative to nitrogen or phosphorus has been correlated with an increase in harmful algal blooms.

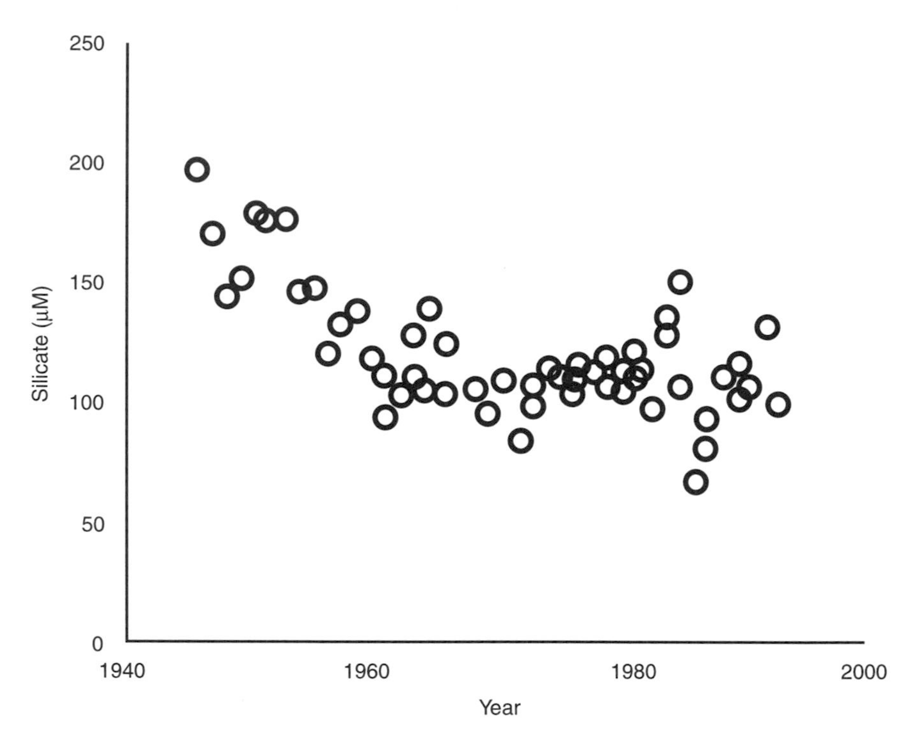

FIGURE 3-8 The concentration of dissolved silicate in waters of the Mississippi River near the Gulf of Mexico from the 1940s into the 1990s. Note the decrease during the 1950s and 1960s, probably in response to eutrophication in upstream freshwater ecosystems trapping silicate (modified from Rabalais et al. 1996).

Decreasing silica availability and the consequent lower abundances of diatoms also lowers organic matter sedimentation and thereby have a partially mitigating influence on low-oxygen events associated with eutrophication. In many coastal systems there may, however, still be sufficient silica to fuel diatom blooms during the critical spring bloom period when the majority of sedimentation often occurs (Conley et al. 1993; Turner et al. 1998). Further, eutrophication can lead to other complex shifts in trophic structure that might either increase or decrease the sedimentation of organic carbon (Turner et al. 1998).

Iron is another element that can affect the community composition of phytoplankton. As discussed in Chapter 4, greater availability of iron may encourage some harmful algal blooms. In some oceanic waters away from shore, iron availability appears to be a major control on rates of primary production (Martin et al. 1994; Coale et al. 1996). However, there is no evidence that iron limits primary production in estuaries and coastal seas (although it may partially limit nitrogen-fixing cyanobacteria in estuaries) (Howarth and Marino 1998; Howarth et al. 1999). Although iron concentrations are lower in estuaries than in freshwater lakes, concentrations in estuaries and coastal seas are far greater than in oceanic waters (Marino et al. 1990; Schlesinger 1997). The solubility of iron in seawater and estuarine waters is low, and complexation with organic matter is critical to keeping iron in solution and maintaining its biological availability. Eutrophication tends to increase the amount of dissolved organic matter in water, and therefore may act to increase iron availability. Furthermore, hypoxia and anoxia accompanying eutrophication may enhance iron availability in the water column due to iron release from sediments as the reducing intensity increases (NRC 1993a).

4

What Are the Effects of Nutrient Over-Enrichment?

KEY POINTS IN CHAPTER 4

This chapter explores the impacts of nutrient over-enrichment and finds:

- The productivity of many coastal marine systems is limited by nutrient availability, and the input of additional nutrients to these systems increased primary productivity.
- In moderation in some systems, nutrient enrichment can have beneficial impacts such as increasing fish production; however, more generally the consequences of nutrient enrichment for coastal marine ecosystems are detrimental. Many of these detrimental consequences are associated with eutrophication.
- The increased productivity from eutrophication increases oxygen consumption in the system and can lead to low-oxygen (hypoxic) or oxygen-free (anoxic) water bodies. This can lead to fish kills as well as more subtle changes in ecological structure and functioning, such as lowered biotic diversity and lowered recruitment of fish populations.
- Eutrophication can also have deleterious consequences on estuaries even when low-oxygen events do not occur. These changes include loss of biotic diversity, and changes in the ecological structure of both planktonic and benthic communities, some of which may be deleterious to fisheries. Seagrass beds and coral reefs are particularly vulnerable to damage from eutrophication and nutrient over-enrichment.
- Harmful algal blooms (HABs) harm fish, shellfish, and marine mammals and pose a direct public health threat to humans. The factors that cause HABs remain poorly known, and some events are entirely natural. However, nutrient over-enrichment of coastal waters leads to blooms of some organisms that are both longer in duration and of more frequent occurrence.
- Although difficult to quantify, the social and economic consequences of nutrient over-enrichment include aesthetic, health, and livelihood impacts.

Nutrient enrichment can have a range of effects on coastal systems. On occasion, in some ecosystems moderate nutrient enrichment can be beneficial because increased primary production can lead to increased fish populations and harvest (Jørgensen and Richardson 1996; Nixon 1998). Far more often, when nutrient enrichment is sufficiently great, the effects are detrimental. In some cases, even small increases in nutrient inputs can be quite damaging to certain types of ecosystems, such as those particularly susceptible to changed conditions (e.g., coral reefs).

Direct and indirect ecological impacts of nutrient enrichment include increased primary productivity, increased phytoplankton biomass, reduction in water clarity, increased incidences of low oxygen events (hypoxia and anoxia), and changes in the trophic structure, trophic interactions, and trophodynamics of phytoplankton, zooplankton, and benthic communities. Harmful algal blooms may become more frequent and extensive. Coral reefs and submerged macrophytic vegetation, such as seagrass beds and kelp beds, may be degraded or destroyed. Fish kills may occur, and more importantly, subtle changes in ecological structure may lead to lowered fishery production. Generally, nutrient over-enrichment leads to ecological changes that decrease the biotic diversity of the ecosystem.

The ecological effects of nutrient over-enrichment can have societal impacts as well, although the economic consequences are generally difficult to quantify. These include aesthetic impacts, such as loss of visually exciting coral reefs and seagrass beds, as well as production of noxious odors and unappealing piles of algal detritus on beaches. Fishery resources can be damaged or lost. Human health is threatened by accumulation of toxins in shellfish. Property can be devalued. This chapter summarizes the societal impacts of nutrient enrichment.

ECOLOGICAL EFFECTS

Increased Primary Productivity

As discussed earlier in this report, eutrophication is a process of increasing organic enrichment of an ecosystem where the increased rate of supply of organic matter causes changes to that system (Nixon 1995). This increased rate of supply is driven by primary productivity. Primary productivity is affected by a variety of factors, including light availability, nutrients, and grazing mortality. The interplay of these factors determines how a coastal marine ecosystem will respond to nutrient additions. (These and other factors that determine an estuary's sensitivity to eutrophication are discussed in detail in Chapter 6.) For many systems, primary productivity is limited largely by nutrient availability, and in these systems increasing the nutrient input increases the primary productivity rate and often the phytoplankton biomass. As explained in Chapter 3, in the majority of coastal systems—at least in the temperate zone—nitrogen is the element most limiting of primary productivity; consequently, rates of primary production and standing stock of phytoplankton biomass are often directly related to nitrogen inputs.

As noted in a previous report from the National Research Council (NRC 1993a), planners and managers now are often at a disadvantage because "no guidelines exist by which to determine whether coastal marine ecosystems are in fact eutrophic." That report goes on to recommend that coastal eutrophication be judged by some measure of the relationship between phytoplankton biomass (as represented by chlorophyll concentrations) and trophic status, the same approach that is generally used by limnologists for freshwater lakes. Adoption of such an approach would lead to the conclusion that "few estuaries are oligotrophic, many are mesotrophic, and many are extremely eutrophic" (NRC 1993a).

Other authors have suggested similar approaches. For instance, Jaworski (1981) has suggested a lake-based framework of nutrient-loading guidelines that, if met, would tend to keep most estuaries from becoming eutrophic. However, demonstrable harm from human-increased nutrient loading to estuaries has occurred in some systems even when the loadings were low enough not to be called eutrophic by these standards (NRC 1993a).

Nixon (1995) suggested another set of guidelines—these based on measured rates of primary production—for determining whether an estuary is eutrophic. In this classification scheme, estuaries with productivity between 300 and 500 g C m^{-2} yr^{-1} would be considered eutrophic, while those with productivities greater than 500 g C m^{-2} yr^{-1} would be considered hypereutrophic. These guidelines, too, lead to the conclusion that many estuaries are eutrophic or even hypereutrophic.

Increased Oxygen Demand and Hypoxia

Eutrophication is accompanied by an increased demand for oxygen. Some of this increased oxygen demand is due to the greater respiration of the increased biomass of plants and animals that are supported in the nutrient-loaded ecosystem. Much of it is often due to respiration of bacteria (in both the water column and sediments) that consume the organic matter produced by the greater plant production. If the loss of oxygen caused by increased respiration is not offset by the direct introduction of additional oxygen by photosynthesis or mixing processes, then hypoxia or anoxia occurs. Biologists generally refer to the situation where some oxygen is present but where dissolved oxygen levels are less than or equal to 2.0 milligrams per liter (mg l^{-1}) as hypoxia. Anoxia is the complete absence of oxygen.

Hypoxia and anoxia are more likely to occur in summer because warming of the water column can lead to stratification and the formation of a barrier that prevents the introduction and mixing of oxygen from surface waters. Also, the solubility of oxygen decreases and oxygen demand (respiration rate) generally increases as temperature increases.

As noted earlier, many studies of the biological impacts of reduced dissolved oxygen concentrations have used 2.0 mg l^{-1} as the cut off for designating conditions as hypoxic (e.g., Pihl et al. 1991, 1992; Schaffner et al. 1992), because below this threshold there are severe declines in the diversity and abundance of species in the systems. There is evidence, however, that 2.0 mg l^{-1} may not be a universal threshold. For example, the results of a study of biological resources in Long Island Sound, New York, revealed that 3.0 mg l^{-1} was the threshold level for finfish and squid (Howell and Simpson 1994). A study of the benthic community in Corpus Christi Bay, Texas, indicated that dissolved oxygen concentrations less than 3.0 mg l^{-1} should be the operational definition of hypoxia in that system (Ritter and Montagna 1999), and that a single value of dissolved oxygen as a water quality standard for estuarine waters may not be appropriate.

Many states have standards for dissolved oxygen levels in aquatic systems that are well above the limits used to define hypoxia. The Florida Department of Environmental Protection mandates that the average level of dissolved oxygen that must be maintained in marine waters designated for the commercial harvest of shellfish, recreation, and for the maintenance of healthy fish and wildlife is greater than 5.0 mg l^{-1} in a 24-hour period and never less than 4.0 mg l^{-1}. Although this level may seem conservative, in the absence of detailed information for a system it may be appropriate.

The occurrence of hypoxic and anoxic bottom waters, particularly in the coastal zone, has become a major concern in recent years because it appears that the frequency, duration, and spatial coverage of such conditions have been increasing, and this increase is thought to be related to human activities (Diaz and Rosenberg 1995). Zones of reduced oxygen can disrupt the migratory patterns of benthic and demersal species, lead to reduced growth and recruitment of species, and cause large kills of commercially important invertebrates and fish (NRC 1993a). Such conditions can also lead to an overall reduction in water quality, thereby affecting other coastal zone activities such as swimming and boating. Reports of a "dead zone," an extensive area of reduced oxygen levels covering an expanse originally of some 9,500 km^2 in the Gulf of Mexico (Rabalais et al. 1991), have focused attention on the problem of coastal zone hypoxia. By the summer of 1999, the hypoxic area in the Gulf of Mexico had grown to an area of 20,000 km^2 (Rabalais personal communication).

Researchers studying the Chesapeake Bay have said since the 1980s that the occurrence of hypoxic and anoxic bottom waters has increased in association with nutrient inputs (Taft et al. 1980; Officer et al. 1984). More recent studies examined pollen distribution, diatom diversity, and the concentration of organic carbon, nitrogen, sulfur, and acid-soluble iron in sediment cores from the mesohaline portion of the bay (Cooper and Brush 1991). The cores represented a 2,000-year history of the bay. Changes in the concentration of organic components and pollen abundance coincided with the new settlement by Europeans in the late 1700s. This period was marked by major land clearing in the watershed, which likely promoted increases rates of sedimentation, mineralization, and nitrification, and an increase in agricultural activity and the use of manures. Analysis of the sediment cores indicated a shift in the phytoplankton community from centric to pennate diatoms for this time period, and this was interpreted as evidence of increased nutrient input to the bay. This historical perspective indicates a role for nutrients in the occurrence of hypoxia in Chesapeake Bay. As discussed in Chapter 5, the input of nutrients to Chesapeake Bay has probably accelerated even more in the last several decades due to increased use of inorganic fertilizer and increased combustion of fossil fuels and the resulting atmospheric deposition of nitrogen.

The northern Adriatic Sea and northern Gulf of Mexico are two other coastal systems that have experienced increasing episodes of hypoxia (Justic et al. 1993; Turner and Rabalais 1994). Both systems are affected by river flow, the Po River in the case of the former and the Mississippi River in the latter. In both systems researchers have documented a seasonal increase in primary productivity in surface waters that was related to nutrients and river flow; this increase was followed by hypoxia in the bottom waters. The hypoxia onset, however, lagged peak river flow in the Gulf of Mexico and Adriatic Sea by two and four months, respectively. This difference in the lag period was ascribed to greater depth of the water column in the Adriatic Sea and differences in the downward flux of organic matter. Again, the evidence showed that the introduction of new nutrients in the river flow contributed to the development of hypoxia in these systems, but stratification of the water column was a necessary condition.

There also is evidence that increased nutrient loading has contributed to the occurrence of hypoxia in Florida Bay and the Florida Keys (Lapointe at al. 1990; Lapointe and Clark 1992). The most severe cases of hypoxia were found in the canal and seagrass systems closest to the discharge areas. Increased nitrogen levels were associated with increased growth of nutrient-limited phytoplankton, whereas high levels of soluble reactive P were associated with increased growth of macroalgae and tropical seagrasses. Lapointe and Matzie (1996) showed that episodic rainfall events led to higher submarine discharge rates that were followed within days by hypoxic oxygen levels.

Shifts in Community Structure Caused by Anoxia and Hypoxia

The occurrence of hypoxic and anoxic bottom waters may also lead to shifts in benthic and pelagic community structure due to the mortality of less mobile or more sensitive taxa, reduction of suitable habitat, and shifts in predator-prey interactions (Diaz and Rosenberg 1995). Hypoxia plays a major role in the structuring of benthic communities because species differ in the sensitivity to oxygen reduction (Diaz and Rosenberg 1995). The response of species to reduced oxygen availability also depends on the frequency and duration of these events. With short bouts of hypoxia, some large or very motile species are able to adjust to or move away from the stress.

Hypoxia tends to shift the benthic community from being dominated by large long-lived species to being dominated by smaller opportunistic short-lived species (Pearson and Rosenberg 1978). In addition, recurring hypoxia may limit successional development to colonizing communities. In such systems more organic matter is available for remineralization by the microbial community. This can decrease the amount of energy available for benthic recruitment when hypoxia and anoxia disappears. Zooplankton that normally vertically migrate into bottom waters during the day may be more susceptible to fish predation if they are forced to restrict their activity to the oxic surficial waters. Roman et al. (1993) concluded that the vertical distribution of copepods in the Chesapeake Bay was altered by the presence of hypoxic bottom waters. Moreover, an hypoxic or anoxic bottom layer may constitute a barrier that de-couples the life cycle of pelagic species (e.g., diatoms, dinoflagellates, and copepods) that have benthic resting stages (Marcus and Boero 1998).

In a controlled eutrophication experiment (Doering et al. 1989), the structure of the zooplankton community was affected by the presence or absence of an intact benthic community. In the absence of an intact benthic community, holoplanktonic forms, especially higher level predators, dominated, whereas meroplanktonic forms were more evident in the presence of an intact benthic community. Although the data did not identify the mechanism behind these shifts, the differences likely reflected alterations in the coupling of the benthic and pelagic environments (nutrient as well as life cycle linkages) (Marcus and Boero 1998).

Changes in predator-prey interactions in the water column can also lead to shifts in energy flow. Increased fish predation on zooplankton can release grazing pressure on the phytoplankton and increase the deposition of organic matter to the sediments. If the duration and severity of the hypoxia is not sufficient to cause mortality of the macrobenthos, the increased supply of organic matter to the benthic system could fuel the growth of benthic fauna and demersal fish populations at the expense of pelagic fisheries. On the other hand, extended hypoxic and anoxic events could lead to the demise of the macrobenthos and the flourishing of bacterial mats. The loss of burrowing benthic organisms that irrigate the sediments and the presence of an extensive bacterial community may alter geochemical cycling and energy flow between the benthic and pelagic systems (Diaz and Rosenberg 1995). For example, the flux of nitrogen out of the sediments is affected by the rates of nitrification and denitrification, and these processes depend on the naturally oxic and anoxic character of the sediments.

Changes in Plankton Community Structure Caused Directly by Nutrient Enrichment

Nutrient over-enrichment can also change ecological structure through mechanisms other than anoxia and hypoxia. Phytoplankton species have wide differences in their requirements for and tolerances of major nutrients and trace elements. Some species are well adapted to low-nutrient conditions where inorganic compounds predominate, whereas others thrive only when major nutrient concentrations are elevated or when organic sources of nitrogen and phosphorus are present. Uptake capabilities of major nutrients differ by an order of magnitude or more, allowing the phytoplankton community to maintain production across a broad range of nutrient regimes. A decrease in silica availability in an estuary and the trapping of silica in upstream eutrophic freshwater ecosystems can occur as a result of eutrophication and thus nitrogen and phosphorus over-enrichment. This decrease in silica often limits the growth of diatoms or causes a shift from heavily silicified to less silicified diatoms (Rabalais et al. 1996). Given these changes in the cycling of nitrogen, phosphorus, and silica, it is no surprise that the phytoplankton community composition is altered by nutrient enrichment (Jørgensen and Richardson 1996).

The consequences of changes in phytoplankton species composition on grazers and predators can be great, but in general these are poorly studied. As noted by Jørgensen and Richardson (1996):

> Any eutrophication induced change in the species composition of the phytoplankton community which leads to a change in size structure of the phytoplankton community will potentially affect energy flow in the entire ecosystem. Thus, eutrophication can, at least in theory, play an

> important role in dictating whether the higher trophic levels in a given system are dominated by marketable fish or by jellyfish. . . . Little is actually known about the effects of eutrophication on the size structure of the phytoplankton community under various conditions, but this is an area that warrants further research.

In particular, a change from diatoms toward flagellates, which may tend to result during eutrophication as the silica supply is diminished, may be deleterious to food webs supporting marketable forms of finfish (Greve and Parsons 1977). On the other hand, if the silica supply remains high enough, moderate eutrophicaton can encourage more growth by diatoms and lead to higher fish production (Doering et al. 1989; Hansson and Rudstam 1990).

Looking beyond the major nutrients, it is also evident that phytoplankton species have variable requirements for nutritional trace elements or have different tolerances for toxic metals (Sunda 1989), and the effects of these elements can be affected by dissolved organic matter (DOM) concentrations. One example of this effect is seen with copper, which is highly toxic to marine organisms and is often significantly elevated in harbors and estuaries due to anthropogenic inputs. Copper is strongly bound to organic chelators in seawater, and this lowers copper's biological availability and consequently its toxicity (Sunda 1989). Moffett et al. (1997) studied copper speciation and cyanobacterial distribution and abundance in four harbors subject to varying degrees of copper contamination from anthropogenic sources. Cell densities of cyanobacteria, one of the most copper-sensitive groups of phytoplankton, declined drastically in high copper waters compared to adjacent unpolluted waters. Because of the variability in the concentrations of the natural organic ligands that bind the copper, relatively small changes in the total copper concentration (7 to 10 times) among the study sites were associated with much larger (greater than 1000 times) changes in the free Cu^{2+} activity, the biologically available form.

The bioavailability of metals such as iron also can be affected by human activities, including nutrient pollution and the resulting eutrophication, and this in turn can affect phytoplankton species composition and thus ecosystem structure and function. Iron is an essential element for algae, and is required for electron transport, oxygen metabolism, nitrogen assimilation, and DNA, RNA, or chlorophyll synthesis. Organic ligands are needed to keep iron in solution at the pH of seawater, as iron hydroxides have extremely low solubility and tend to transform into stable crystalline forms that do not directly support algal growth. DOM plays a critical role in enhancing the bioavailability of iron in seawater. A variety of factors can affect DOM levels in estuaries and coastal systems, but in general eutrophication results in higher DOM levels—due to higher levels of primary production, leakage of DOM from phytoplankton, release as phytoplankton are eaten or decompose—with concomitant changes in iron availability.

Although the potential impacts of nutrient enrichment on phytoplankton community structure in the field seem obvious, there are few well-documented examples. This is because the changes in community structure often are gradual and easily obscured by fluctuations in other controlling factors, such as temperature, light, or physical forcings.

Long-term data sets offer another insight into possible changes since they allow sustained trends to be detected in spite of short-term variability caused by weather or other environmental forcings. For example, a 23-year time series off the German coast documented the general enrichment of coastal waters with nitrogen and phosphorus, as well as a four-fold increase in the nitrogen:silicon and phosphorus:silicon ratios (Radach et al. 1990). This was accompanied by a striking change in the composition of the phytoplankton community, as diatoms decreased and flagellates increased more than ten-fold. Other data from nearby regions showed a change in the phytoplankton species composition accompanying a shift in the nitrogen:phosphorus supply ratio along the Dutch coast (Cadée 1990), as well as increased incidence of summer blooms of the marine haptophyte *Phaeocystis* after a shift from phosphorus-limitation to nitrogen-limitation (Riegman et al. 1992). Nutrient status, particularly phosphorus-limitation, is now believed to be a major factor driving colony formation in this genus. Experiments performed with cultures of *Phaeocystis* demonstrate that free-living solitary cells outcompete the more harmful colonial forms in ammonium- and phosphate-limited conditions, whereas colonies dominate in nitrate-replete cultures. This suggests that free-living *Phaeocystis* cells would be prevalent in environments that are regulated by regenerated nitrogen, whereas colonial forms would require a nitrate supply and thus would be associated with "new" nitrogen such as that supplied by pollution.

Another long-term perspective on nutrient enrichment on phytoplankton community structure is seen in recent data examining the abundance of dinoflagellate cysts in bottom sediments of Oslofjord, Norway (Dale et al. 1999). Dinoflagellate cysts are an important group of microfossils used extensively for studying the biostratigraphy and paleoecology of sediments. In this study, dinoflagellate cyst records were analyzed from sediment cores that covered a period of anthropogenic nutrient enrichment that began in the mid- to late-1800s, was heaviest from 1900 to the 1970s, and then diminished from the mid-1970s to the present. Over the period of nutrient and organic enrichment, cyst abundance in the sediments doubled and a marked increase in one species in particular,

Lingulodinium machaerophorum (=*Gonyaulax polyedra*), from less than 5 percent to around 50 percent of the assemblage was noted. In the core considered most representative of general water quality in the inner fjord (Figure 4-1), these trends reversed back to pre-industrial levels during the 1980s and 1990s when improved sewage treatment took effect. Other changes in the phytoplankton community no doubt occurred that were not revealed with this approach, but the cyst record nevertheless demonstrates substantial changes in the abundance and composition of a major phytoplankton class.

Although at times changes in community structure are directly the result of nutrient enrichment, sometimes they are an indirect result of other changes caused by increased nutrients. For instance, a change in the phytoplankton community in the form of selection for different species can be caused directly by increased nitrogen. On the other hand, a change in phytoplankton community structure can be caused indirectly by increased nitrogen, because higher levels of nitrogen increase productivity, which increases dissolved organic carbon, which in turn causes changes in the community structure. Generally it is difficult to determine whether community structure changes are direct or indirect.

Harmful Algal Blooms

Among the thousands of species of microscopic algae at the base of the marine food web are a few dozen that produce potent toxins or that cause harm to humans, fisheries resources, and coastal ecosystems. These species make their presence known in many ways, ranging from massive blooms of cells that discolor the water, to dilute, inconspicuous concentrations of cells noticed only because of the harm caused by their highly potent toxins. The impacts of these phenomena include mass mortalities of wild and farmed fish and shellfish, human intoxications or even death from contaminated shellfish or fish, alterations of marine trophic structure through adverse effects on larvae and other life history stages of commercial fisheries species, and death of marine mammals, seabirds, and other animals.

"Blooms" of these algae are sometimes called red tides, but are more correctly called HABs, and are characterized by the proliferation and occasional dominance of particular species of toxic or harmful algae. As with most phytoplankton blooms, this proliferation results from a combination of physical, chemical, and biological mechanisms and interactions that are, for the most part, poorly understood. HABs have one feature in common, however; they cause harm either due to their production of toxins or to the way the cells' physical structure or accumulated biomass affect co-occurring organisms and alter food web dynamics. This descrip-

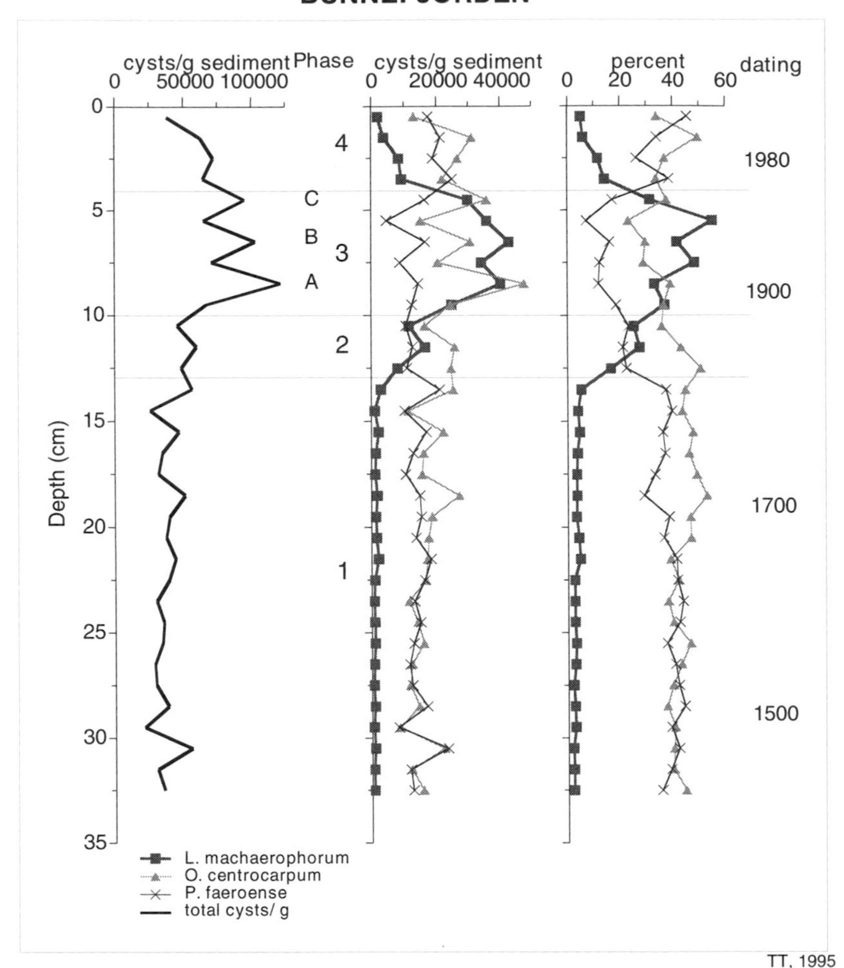

FIGURE 4-1 Water quality in Oslofjord, Norway, as indicated by the changes in the proportions of dinoflagellate cysts in sediment cores. The left panel shows the concentrations of all cysts combined, the middle panel shows the concentration of the three most important species (*Lingulodinium machaerophorum*, *Operculodinium centrocarpum*, and *Peridinium faeroense*), and the right panel shows the relative abundance of those species in percent. Eutrophication of Oslofjord resulted in a doubling of total cyst abundance and by a marked increase in *L. machaerophorum* from less than five percent to nearly 50 percent of the assemblage. These trends reversed with improved water quality in the 1980s and 1990s (modified from Dale et al. 1999).

tor applies not only to microscopic algae but also to benthic or planktonic macroalgae that can proliferate and cause major ecological impacts, such as the displacement of indigenous species, habitat alteration, and oxygen depletion. The causes and effects of macroalgal blooms are similar in many ways to those associated with harmful microscopic phytoplankton species.

HAB phenomena take a variety of forms. One major category of impact occurs when toxic phytoplankton are filtered from the water as food by shellfish that then accumulate the algal toxins to levels that can be lethal to humans or other consumers. These poisoning syndromes have been given the names paralytic, diarrhetic, neurotoxic, and amnesic shellfish poisoning (PSP, DSP, NSP, and ASP). A National Research Council report (NRC 1999b) summarized the myriad human health problems associated with toxic dinoflagellates. In addition to gastrointestinal and neurological problems associated with the ingestion of contaminated seafood, respiratory and other problems may arise from toxins that are released directly into seawater or become incorporated in sea spray. Whales, porpoises, seabirds, and other animals can be victims as well, receiving toxins through the food web from contaminated zooplankton or fish.

Another type of HAB impact occurs when marine fauna are killed by algal species that release toxins and other compounds into the water or that kill without toxins by physically damaging gills. Farmed fish mortalities from HABs have increased considerably in recent years, and are now a major concern to fish farmers and their insurance companies. The list of finfish, shellfish, and wildlife affected by algal toxins is long and diverse (Anderson 1995) and accentuates the magnitude and complexity of the HAB phenomena. In some ways, however, this list does not adequately document the scale of toxic HAB impacts, as adverse effects on viability, growth, fecundity, and recruitment can occur within different trophic levels, either through toxin transmitted directly from the algae to the affected organism or indirectly through food web transfer. This is because algal toxins can move through ecosystems in a manner analogous to the flow of carbon or energy.

Yet another HAB impact is associated with blooms that are of sufficient density to cause dissolved oxygen levels to decrease to harmful levels as large quantities of algal biomass fall to the sediment and decay as the bloom declines. Oxygen levels can also drop to dangerous levels in "healthy" blooms due to algal respiration at night. Estuaries and nearshore waters are particularly vulnerable to low dissolved oxygen problems during warm summer months, especially in areas with restricted flushing.

One of the explanations given for the increased incidence of HAB outbreaks worldwide over the last several decades is that these events are a reflection of pollution and eutrophication in estuarine and coastal waters (Smayda 1990). Some experts argue that this is evidence of a fundamental change in the phytoplankton species composition of coastal marine ecosystems due to the changes in nutrient supply ratios from human activities (Smayda 1990). This is clearly true in certain areas of the world where pollution has increased dramatically. It is perhaps real, but less evident, in areas where coastal pollution is more gradual and unobtrusive. A frequently cited dataset from an area where pollution is a significant factor is from Tolo Harbor in Hong Kong, where population growth in the watershed grew six-fold between 1976 and 1986. During that time, the number of observed red tides increased eight-fold (Lam and Ho 1989). The underlying mechanism is presumed to be increased nutrient loading from pollution that accompanied human population growth. A similar pattern emerged from a long-term study of the Inland Sea of Japan (Box 4-1).

Both the Hong Kong and Island Sea of Japan examples have been criticized, since both could be biased by changes in the numbers of observers through time, and both are tabulations of water discolorations from algal blooms, rather than just toxic or harmful episodes. Nevertheless, the data demonstrate that coastal waters receiving industrial, agricultural, and domestic effluents, which frequently are high in plant nutrients, do in fact experience a general increase in algal growth.

Nutrients can stimulate or enhance the impact of toxic or harmful species in several ways. At the simplest level, toxic phytoplankton may increase in abundance due to nutrient enrichment but remain as the same relative fraction of the total phytoplankton biomass (i.e., all phytoplankton species are stimulated equally by the enrichment). In this case, we would see an increase in HAB incidence, but it would coincide with a general increase in algal biomass. Alternatively, some contend that there has been a *selective* stimulation of HAB species by nutrient pollution. This view is based on the nutrient ratio hypothesis (Smayda 1990), which argues that environmental selection of phytoplankton species has occurred because human activities have altered nutrient supply ratios in ways that favor harmful forms. For example, diatoms, the vast majority of which are harmless, require silicon in their cell walls, whereas most other phytoplankton do not. As discussed in Chapter 3, silica availability is generally decreased by eutrophication. In response to nutrient enrichment with nitrogen and phosphorus, the nitrogen:silicon or phosphorus:silicon ratios in coastal waters have increased over the last several decades.

Diatom growth in these waters will cease when silicon supplies are depleted, but other phytoplankton classes (which have more toxic species) can continue to proliferate using the "excess" nitrogen and phosphorus. The massive blooms of *Phaeocystis* that have occurred with increasing frequency along the coast of western Europe are an example of this phe-

BOX 4-1
Red Tides and Eutrophication in the Inland Sea of Japan

A prominent example of the link between eutrophication and increased HABs is seen in a long-term data set of red tides in the Inland Sea of Japan. As Japanese industrial production grew rapidly in the late 1960s and early 1970s, pollution of coastal waters also increased. Currently, the number of visible red tides increased from 44 per year in 1960 to more than 300 a decade later, matching the pattern of increased nutrient loading from pollution (Figure 4-2). Japanese authorities instituted effluent controls in the early 1970s through the Seto Inland Sea Law, resulting in a 70 percent reduction in the number of red tides that has persisted to this day (Okaichi 1997).

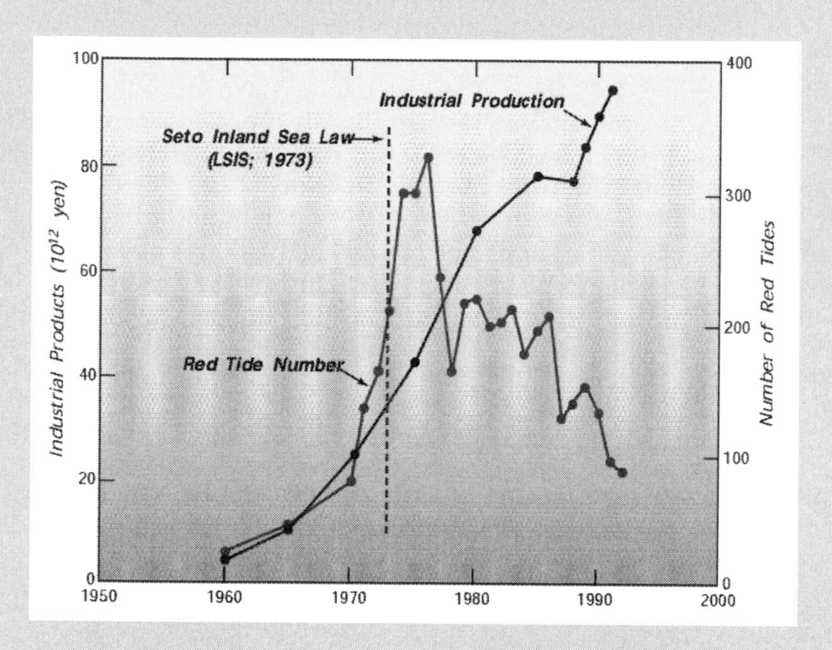

FIGURE 4-2 Changes in the number of visible red tides in the Inland Sea of Japan, 1960-1990 (Okaichi 1997; used with permission from Terra Scientific Publishing).

nomenon (Lancelot et al. 1998). Other examples include the fish killing blooms of *Chattonella* species, which have caused millions of dollars of damage in the Seto Inland Sea, though the frequency and severity of these outbreaks has decreased since pollution loading was reduced and eutrophication abated somewhat (Okaichi 1997).

Another frequently cited example of the potential linkage between HABs and pollution involves the recently discovered "phantom" dinoflagellate *Pfiesteria*. In North Carolina estuaries and in the Chesapeake Bay, this organism has been linked to massive fish kills and to a variety of human health effects, including severe learning and memory problems (Burkholder and Glasgow 1997). A strong argument is being made that nutrient pollution is a major stimulant to outbreaks of *Pfiesteria* or *Pfiesteria*-like organisms because the organism and associated fish kills have occurred in watersheds that are heavily polluted by hog and chicken farms and by municipal sewage. The mechanism for the stimulation appears to be two-fold. First, *Pfiesteria* is able to take up and use some of the dissolved organic nutrients in waste directly (Burkholder and Glasgow 1997). Second, this adaptable organism can consume algae that have grown more abundant from nutrient over-enrichment. Even though the link between *Pfiesteria* outbreaks and nutrient pollution has not been fully proven, the evidence is strong enough that legislation is already in various stages of development and adoption to restrict the operations of hog and chicken farms in order to reduce nutrient loadings in adjacent watersheds. *Pfiesteria* has thus provided the justification needed by some agencies to address serious and long-standing pollution discharges by nonpoint sources, which heretofore have avoided regulation.

Degradation of Seagrass and Algal Beds and Formation of Nuisance Algal Mats

Many coastal waters are shallow enough that benthic plant communities can contribute significantly to autotrophic production if sufficient light penetrates the water column to the seafloor. In areas of low nutrient inputs, dense populations of seagrasses and perennial macroalgae (including kelp beds) can attain rates of net primary production that are as high as the most productive terrestrial ecosystems (Charpy-Roubaud and Sournia 1990). These perennial macrophytes are less dependent on water column nutrient levels than phytoplankton and ephemeral macroalgae, and light availability is usually the most important factor controlling their growth (Sand-Jensen and Borum 1991; Dennison et al. 1993; Duarte 1995). As a result, nutrient enrichment rarely stimulates these macrophyte populations, but instead causes a shift to phytoplankton or bloom-forming benthic macroalgae as the main autotrophs. Fast-growing micro- and macroalgae with rapid nutrient uptake potentials can replace seagrasses as the dominant primary producers in enriched systems (Duarte 1995; Hein et al. 1995). The biotic diversity of the community generally decreases with these nutrient-induced changes (Figure 4-3A&B).

Over the last several decades, nuisance blooms of macroalgae ("seaweeds") in association with nutrient enrichment have been increasing along many of the world's coastlines (Lapointe and O'Connell 1989). Phytoplankton biomass and total suspended particles increase in nutrient-enriched waters and reduce light penetration through the water column to benthic plant communities. Epiphytic microalgae become more abundant on seagrass leaves in eutrophic waters and contribute to light attenuation at the leaf surface, as well as to reduced gas and nutrient exchange (Tomasko and Lapointe 1991; Short et al. 1995; Sand-Jensen 1977). Ephemeral benthic macroalgae have light requirements that are significantly less than either seagrasses or perennial macroalgae, and also can shade perennial macrophytes such as seagrasses and contribute to their decline (Markager and Sand-Jensen 1990; Duarte 1995). These nuisance algae are typically filamentous (sheet-like) forms (e.g., *Ulva*, *Cladophora*, *Chaetomorpha*) that can accumulate in extensive thick mats over the seagrass or sediment surface, and this can lead to destruction of these submerged aquatic seagrass systems. Massive and persistent macroalgal blooms ultimately displace seagrasses and perennial macroalgae through shading effects (Valiela et al. 1997). The nuisance algae also wash up on beaches, creating foul-smelling piles.

In addition to shading, seagrass distribution in eutrophic waters is influenced by increased sediment sulfide concentrations resulting from decomposition in anoxic organic-rich sediments. Elevated sediment sulfide has been shown experimentally to reduce both light-limited and light-saturated photosynthesis, as well as to increase the minimum light requirements for survival (Goodman et al. 1995). Both effects interact with increased light attenuation to decrease the depth penetration of seagrasses in eutrophic waters.

Decreased photosynthetic oxygen production at all light levels also reduces the potential for oxygen translocation and release to the rhizosphere, and creates a positive feedback that reduces sulfide oxidation around the roots, further elevating sediment sulfide levels. In Florida, chronic sediment hypoxia and high sediment sulfide concentrations have been associated with the decline of the tropical seagrass *Thalassia testudinum* (Robblee et al. 1991). Sulfide also may reduce growth and production of seagrasses by decreasing nutrient uptake and plant energy status, as has been shown for salt marsh grasses (Bradley and Morris 1990, Koch et al. 1990).

Declines in seagrass distribution caused by decreased light penetra-

FIGURE 4-3 (A) The bottom-dwelling plants of a marine ecosystem that received natural rates of nitrogen addition. Note the high diversity of these plants and their spacing. (B) The bottom-dwelling plants of a marine ecosystem that received high rates of nitrogen input. Note that there are few plant species, and that the leaves of these are covered with a thick layer of algae (photos by R. Howarth; Vitousek et al. 1997).

tion in deeper waters or changes in community composition prompted by the proliferation of benthic macroalgae in shallower waters will have significant trophic consequences. Seagrass roots and rhizomes stabilize sediments, and their dense leaf canopy promotes sedimentation of fine particles from the water column. Loss of seagrass coverage increases sediment resuspension and causes an efflux of nutrients from the sediment to the overlying water that can promote algal blooms. Seagrasses also provide food and shelter for a rich and diverse fauna, and reduced seagrass depth distribution or replacement by macroalgal blooms will result in marked changes in the associated fauna (Thayer et al. 1975; Norko and Bonsdorff 1996).

In addition, where mass accumulations of macroalgae occur, their characteristic bloom and die-off cycles influence oxygen dynamics in the entire ecosystem. As a result, eutrophic shallow estuaries and lagoons often experience frequent episodic oxygen depletion throughout the water column rather than the seasonal bottom-water anoxia that occurs in stratified, deeper estuaries (Sfriso et al. 1992; D'Avanzo and Kremer 1994). Benthic macroalgae also uncouple sediment mineralization from water column production by intercepting nutrient fluxes at the sediment-water interface (Thybo-Christesen et al. 1993; McGlathery et al. 1997) and can outcompete phytoplankton for nutrients (Fong et al. 1993). Except during seasonal macroalgal die-off events in these shallow systems, phytoplankton production is typically nutrient-limited and water column chlorophyll concentrations are uncharacteristically low despite high nutrient loading (Sfriso et al. 1992).

Coral Reef Destruction

Coral reefs are among the most productive and diverse ecosystems in the world. They grow as a thin veneer of living coral tissue on the outside of the hermatypic (reef-forming) coral skeleton. The world's major coral reef ecosystems are found in nutrient-poor surface waters in the tropics and subtropics. Early references to coral reef ecosystems preferring or "thriving" in areas of upwelling or other nutrient sources have since been shown to be incorrect (Hubbard, D. 1997). Rather, high nutrient levels generally are detrimental to "reef health" (Kinsey and Davies 1979) and lead to phase shifts away from corals and coralline algae toward dominance by algal turf or macroalgae (Lapointe 1999). For example, some offshore bank reefs in the Florida Keys that contained more than 70 percent coral cover in the 1970s (Dustan 1977) now have only about 18 percent coral cover; turf and macroalgae now dominate these reefs, accounting for 48 to 84 percent cover (Chiappone and Sullivan 1997). Reduced herbivory, either caused by disease or overfishing, also can lead to increases in macroalgal cover and reduced coral abundance.

Because the growth rates of macroalgae under the high light intensities and warm temperatures found in coral reef waters are highly dependent on the concentration of the growth-limiting nutrient (typically either nitrogen or phosphorus), even slight increases in ambient dissolved nutrient concentrations can lead to expansion of algae at the expense of coral. Standing stock concentrations are not the best indicators of nutrient availability because they do not take into account turnover times of nutrient pools (Howarth 1988), but in a comparative sense they can provide information on trends in ambient nutrient conditions. In a review of eutrophication on coral reefs in Kaneohe Bay, Hawaii, fringing reefs in Barbados, and the Great Barrier Reef, Bell (1992) found that macroalgae tended to dominate coral communities at reported dissolved inorganic nitrogen (DIN) concentrations above 1.0 μM and 0.2 μM soluble reactive phosphorus (SRP). Macroalgae also have became the dominant space-occupying organisms on some carbonate-rich reefs in the Caribbean (Lapointe et al. 1993) and in localized areas near the Florida Keys (Lapointe and Matzie 1996) even though there were only small increases in measured DIN and SRP concentrations. The growth rate of the coral reef macroalga *Dictyosphaeria cavernosa* , which overgrew Kaneohe Bay in the 1960s due to sewage nutrient enrichment (Banner 1974), has recently been shown to be nitrogen-limited and to achieve maximum growth rates at concentrations as low as 1.0 μM DIN (Larned and Stimson 1996). That nutrient enrichment at these low levels enhances macroalgal growth and triggers such dramatic ecological changes underscores the extreme sensitivity of these oligotrophic ecosystems to even slight increases in nutrient enrichment.

Increased macroalgal cover on reefs inhibits the recruitment of corals and leads to second-order ecological effects. For instance, macroalgal blooms can lead to hypoxia and anoxia of the reef surface, reducing the habitat quality needed to support the high diversity of coral reef organisms and potentially important grazers.

The recent recognition that the "global nitrogen overload problem" (Moffat 1998) now affects remote areas via atmospheric pathways (Vitousek et al. 1997) illustrates how nitrogen enrichment can potentially impact even remote reef locations on earth and contribute to coral "stress." There is some evidence that nitrogen availability, in addition to temperature, light, and other environmental factors, may influence the "coral bleaching" (i.e., loss of zooxanthellae) phenomenon that has expanded globally in recent years (D'Elia et al. 1991). Over a six-year period, Fagoonee et al. (1999) found that the zooxanthellae density of corals in Mauritius varied considerably, correlating most significantly with nitrate concentration of the water column.

Myriad other direct and indirect effects of coastal eutrophication are known to affect coral reefs. Direct effects include decreased calcification associated with elevated nutrients. This is presumably at least in part due to shading by increased macroalgal biomass, since calcification is a direct function of photosynthesis by zooxanthallae. Indirect effects on coral reefs can stem from increased phytoplankton biomass that alters the quality and quantity of particulate matter and optical properties of the water column (Yentsch and Phinney 1989). For instance, reduced light levels associated with turbidity on reefs in Barbados depressed larval development and maturation in the coral *Porites porites* (Tomascik and Sander 1985). Another indirect effect occurs when nutrient enrichment enhances predator species; for example, outbreaks of the "Crown-of-Thorns" starfish in the South Pacific, which preys on living coral tissue, have been related to the stimulatory effects of nutrient runoff on their larval development (Birkeland 1982).

Disease and Pathogen Increases

The occurrence of microbial pathogens in the marine environment that is of concern to human health generally is associated with environmental contamination by human sewage and not nutrients per se (NRC 1993a). However, one group of pathogens, the *Vibrios*, has been identified as autochthanous members of the microbial community in brackish estuarine and coastal waters (Colwell 1983). In laboratory studies, the growth rate of *Vibrio cholerae* has been positively correlated with organic enrichment (Singleton et al. 1982). Another species *V. vulnificus* has been identified as a dominant member of the heterotrophic bacterial community of the Chesapeake Bay (Wright et al. 1996). It is possible, therefore, that eutrophication promotes the growth of these pathogens under field conditions. Research has also revealed an association of *V. cholerae* and *V. parahaemolyticus* with zooplankton, particularly copepods, to whose surfaces the bacteria attach (Kaneko and Colwell 1975; Huq et al. 1983). Colwell (1996) has suggested that phytoplankton blooms may be the ultimate cause of some outbreaks of *V. cholerae*, by fueling the growth of copepods. The increased abundance of copepods with their associated *Vibrio* flora could provide the dose necessary to cause cholera, and it could be worthwhile to determine if higher level predators of copepods (e.g., fish) could become contaminated by *Vibrios* and transmit the pathogens to human consumers.

ECONOMIC IMPACTS

The impacts of nutrient enrichment can be measured not only in ecological terms but also in economic terms. The economic impacts are a measure of either the damages (i.e., lost value) from nutrient enrichment or the benefits of the improvements that result from reducing or reversing this process. To measure value, economists use the concept of "willingness to pay" (WTP) for improvements in, for example, water quality, or "willingness to accept" (WTA) compensation for environmental degradation (Freeman 1993; Smith 1996).[1] Because of the difficulty of measuring WTA, most empirical studies estimate WTP measures of value. WTP represents the amount of money that an individual would be willing to give up or pay to secure an environmental improvement, which reflects how much the improvement is "worth" to the individual.[2]

Note that the amount the individual is willing to pay may differ substantially from the amount actually paid. For example, many recreational opportunities are available free of charge.[3] This does not mean, however, that the opportunity has no value. As long as an individual would have been willing to pay for that opportunity (e.g., for access to the marine resource), it has a positive economic value (i.e., it generates gross benefits for that individual) even if the individual does not pay for it. Likewise, if a water quality improvement increases the amount an individual would have been willing to pay for a recreational opportunity, that improvement has a positive economic benefit even if the individual did not actually pay for the improvement. A primer on the concept of economic value and its application to the valuation of coastal resources is available from National Oceanic and Atmospheric Administration (NOAA) (Lipton and Wellman 1995). This handbook is specifically designed to provide coast resource managers with an introduction to economic valuation.

Types of Economic Value

There are a number of ways to classify economic value (Freeman 1993). A fundamental distinction can be drawn between use value and non-use value (Freeman 1993; Smith 1996). Use value is the value an individual derives from directly using a resource. In the context of marine

[1] These measures are alternatively referred to as "compensating surplus" and "equivalent surplus" measures of value (Freeman 1993, 1995; Smith 1996).

[2] Willingness to pay provides a measure of the gross value or benefits of an improvement in environmental quality. To estimate net benefits, the cost of achieving that improvement must be subtracted. These net benefits represent a measure of the change in social surplus (to producers and consumers) from the improvement (Freeman 1993; Smith 1996).

[3] This implies that there is no access or entrance fee to the recreational site. It may, however, be costly to get to the site or to buy the necessary equipment. In fact, the existence of these related expenses provides an opportunity for inferring WTP. See the discussion of the travel cost method of valuation below.

resources, this includes value derived from the use of the resource for swimming, recreational fishing, commercial fishing, wildlife viewing, boating, and beach uses (Bockstael et al. 1989). In contrast, non-use value is the value derived from the resource even though it is not currently used. Non-use values include (1) existence value, which is the value derived simply from knowing that the resource exists and is maintained, (2) bequest value, the value that the current generation received from knowing that the resource will be available for future generations, and (3) option value, which derives from preserving the resource so that the option of future use is retained (Smith 1996). This categorization encompasses a very broad definition of economic value. In particular, even when a resource is not currently being used by humans, it will still have economic value if individuals are willing to pay to preserve it despite the lack of current use. Thus, the ecological function of an ecosystem will have economic value as long as individuals value that ecological function (for whatever reason).

A second categorization of values that also encompasses this broad definition of value is the distinction between the value of market and non-market goods. Market goods are goods and services that are traded in the market at a given price. For these goods, the price that people actually pay for the good or service serves as a reasonable proxy for how much they would be willing to pay (at the margin) for the good and hence serves as a useful proxy for its value. Market goods are the source of commercial value for many resources (e.g., commercial fisheries).

Non-market goods, on the other hand, are not bought and sold in markets and hence have no observable price. Water quality is a classic example of a non-market good. While individuals value water quality improvements (in the same way that they would value an increase in the consumption of shellfish, for example), they cannot go to the market and "purchase" additional water quality for a given price (in the same way that they can purchase an additional pound of shellfish). Without a price that reveals a minimum amount that individuals would be willing to pay for the water quality improvement, some other mechanism for estimating the value of the improvement must be found. A large amount of research has gone into the development of non-market valuation techniques (Box 4-2).

Alternative Valuation Techniques

A wide variety of techniques[4] exist for estimating the dollar value of

---

[4] For a more detailed discussion and assessment of techniques for valuing natural resources (NRC 1997a). Detailed discussions in the context of water quality benefits are also in Ribaudo and Hellerstein (1992) and Carson and Mitchell (1993).

BOX 4-2
What Do House Prices and Travel Costs Tell Us About the Economic Impacts of Nutrient Over-enrichment?

For goods that are bought and sold in markets (e.g., commercial fish), the price of the good reveals the price people are willing to pay for one more unit of that good (i.e., what they think an additional unit of the good is "worth"). Some goods, such as water quality, do not have a market price, and as a result economists have had to look for other indicators of their value. For example, a study by Boyle et al. (1998) used differences in house prices across lakes of different water clarity to infer a measure of the value of improved water clarity. If people value water clarity and other factors remain the same, they should be willing to pay more for a house on a lake with greater clarity.

Using statistical techniques applied to data on house prices, water clarity, and other house characteristics for sales around 36 lakes in Maine, the researchers found that for a select group of lakes, a 1-meter reduction in visibility, a measure of water quality resulted in a reduction in the average property value between $6,001 and $7,629 (in 1998 dollars; Boyle et al. 1998). They then used the property value information to calculate the impact on property owners.

Similarly, economists have developed methods for inferring the value of water quality improvements from observations about recreational travel behavior. If individuals value water quality, they should be willing to travel farther or more often to sites with better water quality. Since increased travel involves increased costs, this increased travel reveals how many people are willing to pay for the improved water quality. For example, using survey data on visits to 11 public beaches on the western shore of Maryland and estimates of associated travel costs, Bockstael et al. (1989) applied a travel cost model to estimate the amount that individuals are willing to pay for additional visits to the beach. In a second-stage analysis, they then estimated the relationship between the value of additional visits and water quality, measured by the loadings of nitrogen and phosphorus. By varying water quality and calculating the corresponding change in the value of visits, the individual value of a water quality improvement could be estimated. Finally, from the individual estimates for an average user and information on the number of users, aggregate estimates of the value of a water quality improvement could be obtained. Applying this technique to the Chesapeake Bay, Bockstael et al. (1989) found that on average a 20 percent reduction in nitrogen and phosphorus inputs near the beach would generate benefits of $34.6 million (in 1984 dollars) from increased public beach use on the western shore.

improvements in environmental quality. The applicability of these techniques varies. Some are applicable only to improvements that directly affect market goods (e.g., commercial fisheries), while others are designed to estimate the value of improvements in non-market goods (e.g., recreational fishing). While different categorizations exist, valuation techniques are generally classified as either (1) indirect, or revealed preference, approaches or (2) direct, or stated preference, approaches. Revealed preference approaches infer values from observed behavior. Stated preference approaches estimate values based on survey responses to questions about hypothetical scenarios.

There are a number of revealed preference approaches to valuation, including (1) derived demand/production cost estimation techniques, (2) cost-of-illness method, (3) the averting behavior (or avoidance cost) method, (4) hedonic price method, and (5) travel cost method (NRC 1997a). Each is designed to capture a particular component of economic value, and hence is applicable to a particular subset of environmental impacts. Thus, each provides only a partial measure of value; none is capable of providing a measure of the total economic value of a water quality improvement.

The derived demand approach is applicable when water quality serves as an input into the production of a marketed good. For example, ambient water quality or the amount of submerged aquatic vegetation can affect the stocks of commercial fish species. Thus, changes in water quality change the supply conditions for these species and hence the profits derived from related commercial fisheries. Under the derived demand approach, the change in profits is a measure of the value of the environmental improvement. Application of this technique requires documentation of the relationship between water quality and supply. In some cases, analyses simply attempt to estimate the impact of a discrete event, such as a *Pfiesteria* outbreak or HAB, on revenues, using a "before" and "after" comparison. For example, Lipton (1998) estimated that the commercial seafood industry in Maryland lost $43 million in sales in 1997 as a result of the public's concern about *Pfiesteria*. However, this approach can both over- and under-estimate the economic impacts of such events. It over-estimates them because it does not account for cost savings due to lower production levels or for the ability of consumers to substitute other products (which raises revenues for the producers of these substitute products). As with all revealed preference approaches, it under-estimates the impact of such outbreaks because it does not capture losses in non-use value (NRC 1997a).

For more continuous water quality problems such as nutrient enrichment, the impact on supply can be estimated using econometric methods, provided there is sufficient variability in the environmental indicator either over space or over time. Examples of studies of this type include Lynne et al. (1981), who estimated the relationship between marsh characteristics and productivity of the blue crab fishery on the Gulf Coast of Florida, and Kahn and Kemp (1985), who estimated the impact of submerged aquatic vegetation on the striped bass population. More recently, Diaz and Solow (1999) estimated the effect of hypoxia on the brown and white shrimp fisheries in the Gulf of Mexico, but failed to find any significant effect of hypoxia on these fisheries during the study period (1985–1995). It is possible that in this case, the natural variability in the underlying data is sufficiently large that it is difficult to detect changes attributable to hypoxia. Alternatively, it may be that discernible effects do not exist at current hypoxia levels, even though the effects could become significant if conditions worsen (Diaz and Solow 1999).

The derived demand approach is designed to capture the effects of environmentally induced supply shifts on producers. In contrast, the cost-of-illness method can be used to estimate the economic impact of environmental events that affect human health. It measures the benefits of a pollution reduction by estimating the possible savings in direct out-of-pocket expenses resulting from the illness (e.g., medicine, doctor, and hospital bills) and the lost earnings associated with the illness (NRC 1997a). It does not account for any discomfort or other health-related impacts that are not avoided through medical treatment, or for any expenditure undertaken to prevent the illness.

Expenditures undertaken to reduce or avoid the damaging effects of pollution are termed "averting" or "avoidance" costs. Reductions in these costs (i.e., cost savings) attributable to water quality improvements constitute a partial measure of the economic value of that improvement. For example, the Environmental Protection Agency (EPA) estimated the benefits of reduced nitrogen loadings to estuaries by calculating the pollution control costs that could be avoided if loadings were reduced (EPA 1998b). They estimated that in the east coast, Tampa, and Sarasota estuaries, the Regional NO_x State Implementation Plan will save $237.8 million (in 1998 dollars) in pollution control costs.

Hedonic pricing models are most applicable to measuring the benefits of environmental improvements to property owners. The basic principle underlying this approach is that the amount consumers are willing to pay for a house or piece of property depends on the characteristics of that house, including the perceived environmental characteristics of the surrounding areas (i.e., those that are of concern to prospective purchasers) (Rosen 1974). Thus, a house on an estuary with high water quality should sell for a higher price than an otherwise comparable one on an estuary with low water quality. The increase in the price of the house provides an estimate of the value of the difference (or, equivalently, an improvement) in water quality. A recent study of lakefront property owners in Maine found that housing prices were significantly affected by water clarity (Box 4-2). Although this study was based on lakes rather than coastal waters, it suggests that homeowners do value water clarity improvements and are willing to pay for them. However, measures of willingness to pay based on hedonic studies of this type capture only the value to the property owner. In particular, they do not capture the value to other users of the marine resource (e.g., recreationists) who do not own property near the resource.

The most commonly used method for estimating the value of improved water quality to recreationists is the travel cost method. Although recreationists may not pay an access fee for use of a recreational resource, they incur costs in the form of travel costs (including the opportunity cost of their time) and other out-of-pocket expenses. These costs can be viewed as the price paid for the recreation trip. Individuals who live farther from a site must pay a higher price (in the form of higher travel costs) and hence would be expected to demand less (i.e., take fewer trips). Thus, by relating travel costs to the number of trips taken, the demand for trips and the associated willingness to pay for them can be estimated. Improved water quality should increase the value of a recreational trip, which should in turn result in more frequent trips or trips from farther away. The changes in demand that result from a water quality improvement can be used to infer the economic value of the improvement.

There is a large theoretical and empirical literature on application of travel cost methods to the valuation of changes in environmental quality, particularly improvements in water quality. In a study of the benefits of improvements in water quality in the Chesapeake Bay, Bockstael et al. (1989) used the travel cost method to examine the impacts on three activities: beach use, boating, and fishing. For beach use and boating, water quality was represented by the level of nutrient enrichment as measured by the total input of nitrogen and phosphorus. For fishing, water quality was proxied by the catch rates for striped bass. From their travel cost study, they estimated the fishing, boating, and swimming benefits of a hypothetical 20 percent improvement in water quality in the bay to be in the range of \$18 to \$55 million per year in 1987 dollars. Bockstael et al. (1989) also used contingent valuation to estimate the benefits of water quality improvements in the bay (see discussion below).

Single-site travel cost models do not account explicitly for the possibility of substitutability across sites. As water quality at one site changes, users may switch to other sites, even if switching leads to increased travel costs. Random utility models are designed to model the choices that individuals make among alternative sites, as determined by environmental quality, distance, and other site characteristics. By examining the tradeoffs that individuals are willing to make between environmental quality and travel costs, estimates of the value of environmental improvements can be derived. In one study, Kaoru et al. (1995) applied a random utility model to the estimation of the value of reductions in nitrogen loadings in the Albemarle and Pamlico Sounds of North Carolina. Their estimates of the benefits to an individual of a 36 percent decrease in nitrogen loadings ranged from $0.12 to $11, depending on the assumptions made. While this study does not provide aggregate measures of value, it does highlight the sensitivity of estimates to individual characteristics and the level of model aggregation.

As noted above, by focusing on a particular user population, each of the revealed preference approaches provides a partial measure of the value of improved water quality. In particular, by their nature, they can capture only the use value of an environmental resource. In some cases, non-use value may constitute a significant portion of total economic value. The only techniques currently available for estimating total economic value, including non-use value, are stated preference approaches. The most commonly used stated preference approach is the contingent valuation method, although alternative methods such as conjoint analysis are being increasingly applied to environmental valuation (NRC 1997a).

The basic instrument used in contingent valuation method is a survey asking hypothetical questions about consumers' willingness to pay. Although open-ended questions have been used, there is a consensus that a dichotomous choice format is preferable (Arrow et al. 1993; NRC 1997a). With a dichotomous choice format, individuals are asked whether they would be willing to pay a specific amount (say, $20) for a given improvement in environmental quality. If they answer "yes," then their WTP exceeds $20. If they answer "no," then their WTP is less than $20. By varying the specified amount and using statistical analysis, an estimate of mean WTP can be derived (Freeman 1993; Smith 1996). The mean values can then be aggregated to the population level to provide estimates of aggregate value. Bockstael et al. (1989) use a contingent valuation survey to estimate the willingness to pay to make water quality in the Chesapeake Bay suitable for swimming. They surveyed both users and non-users of the Chesapeake Bay. The average willingness to pay was $121 (in 1984 dollars) for users and $38 for non-users. For the total population in the District of Columbia and Baltimore standard metropolitan statistical area, the aggregate value of the improvement is estimated to be between $65.7 and $116.6 million (in 1987 dollars).

Although the contingent valuation method has the advantage of allowing estimation of total economic value, it has been the subject of much controversy (Diamond and Hausman 1994; Hanemann 1994). Concerns about its use stem from the potential for biased estimates as a result of the hypothetical nature of the survey responses. Much research has been devoted to testing or correcting for, or designing methodologies that eliminate, potential biases (Carson and Mitchell 1993; Smith 1996). The potential bias can be reduced through careful survey design and pre-testing.

Because neither revealed nor stated preference approaches provide an ideal technique for measuring environmental values, economists have

BOX 4-3
The Challenge of Estimating Economic Impacts

Each of the various techniques available to estimate economic impacts provides only a partial estimate of the benefits of water quality improvement. Aggregating benefits across categories of use to get a total measure of value is difficult for at least two reasons. Aggregation requires that the various studies consider identical water quality changes. This is rarely the case if the studies have been done independently, and maintaining comparability can be difficult even within a given study. For example, in estimating the benefits of water quality improvements using the travel cost method, Bockstael et al. (1989) considered a 20 percent reduction in nitrogen and phosphorus loadings when estimating values relating to beach use and a 20 percent increase in catch rates when estimating values derived from sport fishing. Since the "goods" that are being valued are not necessarily the same (i.e., a 20 percent reduction in nutrient input does not necessarily lead to a 20 percent increase in catch rates), combining the value estimates across these categories is problematic. In addition, in aggregating estimates across categories, care must be taken to avoid double counting. For example, the contingent valuation estimates of Bockstael et al. (1989) should capture both use and non-use values. These cannot be combined with other partial measures of use value without double counting.

Despite these difficulties, a recent NOAA study (Doering et al. 1999) attempted to combine the results of a number of valuation studies to estimate the wetlands benefits of reduced nutrient loadings to the Gulf of Mexico. The total benefit of wetlands restoration from reduced loadings are derived from both market goods (commercial fishing and fur trapping) and non-market goods. The non-market goods generated both use value (from recreation, sport fishing and waterfowl hunting) and non-use value (existence and bequest value). Combining estimates of these various components of value from previous valuation studies, the researchers estimated that the value to residents of the Mississippi drainage basin of restoring 100,000 acres of coastal wetlands would range between $11.8 and $40 billion (in 1999 dollars). The largest benefits are derived from non-use values, followed by the value from recreational fishing and ecological services provided by wetlands. The NOAA study also considered erosion control benefits of reduced nutrient loadings. Because of limited data, however, the study makes no attempt to put a dollar value on the other economic impacts of eutrophication (e.g., impacts on swimming, boating, or commercial fishing).

Even though there has been considerable research on the ecological impacts of nutrient pollution, to date research on the economic impacts has been very limited. There have been a few key studies, using a variety of valuation techniques, but each captures only a subset of benefits relating to a specific use or user group. Aggregating these across uses, as well as across locations, is difficult, and neither national nor regional estimates of total economic impact exist. Some of the studies that have been done show large economic impacts in some cases or under some assumptions, and rather modest impacts in other cases.

The literature clearly shows that people value improvements in water quality and are willing to pay for those improvements, even though market-based measures of value significantly under-estimate total value since a large component of

continued

BOX 4-3 Continued

total value stems from non-market uses of the marine resources (e.g., for swimming, recreational fishing, boating, wildlife viewing). Furthermore, the studies that have been done have consistently concluded that non-use value comprises a significant share of total value. In other words, people are willing to pay significant amounts for water quality improvements even when they do not directly use the waterbody. Although non-use values are particularly difficult to measure and the techniques used to measure them are controversial, available data is consistent in showing that non-use values are important. Thus, estimates of economic impacts that fail to include non-market benefits, including non-use values, are likely to significantly underestimate total economic impacts.

recently developed methodologies that combine information from both types of approaches in an effort to improve benefit estimates (Cameron 1992; Adamowicz et al. 1994, 1997; Englin and Cameron 1996; Huang et al. 1997). Combining the two approaches has the potential to increase the reliability of the estimates, although a possible inconsistency in the use of joint estimation exists (Huang et al. 1997). It also allows a decomposition of total value into use and non-use values. A study by Huang et al. (1997) combines stated and revealed preference models in the estimation of the willingness to pay for improvements in water quality in the Albemarle and Pamlico Sounds of North Carolina. While they do not provide measures of aggregate value, their results do show that non-use value constitutes a significant portion (over half) of total value. Bockstael et al. (1989) also found significant non-use values in their study. This suggests that ignoring non-use values will lead to significant under estimation of the total value of a water quality improvement (Box 4-3).

5

Sources of Nutrient Inputs to Estuaries and Coastal Waters

KEY POINTS IN CHAPTER 5

This chapter reviews the sources and amounts of nutrients supplied to coastal water bodies and finds:

- Globally, human activity has dramatically increased the flux of phosphorus (by a factor of almost 3) to the world's oceans. There has been an even more dramatic increase in nitrogen flux, especially in the last 40 years, with the greatest flux adjacent to areas of highest population density. Human activity has increased the flux of nitrogen in the Mississippi River by some 4-fold, in the rivers in the northeastern United States by some 8-fold, and in the rivers draining to the North Sea by more than 10-fold.
- Although point source nutrients are the major problem for small watersheds adjacent to major population centers, these inputs are relatively easy to minimize with tertiary wastewater treatment processes. In contrast, nutrients from nonpoint sources have become the dominant and least easily controlled component of nutrients transported into coastal waters from large watersheds, and especially from watersheds with extensive agricultural activity or atmospheric nitrogen pollution.
- Phosphorus flux to estuaries is dominantly derived from agricultural activities as particle-bound forms mobilized in runoff. In some areas, groundwater transported phosphorus is also important.
- Nitrogen input to estuaries is derived from both agricultural activity (e.g., dominant in the Mississippi River) and fossil-fuel combustion (e.g., dominant in the northeastern United States). Animal feeding operations have become a major contributor to nitrogen exports.
- It is likely that the atmospheric component of nitrogen flux into estuaries has previously been under-estimated. This component is derived from fossil-fuel

combustion and from animal feedlots and other agricultural sources, and is both deposited directly into estuaries and also deposited initially onto the land surface and then carried into estuaries by runoff.

Human activity has an enormous influence on the global cycling of nutrients, especially on the movement of nutrients to estuaries and other coastal waters. For phosphorus, global fluxes are dominated by the essentially one-way flow of phosphorus carried in eroded materials and wastewater from the land to the oceans, where it is ultimately buried in ocean sediments (Hedley and Sharpley 1998). The size of this flux is currently estimated at 22 Tg P yr^{-1} (Howarth et al. 1995). Prior to increased human agricultural and industrial activity, the flow is estimated to have been around 8 Tg P yr^{-1} (Howarth et al. 1995). Thus, current human activities cause an extra 14 Tg of phosphorus to flow into the ocean sediment sink each year, or approximately the same as the amount of phosphorus fertilizer (16 Tg P) applied to agricultural land each year.

The effect of human activity on the global cycling of nitrogen is equally immense, and furthermore, the rate of change in the pattern of use is much greater (Galloway et al. 1995). The single largest global change in the nitrogen cycle comes from increased reliance on synthetic inorganic fertilizers, which accounts for more than half of the human alteration of the nitrogen cycle (Vitousek et al. 1997). The process for making inorganic nitrogen fertilizer was invented during World War I, but was not widely used until the 1950s. The rate of use increased steadily until the late 1980s, when the collapse of the former Soviet Union led to great disruptions in agriculture and fertilizer use in Russia and much of eastern Europe. These disruptions resulted in a slight decline in global nitrogen fertilizer use for a few years (Matson et al. 1997). By 1995, the global use of inorganic nitrogen fertilizer was again growing rapidly, with much of the growth driven by increased use in China (Figure 5-1). Use as of 1996 was approximately 83 Tg N yr^{-1}. Approximately half of the inorganic nitrogen fertilizer that was ever used on Earth has been applied during the last 15 years.

Production of nitrogen fertilizer is the largest process whereby human activity mobilizes nitrogen globally (Box 5-1). However, other human-controlled processes, such as combustion of fossil fuels and production of nitrogen-fixing crops in agriculture, convert atmospheric nitrogen into biologically available forms of nitrogen. Overall, human fixation of nitrogen (including production of fertilizer, combustion of fossil fuel, and production of nitrogen-fixing agricultural crops) increased globally some two- to three-fold between 1960 to 1990 and continues to grow (Galloway et al. 1995). By the mid 1990s, human activities made new nitrogen avail-

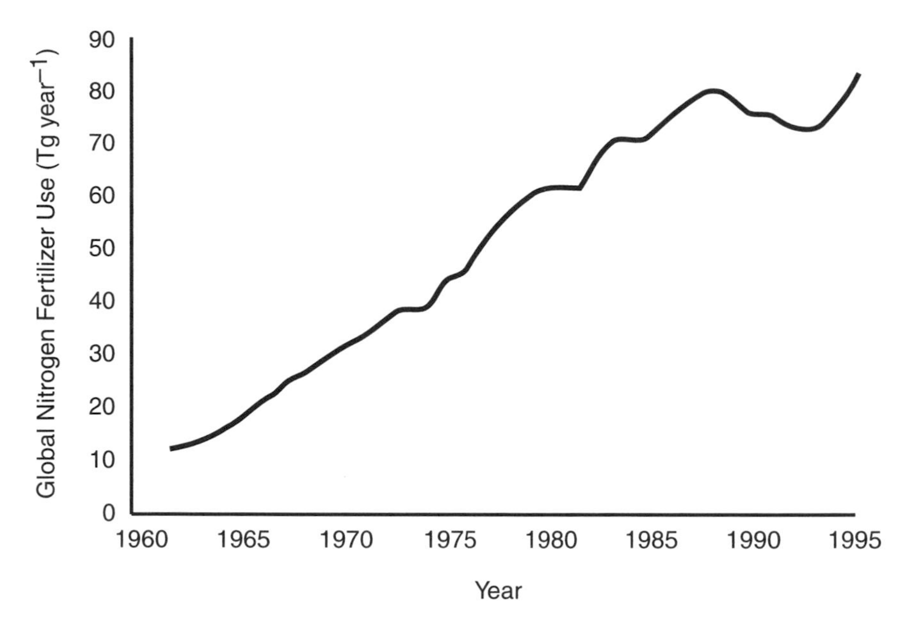

FIGURE 5-1 Annual global nitrogen fertilizer consumption for 1960-1995 (1 Tg = 10^{12} g; data from FAO 1999). The rate of increase was relatively steady until the late 1980s, when collapse of the former Soviet Union reduced fertilizer use in Russia. Fertilizer use is growing again, driven in large part by use in China (modified from Matson et al. 1997).

BOX 5-1
The Fate of Nitrogen Fertilizer in North America

When nitrogen fertilizer is applied to a field, it can move through a variety of flow paths to downstream aquatic ecosystems (Figure 5-2). Some of the fertilizer leaches directly to groundwater and surface waters, with the range varying from 3 percent to 80 percent of the fertilizer applied, depending upon soil characteristics, climate, and crop type (Howarth et al. 1996). On average for North America, some 20 percent is leached directly to surface waters (NRC 1993a; Howarth et al. 1996). Some fertilizer is volatilized directly to the atmosphere; in the United States, this averages 2 percent of the application, but the value is higher in tropical countries and also in countries that use more ammonium-based fertilizers, such as China (Bouwman et al. 1997). Much of the nitrogen from fertilizer is incorporated into crops and is removed from the field in the crops when they are harvested, which is of course the objective of the farmer. A recent National Research Council report

continued

BOX 5-1 Continued

(NRC 1993a) suggests that on average 65 percent of the nitrogen applied to croplands in the United States is harvested, although other estimates are somewhat lower (Howarth et al. 1996). By difference, on average approximately 13 percent of the nitrogen applied must be building up in soils or denitrified to nitrogen gas.

Since much of the nitrogen is harvested in crops, it is important to trace its eventual fate. The majority of the nitrogen is fed to animals (an amount equivalent to 45 percent of the amount of fertilizer originally applied, if 65 percent of the nitrogen is actually harvested in crops; Bouwman and Booij 1998). Some of the

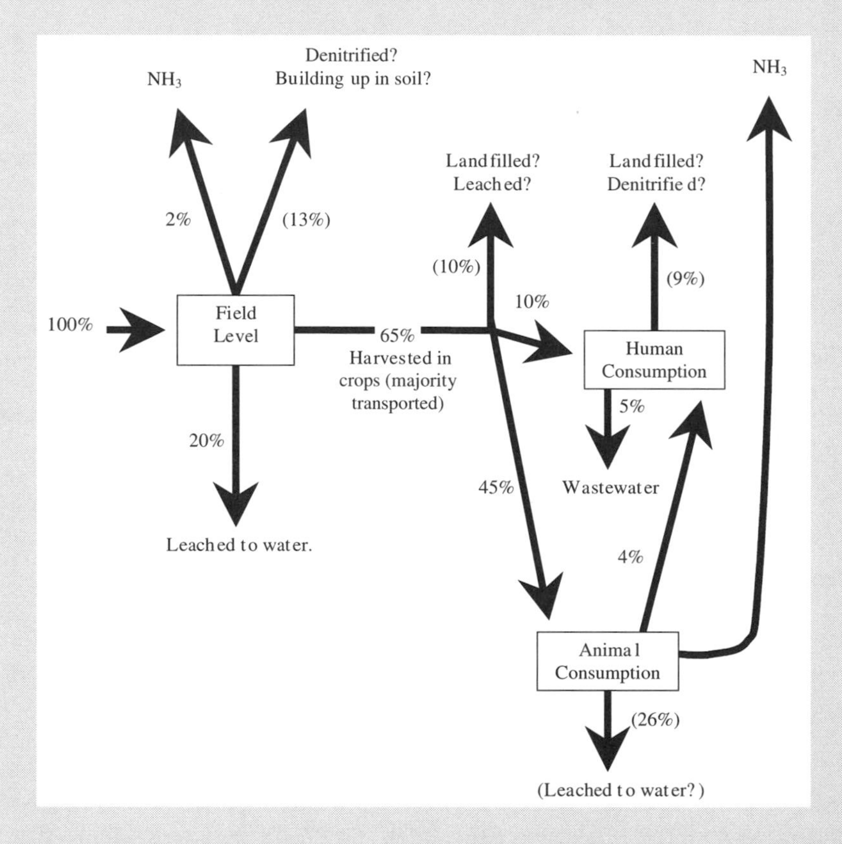

FIGURE 5-2 The average fate of nitrogen fertilizer applied to agricultural fields for North America. The numbers in parentheses are calculated by difference, and the other numbers are direct estimates (unpublished figure by R. Howarth).

continued

BOX 5-1 Continued

nitrogen is directly consumed by humans eating vegetable crops—in North America perhaps 10 percent of the amount of nitrogen originally applied to the fields (Bouwman and Booij 1998). By difference, perhaps 10 percent of the amount of nitrogen originally applied to fields is lost during food processing, being placed in landfills or released to surface waters from food-processing plants.

Of the nitrogen that is consumed by animals, much is volatilized from animal wastes to the atmosphere as ammonia. In North America, this volatilization is roughly one-third of the nitrogen fed to animals (Bouwman et al. 1997), or 15 percent of the amount of nitrogen originally placed on the fields. This ammonia is deposited back onto the landscape, often near the source of volatilization, although some of it first travels for long distances through the atmosphere (Holland et al. 1999). Some of the nitrogen in animals is consumed by humans, an amount roughly equivalent to 10 percent of the amount of nitrogen fed to the animals, or 4 percent of the nitrogen originally applied to fields. By difference, the remainder of the nitrogen—over 25 percent of the amount of nitrogen originally applied to the fields—is contained in animals wastes that are building up somewhere in the environment. Most of this may be leached to surface waters.

Of the nitrogen consumed by humans, either through vegetable crops or meat, some is released through wastewater treatment plants and from septic tanks. In North America, this is an amount equivalent to approximately 5 percent of the amount of nitrogen originally applied to fields (Howarth et al. 1996). By difference, the rest of the nitrogen is placed as food wastes in landfills or is denitrified to nitrogen in wastewater treatment plants and septic tanks.

The conclusion is that fertilizer leaching from fields is only a portion of the nitrogen that potentially reaches estuaries and coastal waters. Probably of greater importance for North America as a whole is the nitrogen that is volatilized to the atmosphere or released to surface waters from animal wastes and landfills. Since food is often shipped over long distances in the United States, the environmental effect of the nitrogen can occur well away from the original site of fertilizer application.

able at a rate of some 140 Tg N yr^{-1} (Vitousek et al. 1997), matching the natural rate of biological nitrogen fixation on all the land surfaces of the world (Vitousek et al. 1997; Cleveland et al. 1999). Thus, the rate at which humans have altered nitrogen availability globally far exceeds the rate at which humans have altered the global carbon cycle (Figure 5-3).

The human alteration of nutrient cycles is not uniform over the earth, and the greatest changes are concentrated in the areas of greatest popula-

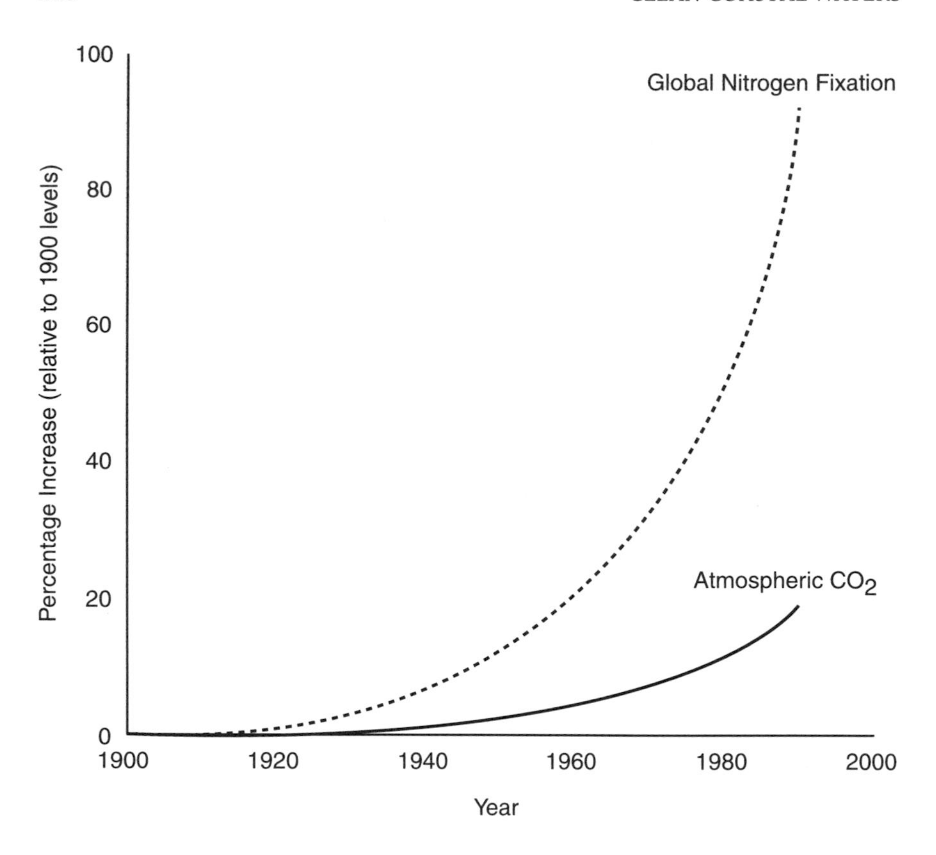

FIGURE 5-3 The relative change in nitrogen fixation caused by human activities globally compared to the increase in carbon dioxide in the atmosphere since 1900. Note that humans are having a much greater influence on nitrogen availability than they are on the production of carbon dioxide, an important greenhouse gas (modified from Vitousek et al. 1997).

tion density and greatest agricultural production. Some regions of the world have seen very little change in the flux of either nitrogen or phosphorus to the coast (Howarth et al. 1995, 1996), while in other places the change has been tremendous. Human activity is estimated to have increased nitrogen inputs to the coastal waters of the northeastern United States generally, and to Chesapeake Bay specifically, by some six- to eightfold (Boynton et al. 1995; Howarth et al. 1996; Howarth 1998). Atmospheric deposition of nitrogen has increased even more than this in the northeast (Holland et al. 1999). The time trends in human perturbation of nutrient cycles can also vary among regions. For example, while the global use of inorganic nitrogen fertilizer continues to increase, the use of nitrogen fertilizer in the United States has increased relatively little since 1985 (Figure 5-4; Evans et al. 1996).

Note, however, that the use of nitrogen fertilizer in the United States in the next century may again increase to support greater exports of food to developing countries. Countries such as China have been largely self sufficient in food production for the past two decades, in part because of increased use of nitrogen fertilizer. The use of fertilizer in China is now very high—almost 10-fold greater than in the United States—and further increases in fertilizer use are less likely to lead to huge increases in food production as they have in the past. Therefore, if China's population continues to grow it may once again be forced to import food from the United States and other developed countries, leading to more use of nitrogen fertilizer here.

WASTEWATER AND NONPOINT SOURCE INPUTS

Traditionally, most water quality management emphasizes control of discharges from wastewater treatment plants and other point sources.

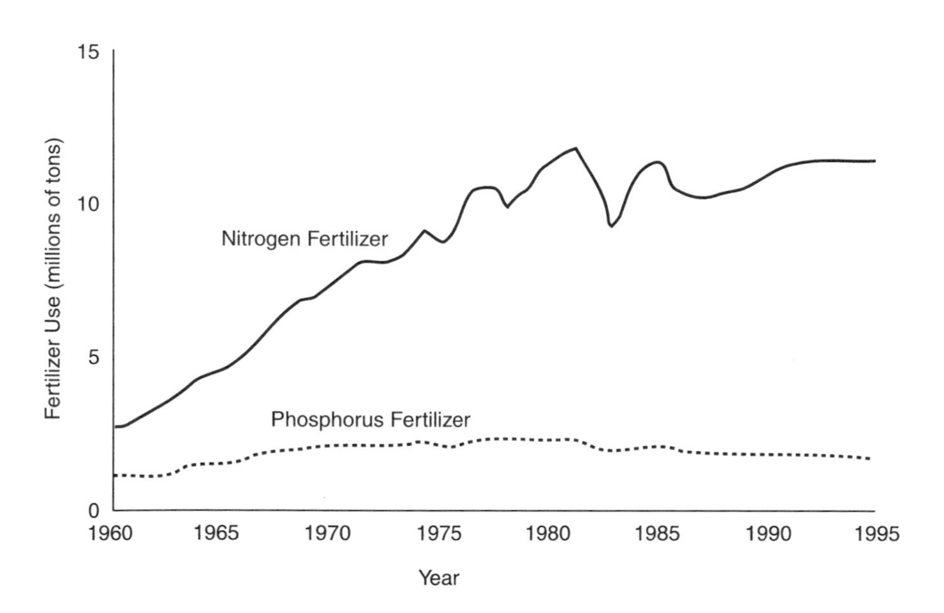

FIGURE 5-4 U.S. commercial fertilizer use (modified from Evans et al. 1996).

However, generally of greater concern for nutrients and coastal eutrophication are "nonpoint sources" of nutrients (NRC 1993a). A regional-scale analysis of fluxes of nitrogen from the landscape to the coast of the North Atlantic Ocean demonstrated that nonpoint sources of nitrogen exceeded sewage inputs for all regions in both Europe and North America (Howarth et al. 1996). Overall, sewage contributed only 12 percent of the flux of nitrogen from the North American landscape to the North Atlantic Ocean (Howarth et al. 1996). Nonpoint sources also dominate for phosphorus inputs to surface waters in the United States (Sharpley and Rekolainen 1997; Carpenter et al. 1998), and because of an effort to control phosphorus point source pollution, nonpoint sources of phosphorus have grown in relative importance since 1980 (Jaworski 1990; Sharpley et al. 1994; Litke 1999).

Wastewater inputs can sometimes be a major source of nitrogen to an estuary when the watershed is heavily populated and small relative to the surface area of the estuary itself (Nixon and Pilson 1983). Even in some estuaries fed by larger watersheds, wastewater can be the largest source of nitrogen if the watershed is heavily populated. For example, wastewater contributes an estimated 60 percent of the nitrogen inputs to Long Island Sound, largely due to sewage from New York City (CDEP and NYSDEC 1998). However, nitrogen and phosphorus inputs from nonpoint sources in most estuaries are greater than are inputs from wastewater, particularly in estuaries that have relatively large watersheds (NRC 1993a). For example, only one-quarter of the nitrogen and phosphorus inputs to Chesapeake Bay come from wastewater treatment plants and other such point sources (Boynton et al. 1995; Nixon et al. 1996). For the Mississippi River, sewage and industrial point sources contribute an estimated 10 percent (Howarth et al. 1996) to 20 percent (Goolsby et al. 1999) of the total nitrogen flux (organic and inorganic nitrogen) and 40 percent of the total phosphorus flux (Goolsby et al. 1999).

As discussed in more detail in Chapter 9, many technologies exist for reducing nutrient discharges from wastewater treatment plants. The relatively standard approaches of using primary and secondary sewage treatment lower phosphorus and nitrogen discharges on average by approximately 20 percent to 25 percent, although there is a significant variation among plants (Viessman and Hammer 1998; NRC 1993a). Additional tertiary treatment for nutrient removal can lower nitrogen discharges by 80 percent to 88 percent and phosphorus discharges by 95 percent to 99 percent (NRC 1993a). However, most wastewater treatment plants in the United States do not have adequate nitrogen removal capabilities. In Tampa Bay, wastewater treatment plants were a major source of nitrogen prior to the institution of tertiary nitrogen removal, and this treatment has successfully reversed the trend in eutrophication there (Johansson and Greening 2000).

Reduction in the eutrophication of most estuaries requires the management of nutrient inputs from nonpoint sources in addition to those of wastewater treatment plants and industrial sources (NRC 1993a). The nature of these sources is described in the remainder of this chapter.

DISTURBANCE, NONPOINT NUTRIENT FLUXES, AND BASELINES FOR NUTRIENT EXPORTS FROM PRISTINE SYSTEMS

In a landscape that is completely undisturbed by humans, export of nitrogen and phosphorus to downstream aquatic ecosystems tends to be small, particularly in the temperate zone (Hobbie and Likens 1973; Omernik 1977; Rast and Lee 1978; Howarth et al. 1996). Assuming that the landscape is in an ecological steady state, the export of nutrients cannot exceed the inputs. For nitrogen, these inputs are biological nitrogen fixation and deposition of nitrogen compounds from the atmosphere; in the temperate zone both tend to be small in the absence of human disturbance (Howarth et al. 1996; Cleveland et al. 1999; Holland et al. 1999). Thus, the export of nitrogen from undisturbed temperate landscapes must also be low, in fact lower than the input because there is some accumulation of nitrogen in the system and some loss of nitrogen through denitrification (the bacterial conversion of reactive nitrate into nonreactive molecular nitrogen). For tropical regions, rates of biological nitrogen fixation and natural deposition of nitrogen from the atmosphere are far higher, and so nitrogen export to downstream ecosystems from undisturbed ecosystems may also be greater (Howarth et al. 1996; Cleveland et al. 1999; Holland et al. 1999; Lewis et al. 1999).

Unfortunately, it is difficult to determine with any precision the magnitude of the natural flux of nitrogen from a temperate landscape like the United States. Atmospheric pollution and the resulting elevated nitrogen deposition are widespread, providing some level of disturbance virtually everywhere in the country, and in fact in most of the world's temperate ecosystems (Holland et al. 1999). There are a few remaining temperate forests that do not receive elevated nitrogen deposition from pollution sources, such as some remote forests in Chile (Hedin et al. 1995). However, these are poor models for most of the temperate systems of the United States as the Chilean forests receive high precipitation and runoff, and have vastly different ecological histories.

An expert panel under the auspices of the International SCOPE (Scientific Committee on Problems of the Environment) Nitrogen Project estimated that pristine temperate-zone ecosystems, such as those that had characterized much of North America and Europe prior to human disturbance, would export between 75 and 230 kg N km^{-2} yr^{-1} to downstream aquatic ecosystems, with the median estimate being 133 kg N km^{-2} yr^{-1} (Howarth et al. 1996; Howarth 1998). This provides the best estimate available for the natural, background load of nitrogen from the landscapes in the continental United States. Valigura et al. (2000) estimate that for estuaries with small watersheds, the nitrogen flux off the pristine landscape prior to European colonization was at the low end of this range, perhaps 78 to 108 kg N km^{-2} yr^{-1}. Assuming a baseline flux of 133 kg N km^{-2} yr^{-1} for an undisturbed temperate landscape, human activity has increased the nitrogen flux in the Mississippi River by more than 4-fold, in the rivers of the northeastern United States by 8-fold, and in the rivers draining to the North Sea by 11-fold (Howarth 1998). In an independent analysis for Chesapeake Bay, Boynton et al. (1995) estimated that nitrogen fluxes have increased some 6- to 8-fold since pre-colonial times, a value consistent with the conclusion from the International SCOPE Nitrogen Project.

In an undisturbed landscape, the major source of phosphorus to a terrestrial ecosystem is the weathering of the soil and parent-rock material, which tends to be relatively slow and therefore sets a low limit on the export of phosphorus. As a global average, the export of phosphorus from the terrestrial landscape prior to human disturbance can be estimated from the oceanic sedimentary record and was somewhat greater than 50 kg P km^{-2} yr^{-1}, expressed per area of land surface (Howarth et al. 1995). However, this clearly depends on the phosphorus content of the parent-rock material, the rate of weathering, and other environmental conditions, including the rate of erosion. The current flux of phosphorus from the landscape is in fact less than 50 kg P km^{-2} yr^{-1} for more than half of the area in the Mississippi River basin (Goolsby et al. 1999), and is only 5 kg P km^{-2} yr^{-1} for the watersheds of Hudson's Bay, Canada (Howarth et al. 1996). On the other hand, the rather large export of phosphorus from the Amazon River basin of over 230 kg P km^{-2} yr^{-1} appears to be a largely natural phenomenon (Howarth et al. 1996). Given the site-specific nature of phosphorus export and the paucity of information on background phosphorus losses from a given location prior to cultivation, no baseline for the natural rate of phosphorus export exists.

Disturbance of the landscape increases the export of both nitrogen and phosphorus, although there are some major differences in the responses of these two nutrients. As a general rule, most export of phosphorus from disturbed systems occurs as phosphorus bound to particles, so factors regulating erosion and sedimentation are critical in controlling phosphorus fluxes. An important exception can occur in sandy soils with low phosphorus adsorption capacities (Sharpley et al. 1998; Sims et al. 1998); such soils can be important to consider when managing eutrophication in portions of the Atlantic coastal plain. Some phosphorus moves through the atmosphere as dust particles, and this can contribute greatly to the phosphorus economies of some remote oceanic waters and forests. In general, such inputs of phosphorus to estuaries and coastal waters are not as important as inputs in surface waters.

For nitrogen, some export also occurs in particle-bound forms, but nitrogen tends to be much more mobile through soils in dissolved form than phosphorus, so significant exports can occur in groundwater (Paerl 1997) or as dissolved nitrogen in surface waters. In addition (and also unlike phosphorus), reactive nitrogen compounds can be quite mobile in the atmosphere. For example, significant amounts of ammonia gas from agricultural sources (particularly urea- and ammonia-based fertilizers, manures, and animal feedlot wastes) volatilize to the atmosphere and are deposited elsewhere in the landscape (Bouwman et al. 1997; Holland et al. 1999). Globally, of the 60 to 80 Tg N yr^{-1} applied as inorganic nitrogen fertilizer, 21 to 52 Tg N yr^{-1} are estimated to be volatilized to the atmosphere as ammonia, either directly from the fertilizer or from animal waste (Holland et al. 1999). That is, on average some 40 percent of the inorganic nitrogen fertilizer that is applied cycles through the atmosphere and is redeposited. In the United States, the value is somewhat lower, but still 25 percent of the inorganic nitrogen fertilizer that is used is volatilized to the atmosphere (Holland et al. 1999).

For phosphorus, agriculture is the largest disturbance controlling nonpoint fluxes of phosphorus in the landscape (Carpenter et al. 1998). For nitrogen, both agriculture and fossil-fuel combustion contribute significantly to nonpoint source flows to estuaries and coastal waters (Howarth et al. 1996). Some of this nitrogen export comes directly from agricultural fields, but because of both substantial nitrogen transport in the atmosphere and nitrogen mobility in dissolved forms, the nitrogen export from other types of ecosystems, including forests, can be substantial. Since agriculture dominates the nonpoint source flux of phosphorus and contributes significantly to nonpoint sources of nitrogen (often dominating it as well), changes in agricultural practices over the last few decades contribute to these nutrient fluxes. Industrial and fossil fuel sources of nitrogen and the mechanisms that control both nitrogen and phosphorus fluxes in the landscape will be discussed later in this chapter.

CHANGES IN AGRICULTURAL PRODUCTION AND NONPOINT SOURCE NUTRIENT POLLUTION

One of the greatest changes in agriculture has been the use of inorganic fertilizers, which expanded dramatically after World War II in response to the demand for increased agricultural output. In the developed countries, large processing plants were built to manufacture nitrogenous fertilizers and convert imported rock phosphate into a variety of water-soluble and partially water-soluble phosphorus fertilizer products. Basic slag, a by-product from the steel industry, also became widely used in the manufacturing of phosphorus fertilizer. In the United States, the use of inorganic phosphorus fertilizer rose rapidly in the 1940s and 1950s, but has been relatively constant since 1960. The rate of use of inorganic nitrogen fertilizer, on the other hand, continued to rise rapidly until the early 1980s (Figure 5-4). This relative gain in nitrogen use over phosphorus use resulted primarily from favorable crop yield responses, especially corn, to nitrogen fertilizers.

Over the last 30 years, agricultural production systems in the United States have become more specialized and concentrated. During this time, overall agricultural production has more than doubled (Evans et al. 1996), and is occurring on less agricultural land and on fewer but larger farms (Evans et al. 1996). Since 1950, U.S. farmland has decreased from 1,200 to 970 million acres (20 percent) and the number of farms has dropped from 5.6 to 2.1 million (63 percent), while average farm size has increased from 213 to 469 acres (120 percent).

In many states, animal feeding operations (AFOs) are now a major source of agricultural income. The rapid growth of the animal industry in certain areas of the United States has been coupled with an intensification of operations. For example, current census information shows an 18 percent increase in the numbers of hogs in the United States over the last 10 years, at the same time as a 72 percent decrease in numbers of hog farms. Over the same 10 years, the number of dairy farms decreased 40 percent, but herd size increased 50 percent. A similar intensification of the poultry and beef industries has also occurred, with 97 percent of poultry production in the United States coming from operations with more than 100,000 birds and over a third of beef production coming from just under 2 percent of the feedlots (Gardner 1998). Driving this intensification is an increased demand for animal products and improved profitability because of advances in transportation, processing, and marketing. But animal feeding operations pose significant challenges with the management of wastes produced.

Prior to World War II, farming communities tended to be self-sufficient, in that enough feed was produced locally to meet animal requirements and the manure nutrients could be effectively recycled to meet crop needs. After World War II, increased fertilizer use in crop production fragmented farming systems and created specialized crop and animal operations that efficiently coexisted in different regions. Since farmers did not need to rely on manures as fertilizers (the primary source until fertilizer production and distribution became cheap), grain and animal production could be spatially separated. By 1995 the major animal producing states imported over 80 percent of their grain for feed (Lanyon and Thompson 1996). In fact, less than a third of the grain produced on farms today is fed to animals on the farm where it is grown (USDA 1989).

This evolution of agricultural systems is resulting in a major transfer of nutrients from grain-producing areas to animal-producing areas and, consequently, accumulation of nitrogen and phosphorus in soils of the animal-producing areas. For example, the potential for nitrogen and phosphorus surplus at the farm scale can be much greater in AFOs than in cropping systems, because nutrient inputs become dominated by feed rather than fertilizer (Isermann 1990; NRC 1993b). Thus, many water quality concerns are a result of this imbalance in system inputs and outputs of nitrogen and phosphorus, which have been brought about by an increase in AFOs. Lander et al. (1998) calculated the amounts of nitrogen and phosphorus produced by manure in confined AFOs on a countywide basis (Figure 5-5). From this and crop yield information, Lander et al. (1998) were able to identify those counties where more than 100 percent of the nitrogen and phosphorus needed for crop production was available from livestock manure. (In other words, counties where manure production exceeded crop need—assuming that manure was applied only to non-legumes and harvested crop land and hay land; Figure 5-6).

The number of U.S. counties where manure nitrogen and phosphorus exceeds the potential crop uptake and removal has been steadily increasing since 1950 (Figure 5-7). This increase has been greater for phosphorus than nitrogen (Kellogg and Lander 1999). In those areas with an excess of nitrogen and phosphorus relative to crop needs, there is a greater risk of nutrient export from agricultural watersheds to surface and ground waters (Figure 5-8). This excess of nutrients in manure tends to occur in areas where downstream export is likely due to relatively wet climates, since high water availability is conducive to animal feeding operations.

The limited large-scale geographic information available to summarize phosphorus soil test results shows trends in soil phosphorus build-up to very high levels in some areas. These areas of phosphorus build up and often coincide with areas of intensive animal production (Fixen 1998; Figure 5-9). Soils in this category require little or no input of phosphorus, either from fertilizer or organic by-products, for economically optimum crop production. In many of these areas, this build-up of soil phosphorus

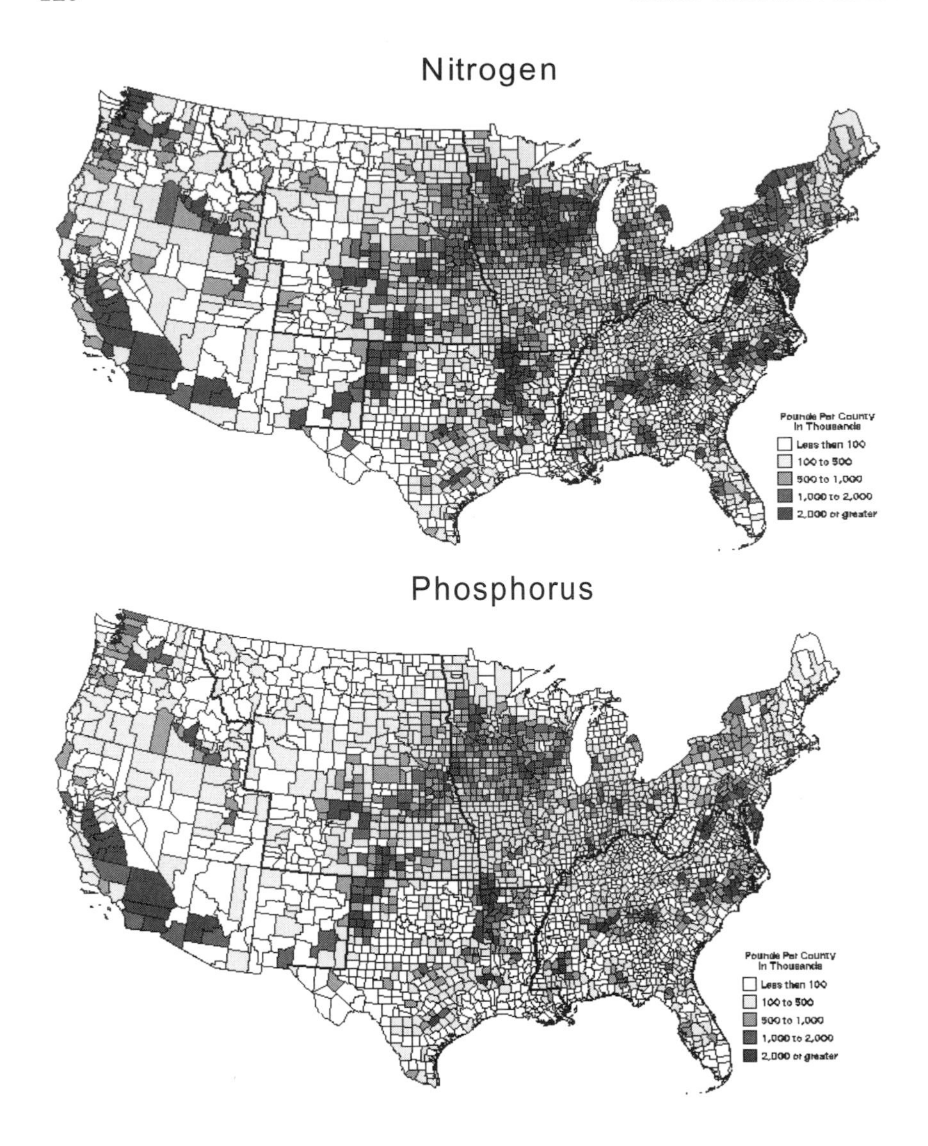

FIGURE 5-5 Estimated manure nitrogen and phosphorus production from confined livestock (modified from Lander et al. 1998).

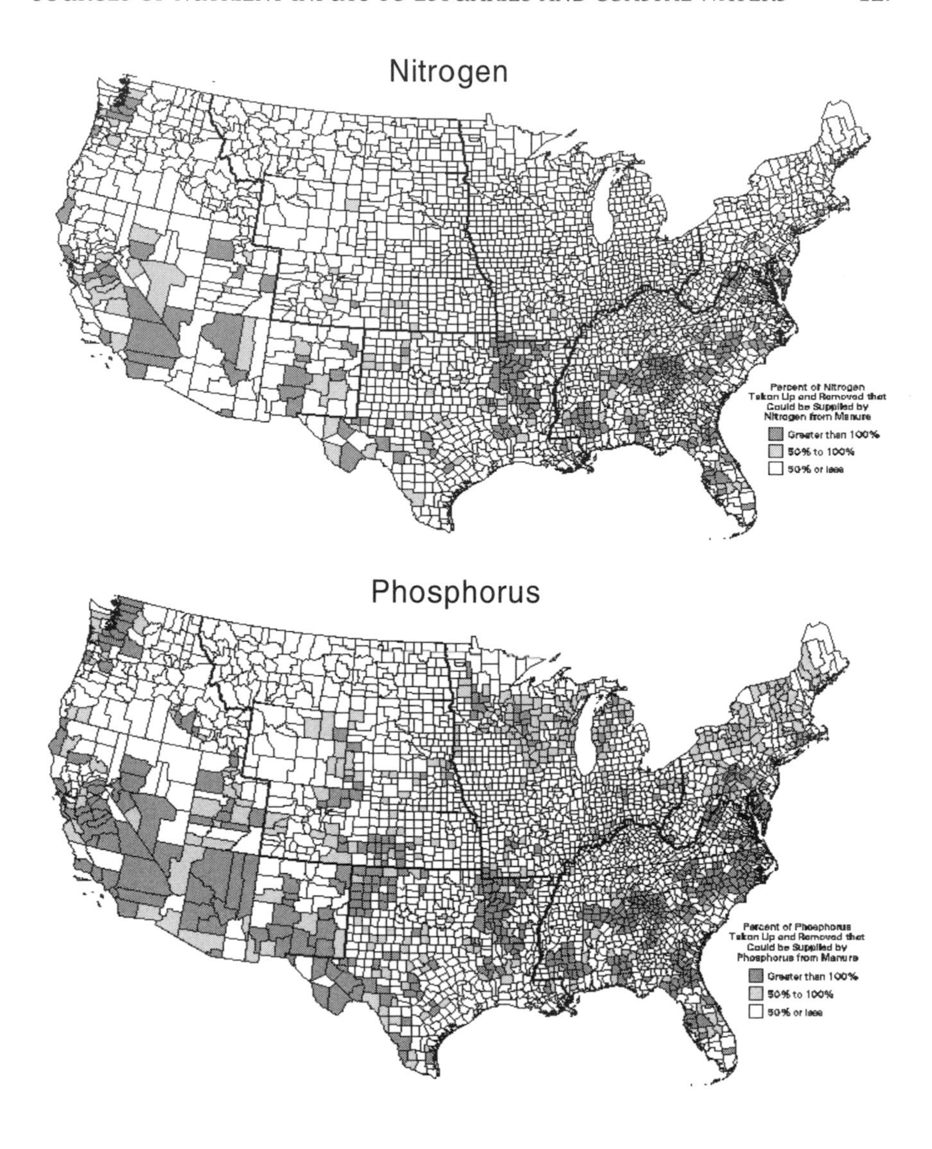

FIGURE 5-6 Potential for nitrogen and phosphorus available from animal manure to meet or exceed plant uptake and removal on non-legume, harvested cropland, and hayland (modified from Lander at al. 1998).

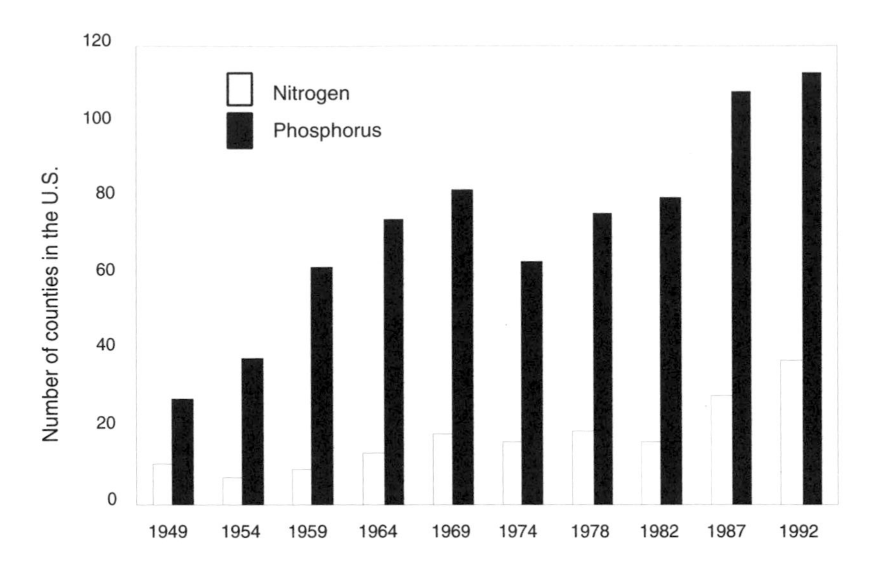

FIGURE 5-7 Number of counties where manure nutrients exceed the potential plant uptake and removal, including pastureland application (modified from Kellogg and Lander 1999).

has increased the risk for phosphorus movement in surface runoff and in some cases (notably Florida and the Delmarva Peninsula) into shallow groundwater aquifers. Unfortunately, many of these areas of soil phosphorus build-up tend to occur in areas with phosphorus-sensitive water resources or major drainage ways, such as the inland waters of the Carolinas, Florida Everglades, Lake Okeechobee, Great Lakes, and the Mississippi River basin (Figure 5-9). Although this survey of soil phosphorus includes only samples sent for analysis and does not represent a complete survey of all soils in the United States, it does highlight some of the effects of long-term changes on agricultural production systems.

How has this come about? Using the Chesapeake Bay drainage basin as an example, if all the manure produced within the basin in 1939 were made available for application to corn, large areas of the corn cropland would not have received adequate amounts (3,000 kg km^{-2} yr^{-1}; Figure 5-10; Lanyon 1999). Some areas of the basin with higher potential applications, such as New York and western Virginia, probably had limited areas of corn production at that time. In other words, without importation of fertilizer from outside the basin, the availability of manure limited crop

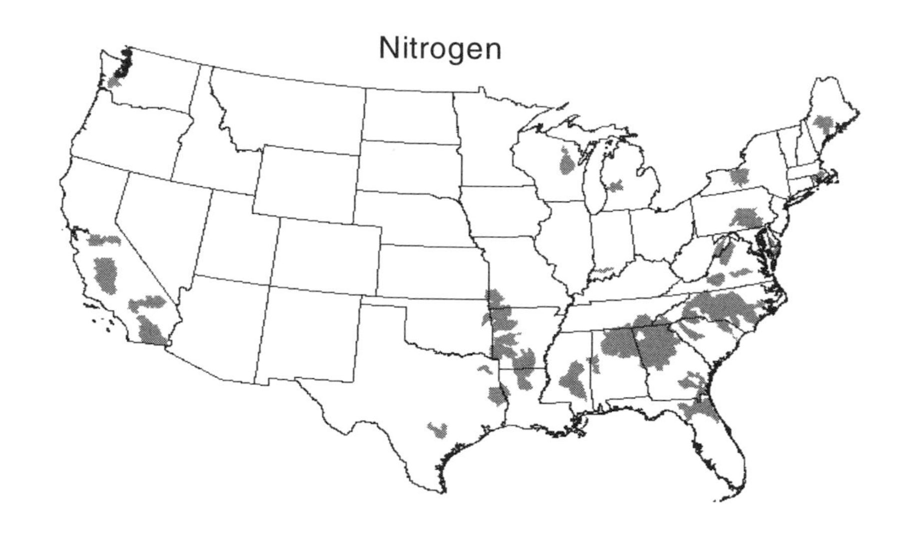

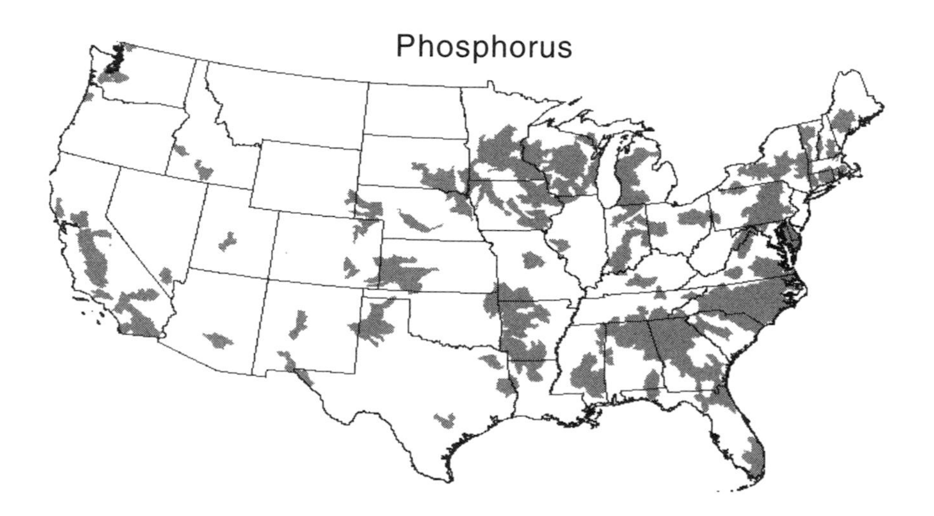

FIGURE 5-8 Watersheds with a high potential for soil and water degradation from manure nitrogen and phosphorus (modified from Kellogg and Lander 1999).

production. However, by 1992 manure production exceeded the corn requirements in large areas of Virginia, West Virginia, Delaware, parts of Pennsylvania, and the New York area of the drainage basin (Figure 5-10), and the need to dispose of excess manure (rather than crop need) began to shape patterns of fertilizer application. These patterns of nutrient distri-

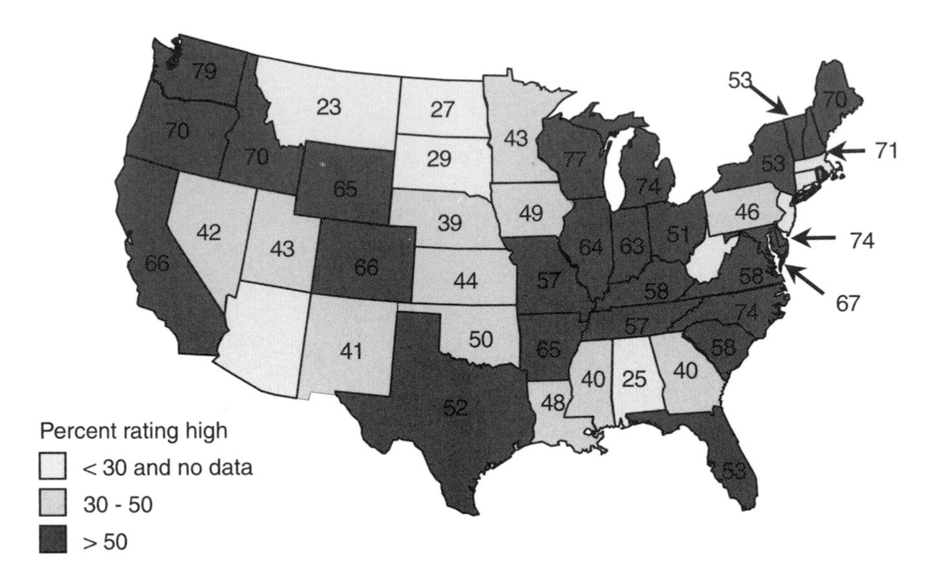

FIGURE 5-9 Percent of soil samples analyzed in state laboratories that tested in the "high or above" range for phosphorus in 1997. Highlighted states had more than 50 percent of soil samples testing in the "high or above" range (modified from Fixen 1998).

bution and accumulation have come about by a number of complex and interrelated factors, not merely independent farmer decisions (Lanyon 1999). Farmers could do little to increase nutrient supplies on their farms when nitrogen and phosphorus fertilizers were scarce. It was only after the emergence of the fertilizer industry and the associated pattern of intensive animal feed operations that nitrogen and phosphorus supplies on farms could be increased to exceed farm nutrient requirements.

Export of Phosphorus from Agricultural Systems

Several surveys of U.S. watersheds have clearly shown that phosphorus loss in runoff increases as the forested portion of the watershed decreases and agriculture increases (Omernik 1977; Rast and Lee 1978). In general, forested watersheds conserve phosphorus, with phosphorus input in dust and in rainfall usually exceeding outputs in stream flow (Taylor et al. 1971; Hobbie and Likens 1973; Schreiber et al. 1976). Surface runoff from forests, grasslands, and other noncultivated soils carries little sediment,

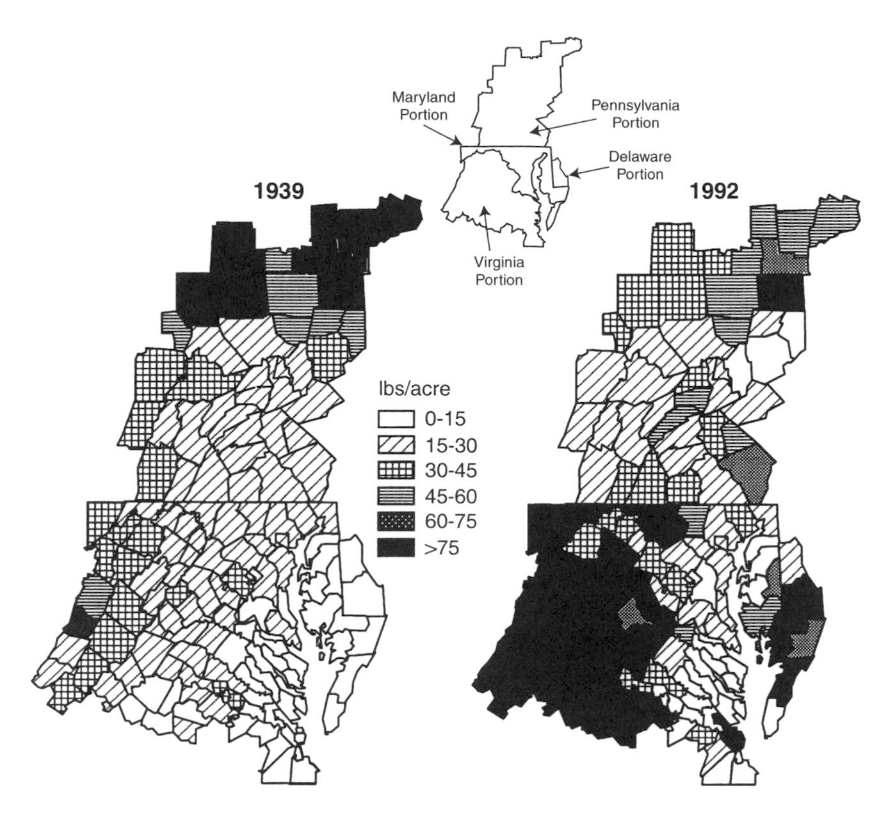

FIGURE 5-10 Available manure phosphorus per acre of corn in the Chesapeake Bay drainage basin before and after World War II (1939 and 1992, respectively) (modified from Lanyon 1999).

so phosphorus fluxes are low and the export that occurs is generally dominated by dissolved phosphorus. This loss of phosphorus from forested land tends to be similar to that found in subsurface or dissolved base flow from agricultural land (Ryden et al. 1973; House and Casey 1988). The cultivation of land in agriculture greatly increases erosion and with it the export of particle-bound phosphorus. Typically, particulate fluxes constitute 60 to 90 percent of phosphorus exported from most cultivated land (Sharpley et al. 1995). In the eastern United States, conversion of land from forests to agriculture between 1700 and 1900 resulted in a 10-fold increase in soil erosion and a presumed similar increase in phosphorus export to coastal waters, even without any addition of phosphorus fertilizer (Meade 1988; Howarth et al. 1996). The soil-bound phosphorus includes both inorganic phosphorus associated with soil particles and phosphorus bound in organic material eroded during flow events. Some of the sediment-bound phosphorus is not readily available (Howarth et al. 1995), but much of it can be a long-term source of phosphorus for aquatic biota (Sharpley 1993; Ekholm 1994).

Increases in phosphorus export from agricultural landscapes have been measured after the application of phosphorus (Sharpley and Rekolainen 1997). Phosphorus export is influenced by the rate, time, and method of phosphorus application, form of fertilizer or manure applied, amount and time of rainfall after application, and land cover. These losses are often small from the standpoint of farmers (generally less than 200 kg P km^{-2}), and represent a minor proportion of fertilizer or manure phosphorus applied (generally less than 5 percent). Thus, these losses are not of economic importance to farmers in terms of irreplaceable fertility. However, they can contribute to eutrophication of downstream aquatic ecosystems.

While phosphorus export from agricultural systems is usually dominated by surface runoff, important exceptions occur in sandy, acid organic, or peaty soils that have low phosphorus adsorption capacities and in soils where the preferential flow of water can occur rapidly through macropores (Sharpley et al. 1998; Sims et al. 1998). Soils that allow substantial subsurface export of dissolved phosphorus are common on parts of the Atlantic coastal plain and in Florida, and are thus important to consider in the management of coastal eutrophication in these regions.

Although there exists a good understanding of the chemistry of phosphorus in soil—water systems, the hydrologic pathways linking spatially variable phosphorus sources, sinks, temporary storages, and transport processes in landscapes are less well understood. This information is critical to the development of effective management programs that address the reduction of phosphorus export from agricultural watersheds.

Runoff production in many watersheds in humid climates is controlled by the variable source area concept of watershed hydrology (Ward 1984). Here, surface runoff is usually generated only from limited source areas in a watershed. These source areas vary over time, expanding and contracting rapidly during a storm as a function of precipitation, temperature, soils, topography, ground water, and moisture status over the watershed (Gburek and Sharpley 1998). Surface runoff from these areas is limited by soil-water storage rather than infiltration capacity. This situation usually results from high water tables or soil moisture contents in near-stream areas.

The boundaries of surface runoff-producing areas will be dynamic both in and between rainfalls (Gburek and Sharpley 1998). During a rainfall, area boundaries will migrate upslope as rainwater input increases. In dry summer months, the runoff-producing area will be closer to the stream than during wetter winter months, when the boundaries expand away from the stream channel.

Soil structure, geologic strata, and topography influence the location and movement of variable source areas of surface runoff in a watershed. Fragipans or other layers, such as clay pans of distinct permeability changes, can determine when and where perched water tables occur. Shale or sandstone strata also influence soil moisture content and location of saturated zones. For example, water will perch on less permeable layers in the subsurface profile and become evident as surface flow or springs at specific locations in a watershed. Converging topography in vertical or horizontal planes, slope breaks, and hill slope depressions or spurs, also influence variable source area hydrology in watersheds. Net precipitation (precipitation minus evapotranspiration) governs watershed discharge and thus total phosphorus loads to surface waters. This should be taken into account when comparing the load estimates from different regions. It is also one reason why there seems to be more concern with phosphorus in humid regions than in more arid regions.

In watersheds where surface runoff is limited by infiltration rate rather than soil-water storage capacity, areas of the watershed can alternate between sources and sinks of surface flow. This again will be a function of soil properties, rainfall intensity and duration, and antecedent moisture condition. As surface runoff is the main mechanism by which phosphorus is exported from most watersheds (Sharpley and Syers 1979), it is clear that, if surface runoff does not occur, phosphorus export can be small.

Export of Nitrogen from Agricultural Systems

The fate of nitrogen applied as fertilizer to agricultural fields has received extensive study. Overall, nitrogen use in agriculture tends to be relatively inefficient (less than 25 percent of that applied), with animal uptake particularly small (less than 20 percent) compared with crop production systems (NRC 1993b). Generally for the United States, 45 percent to 75 percent of the nitrogen in fertilizer is removed in crop harvest (Bock 1984; Nelson 1985; NRC 1993b). Of the remainder, some is stored as organic nitrogen in the soil, some is volatilized to the atmosphere, and some leaches to ground and surface waters. A variety of factors, including soil type, climate, fertilizer type, and farming practices, influence the fate of fertilizer use (Howarth et al. 1996). For typical farming practices in the United States, the percentage of fertilizer that leaches to ground and surface waters varies between 10 and 40 percent for loam and clay soils, and 25 and 80 percent for sandy soils (Howarth et al. 1996). Overall in North America, it is estimated that 20 percent of the fertilizer nitrogen applied to agricultural fields leaches into ground and surface waters (Howarth et al. 1996), although much of that is lost to denitrification in downstream wetlands, streams, and rivers before reaching estuaries or coastal waters.

A variety of factors affect the volatilization of nitrogen from fertilizer to the atmosphere, including soil type, climate, farming practices, and type of fertilizer (Bouwman et al. 1997). For example, when ammonium sulfate is applied to a soil with a pH below 5.5, less than 2 percent of the ammonium is volatilized (in the form of ammonia) to the atmosphere. Conversely, when ammonium sulfate is applied to calcareous soil (which has a higher pH), up to 50 percent of the nitrogen can be volatilized as ammonia gas to the atmosphere (Whitehead and Raistrick 1990; Bouwman et al. 1997). For typical farming practices, climate, and soils in the United States and Europe, Bouwman et al. (1997) estimated that on average 8 percent of the nitrogen in ammonium sulfate and 15 percent of the nitrogen in urea is volatilized to the atmosphere. The percentages are greater in tropical countries, and the volatilization from nitrate-based fertilizers is much less. While emissions of nitric oxide to the atmosphere are an important nitrogen loss from fertilized fields in tropical areas, this is generally a very small flux in temperate regions, including the United States (Holland et al. 1999). Virtually all the nitrogen volatilized from agricultural fields is eventually redeposited back onto the landscape and can reach estuaries and coastal waters (Howarth et al. 1996). Generally, this nitrogen is redeposited quite close to the point of emission (Holland et al. 1999).

Since 45 to 75 percent of the nitrogen applied as fertilizer is harvested in crops, tracing the fate of nitrogen in food and feedstock is important for understanding nitrogen inputs to natural waters (Howarth et al. 1996). The nitrogen in foods that are consumed by humans becomes sewage and is released in sewage effluent, where it is volatilized to the atmosphere as ammonia from sewage treatment plants or is denitrified (converted to plant-unavailable nitrogen) in the sewage treatment plants. However, in the United States most crops are fed to animals (Bouwman and Booij 1998). Thus, most of the nitrogen in harvested crops is excreted by animals. For animals such as poultry, hogs, and cows kept in barns or sheds, 36 percent of the excreted nitrogen on average is volatilized to the atmosphere as ammonia; keeping cows in meadows instead of barns reduces the atmospheric volatilization by more than 50 percent (Bouwman et al. 1997).

Assuming that (1) 65 percent of the nitrogen applied as fertilizer is removed in crops (NRC 1993b); (2) two-thirds of the crop production in the United States is fed to animals (Bouwman and Booij 1998); (3) the nitrogen growth efficiency for animals is 10 percent (Bouwman and Booij

1998); and (4) 36 percent of the nitrogen excreted by animals is volatilized to the atmosphere (Bouwman et al. 1997), then some 14 percent of all nitrogen applied in fertilizer is eventually volatilized to the atmosphere as ammonia after being consumed by animals. This is in addition to direct volatilization of ammonia from fertilizers and from sewage treatment plants.

Stated another way, ammonia volatilization to the atmosphere from agricultural systems in the United States is of the same order of magnitude as nitrate leaching from agricultural fields into surface waters. In addition, although losses are poorly documented, animal wastes also contribute nitrogen directly to surface waters (Howarth 1998). In a regional comparison of nitrogen cycling in major regions of the United States and Europe, Howarth (1998) found that estimates of nitrogen consumption by domestic animals were far better as predictors of nonpoint source nitrogen fluxes in rivers than were rates of application of inorganic nitrogen fertilizer.

Fate of Nitrogen in Atmospheric Deposition

Reactive nitrogen in the atmosphere includes both reduced compounds (NH_y) and oxidized compounds (NO_y). These come from a variety of sources, including fossil-fuel combustion, biomass burning, lightning, and emissions from soils. In the United States, most NO_y comes from fossil-fuel combustion and most NH_y comes from emissions from agricultural sources (Howarth et al. 1996; Prospero et al. 1996; Bouwman et al. 1997; Holland et al. 1999). The lifetime in the atmosphere for many of these reactive nitrogen compounds is short—from hours to a few days—and a large portion of the nitrogen is deposited near its source (Holland et al. 1999). NO_y contributes to "acid rain," but estuarine waters are well buffered and are not directly susceptible to acidification. Thus, the threat from NO_y discussed here is its role as a contributor of nitrogen for coastal eutrophication.

Nitrogen deposition directly onto the water surfaces of estuaries and coastal waters can be substantial, although this is difficult to measure. Monitoring stations for atmospheric input of nitrogen tend to be scarce in coastal areas (Chapter 8). Where monitoring stations exist, they tend to measure only the nitrogen deposited in precipitation (wet deposition). Dry deposition of nitrogen (the impaction of particles and gases of nitrogen onto water, plant, or land surfaces) has proven difficult to measure in any type of ecosystem, and usually only wet deposition or at best some portion of dry deposition are measured at monitoring sites (Holland et al. 1999).

Evidence indicates that deposition directly onto the water surfaces of estuaries tends to contribute from 1 percent to 40 percent of the total nitrogen inputs (Nixon et al. 1996; Paerl 1997; Paerl and Whitall 1999; Valigura et al. 2000), with estuaries such as the Baltic Sea (Nixon et al. 1996) and Tampa Bay (Zarbock et al. 1996) at the upper end of this range. Furthermore, evidence suggests a significant movement of nitrogen in the atmosphere from the eastern United States to the coastal and even offshore waters of the North Atlantic Ocean where it is deposited (Prospero et al. 1996; Holland et al. 1999); this flux could be as large as half the entire amount of reactive nitrogen emitted into the Earth's atmosphere from the United States. However, because of the large natural flux of nitrogen from the deepwater of the North Altantic Ocean onto the continental shelf off the eastern U.S., this atmospheric deposition probably contributes less than 10 percent of the total input of nitrogen to the surface waters of the continental shelf (Howarth 1998).

Much of the reactive nitrogen deposited from the atmosphere falls onto terrestrial ecosystems. This can affect estuaries and coastal waters to the extent that it is exported from land. The fate of nitrogen deposition in forests has received extensive study. Productivity of most U.S. forests in their natural state is limited by the supply of nitrogen (Vitousek and Howarth 1991). As more nitrogen is made available to these forests from atmospheric deposition, production and storage of nitrogen in organic matter can be expected to increase temporarily. However, this ability of forests to store nitrogen is limited. Forests become "saturated" with respect to nitrogen when inputs exceed the total amount needed by trees and the assimilation capacity (through microbial and abiotic processes) of soil organic matter (Aber et al. 1989; Gundersen and Bashkin 1994; Magill et al. 1997; Emmett et al. 1998). Once a forest is saturated with respect to nitrogen, losses both to the atmosphere and to downstream ecosystems can increase rapidly. In European forests that have received high levels of nitrogen deposition for some time, the downstream export of nitrogen can be high, often greater than 500 kg N km^{-2} yr^{-1} (van Breement et al. 1982; Hauhs et al. 1989; Schulze et al. 1989; Durka et al. 1994). Some evidence indicates that the process whereby forests switch from retaining nitrogen to exporting nitrogen as they become nitrogen saturated can be self-accelerating due to related changes in biogeochemical cycling and ecosystem decline (Schulze et al. 1989; Howarth et al. 1996).

Ecological theory suggests that young aggrading forests tend to retain more nitrogen and be less likely to become nitrogen saturated than old-growth mature forests (Vitousek and Reiners 1975; Aber et al. 1989). Therefore, forests that have been logged or burned within the past several decades to a century can be expected to retain more nitrogen from deposition. However, a variety of factors, in addition to land-use history, can affect the ability of a forest to retain nitrogen, including the species composition of trees, climate, and soil type (Howarth et al. 1996; Aber and Driscoll 1997; Aber et al. 1997; Magill et al. 1997; Emmett et al. 1998). For example, Lajtha et al. (1995) found that only about half the nitrogen input from atmospheric deposition was retained by forests at Cape Cod, Massachusetts, whether the forests were young or mature, apparently because the sandy soils there allow nitrogen to pass through so quickly. On the other hand, forests on stony and sandy loam soils in western Massachusetts retained over 85 percent of nitrogen inputs even when heavily fertilized with nitrogen over a six-year period (Magill et al. 1997). In this fertilization experiment, there was some evidence that nitrogen retention decreased over time as the nitrogen content of the forest increased and that nitrogen saturation occurs more rapidly in pine forests than in hardwood forests (Magill et al. 1997).

In a review of nitrogen retention in U.S. forests, Johnson (1992) found no relationship between nitrogen inputs and nitrogen losses to downstream ecosystems—the percentage of nitrogen deposition that was retained varied among forests from nearly none to virtually all. Much of this variation could have been caused by differences in land use, soil type, and dominant tree species (Lajtha et al. 1995; Aber and Driscoll 1997; Magill et al. 1997; Emmett et al. 1998). Some of the variation, however, could have been due to the exclusion of dissolved organic nitrogen fluxes in the budgets considered by Johnson (1992), all of which included losses only of inorganic nitrogen; losses of organic nitrogen can be considerable from some forests (Hedin et al. 1995; Lewis et al. 1999). In fact, even in northern New England where atmospheric deposition of nitrogen is moderately high, most of the dissolved nitrogen leaving forests is organic nitrogen, rather than inorganic nitrogen (Campbell et al. 2000). Further, many of the budgets reviewed by Johnson (1992) were based on short-term studies, and losses of nitrogen from forests can show considerable year-to-year variation in response to climatic variation (Aber and Driscoll 1997). Finally, dry deposition of nitrogen is difficult to estimate precisely (Howarth et al. 1996; Holland et al. 1999; Valigura et al. 2000), and may not have been correctly characterized in some of the budgets summarized by Johnson (1992).

Emmett et al. (1998) proposed that the extent of nitrogen leaching from a forest can be easily predicted from the "nitrogen status" of the forest, as measured by the ratio of organic carbon to nitrogen in the forest floor. They experimentally illustrated that forests with a low nitrogen status (forest floor carbon:nitrogen greater than 30:1) retain most of the nitrogen added (well over 90 percent), whereas forests with a high nitrogen status (forest floor carbon:nitrogen less than 25:1) retain less than half the nitrogen added through deposition and fertilizer. Similarly, Campbell et al. (2000) have demonstrated that the ratio of organic carbon to organic nitrogen in streamwater draining forests in the northeastern United States is a good predictor of export of inorganic nitrogen from those forests. For example, export of inorganic nitrogen increased dramatically in systems where the organic carbon to organic nitrogen ratio of the streamwater was below 20:1 to 25:1 (Campbell et al. 2000). A variety of factors control whether a forest acts as a sink or source of nitrogen, but forests experiencing long periods of high nitrogen inputs through atmospheric deposition will tend to become saturated with respect to nitrogen (Emmett et al. 1998). Thus, over time a forested watershed that experiences high inputs of nitrogen deposition will reach its capacity to store nitrogen and will begin to act as a source of nitrogen to the streams that drain it. Thus, its not surprising that experiments showed that nitrogen leaching from a forest slowed quickly after deposition was reduced through the use of roof exclosures (Bredemeier et al. 1998). This lead Emmett et al. (1998) to suggest that "immediate benefits in water quality could be expected following any reduction in nitrogen deposition loading."

For management of lake acidification in some areas of Europe, managers have adopted the "critical load" concept (Bashkin 1997). This approach sets a goal of keeping atmospheric deposition below some level where it is thought that downstream release will be kept small enough to keep any ecological damage at an acceptable level. Research supports the conclusion that downstream release of nitrogen (and associated acid) can be expected to occur when a critical load value of 1,000 kg N km^{-2} yr^{-1} from atmospheric deposition is reached (Schulze et al. 1989; Pardo and Driscoll 1993; Emmett and Reynolds 1996; Williams et al. 1996; Skeffington 1999; Figure 5-11). Average levels of nitrogen deposition (wet plus dry) currently exceed 1,000 kg N km^{-2} yr^{-1} for the northeastern United States and for much of Europe (Howarth et al. 1996; Prospero et al. 1996; Holland et al. 1999; Valigura et al. 2000).

The export of nitrogen following deposition onto terrestrial ecosystems other than forests has received less study. Some evidence indicates that grasslands are as retentive of nitrogen as forests, or even more so (Dodds et al. 1996). When nitrogen from the atmosphere is deposited onto agricultural fields, its fate is similar to the fate of nitrogen fertilizer applied to such fields, although nitrogen deposited during the non-growing season could be prone to greater loss to downstream ecosystems. The fate of atmospheric nitrogen deposited onto urban and suburban landscapes appears to be virtually unstudied, although the nitrogen content of stormwater runoff from urban environments is high (EPA 1983). Deposition onto urban landscapes is high, as expected since much of the reactive nitrogen in the atmosphere is deposited near sources, and it is reasonable to expect that the export of this deposition to coastal waters is also high. However, nitrogen deposition (wet or dry) in urban environments is

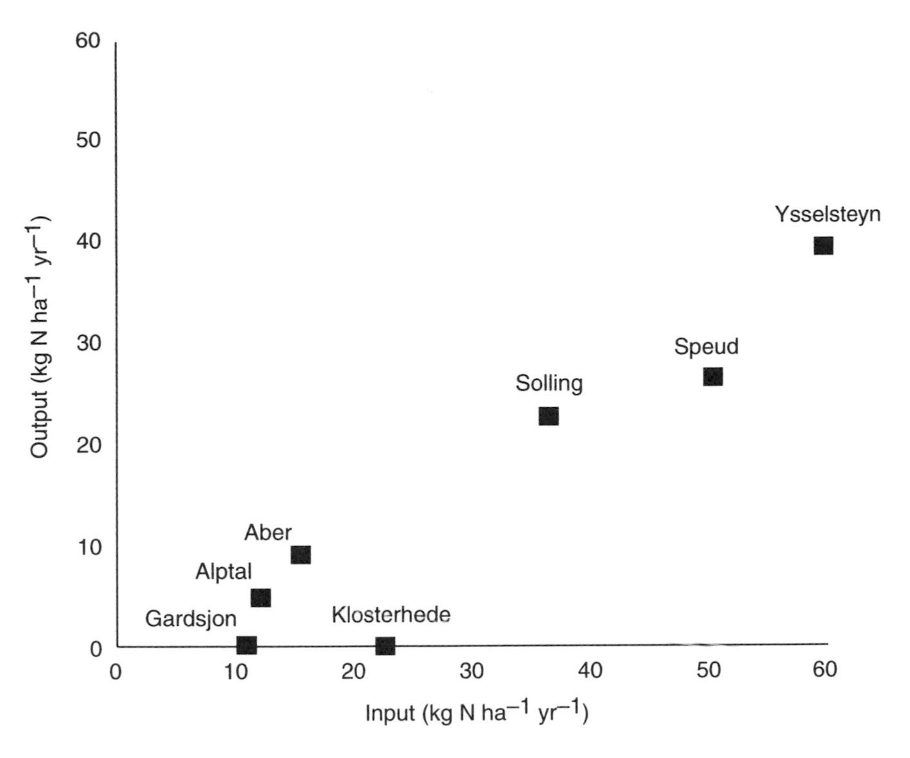

FIGURE 5-11 Ambient inputs (throughfall) and leaching losses at the Nitrogen Saturation Experiment sites (modified from Emmett et al. 1998).

poorly measured since most deposition monitoring sites are in rural environments (Holland et al. 1999). Uncertainty over the extent of nitrogen deposition in urban environments is one of the greatest uncertainties in the nitrogen budget for the United States (Holland et al. 1999).

PROCESSING OF NITROGEN AND PHOSPHORUS IN WETLANDS, STREAMS, AND RIVERS

Not all the nitrogen or phosphorus that is exported from a forest, a corn field, or an animal feedlot will reach an estuary or coastal waters because significant processing of nutrient flows can occur in wetlands, streams, lakes, reservoirs, and rivers that lie between the terrestrial systems and coastal waters (Kirchner and Dillon 1975; Kelly et al. 1987; Howarth et al. 1995, 1996; Rigler and Peters 1995). The sediments in wetlands and aquatic systems sometimes retain phosphorus and sometimes they do not, depending on the presence of such phosphorus-sorptive phases as iron (III) hydroxides and oxides and calcium carbonate minerals (Howarth et al. 1995). Where iron compounds are the dominant phosphorus-adsorbing minerals, the oxidation state of the minerals is important, as oxidized iron compounds sorb phosphorus and reduced iron compounds do not (Theis and McCabe 1978; Howarth et al. 1995). Biological controls on the pH of the immediate surface sediment can also be important (Knuuttila et al. 1994). Despite these details, the retention of phosphorus in lakes can often be predicted from the residence time of water in the lakes, and lakes often retain 80 percent or more of the entering phosphorus (Kirchner and Dillon 1975; Rigler and Peters 1995; Nixon et al. 1996). Wetlands also frequently retain significant quantities of phosphorus, so buffer or riparian zones around streams or waterbodies can reduce inputs from agricultural land (Lowrance et al. 1984a, b, 1985).

Wetlands and aquatic systems can also remove significant quantities of nitrogen, although the mechanism is different. Generally, most removal of nitrogen in these systems is by denitrification. The process occurs largely in anoxic environments, and often occurs at high rates in the sediments of wetlands, lakes, and rivers. Most studies show a large removal of nitrate in groundwater that flows through wetlands, with most of this presumed to be denitrified (Peterjohn and Correll 1984; Lowrance et al. 1984b; Correll et al. 1992; Jordan et al. 1993; Jansson et al. 1994; Vought et al. 1994; Howarth et al. 1996), although some studies show a conversion to organic nitrogen (Devito et al. 1989; Brunet et al. 1994). Riparian wetlands that intercept waters flowing from agricultural fields before they enter streams can be effective at lowering nitrogen loads. However, in many areas these riparian wetlands were drained or otherwise destroyed as land was converted for agriculture (Vought et al. 1994). Krug (1993) has shown that in southern Sweden, the conversion of the last 10 percent to 15 percent of land into agricultural use disproportionately destroyed fringing wetlands and therefore doubled nitrogen inputs to streams. Wetland restoration has been suggested as both the cheapest and most effective approach for lowering nitrogen fluxes through the landscape to rivers (Rosenberg et al. 1990; Haycock et al. 1993).

Denitrification occurs in the water column of streams and rivers when the water is anoxic or extremely hypoxic. However, with the general improvement in river water quality that accompanied the widespread use of secondary sewage treatment for removal of biological oxygen demand, few rivers in the United States now have such low-oxygen events (NRC 1993a). On the other hand, the sediments of streams and rivers are frequently anoxic even when water quality is high, and this provides an ideal location for denitrification. Similarly, the sediments of lakes are almost always anoxic below the first few meters of water. For lakes, streams, and rivers with an oxic water column, the extent of removal of nitrogen by denitrification can be modeled as a function of depth and mean residence time (Kelly et al. 1987; Howarth et al. 1996); denitrification is greater in shallower systems and in systems with a longer water residence time, since both of these lead to greater contact of nitrogen in the water with anoxic sediments. Using this model, Howarth et al. (1996) suggested that denitrification in river systems in the United States are in general unlikely to denitrify more than 20 percent of the nitrogen that flows into them.

NUTRIENT FLUXES TO THE COAST

Insights from a Regional Analysis

At the scale of individual estuaries, it has proven exceedingly difficult to determine the ultimate source for nitrogen inputs and the magnitude of the load each source contributes. Numerous obstacles exist to understanding nutrient fluxes to the coast; including: (1) the existence of multiple sources (fossil-fuel combustion from both mobile and stationary sources plus agricultural sources); (2) the difficulty in estimating dry deposition at the scale of whole watersheds; (3) the difficulty in measuring gaseous losses from ecosystems (including denitrification of nitrate to nitrogen as well as volatilization of ammonia and NO_y compounds), and (4) the multiple pathways for nitrogen flows (surface waters, groundwaters, and atmosphere). (Generally, these problems are far less significant when dealing with phosphorus fluxes.) However, much insight into nitrogen fluxes to coastal waters has been gained recently by analyzing fluxes at relatively large spatial scales.

Over the past six years, the International SCOPE Nitrogen Project has been analyzing nitrogen fluxes at the scale of large regions, such as the combined watersheds of the North Sea, the combined watersheds of the northeastern United States from Maine through the Chesapeake Bay, and the Mississippi River basin. The International SCOPE Nitrogen Project was authorized by the International Council of Scientific Unions and has worked with other international efforts, including the International Geosphere-Biosphere Program, the United Nations' Environmental Program, and the World Meteorological Organization. The motivation was to see what insights on nitrogen pollution could be gained by studying nitrogen biogeochemistry at a scale smaller than global but larger than small watersheds (Howarth 1996; Townsend 1999).

As one of its first activities, the International SCOPE Nitrogen Project evaluated nitrogen exports to the North Atlantic Ocean from the terrestrial landscape (both in Europe and in America) at the scale of large regions (Howarth et al. 1996). At this scale, it is possible to evaluate the net influences of all the processes acting in the region at the scale of individual fields, feedlots, cities, forests, wetlands, and rivers. Some nitrogen inputs are actually easier to estimate at large spatial scales than at the scale of small watersheds and fields. For instance, dry deposition of nitrogen is difficult to estimate at the scale of individual watersheds and can be variable in space. However, dry deposition at the scale of large regions can be estimated with greater accuracy by using mass-balance constraints and knowledge of broad-scale atmospheric transport (Howarth et al. 1996; Prospero et al. 1996; Holland et al. 1999).

The International SCOPE Nitrogen Project showed large variations in the export of nitrogen to the North Atlantic from regions in the temperate zone, with fluxes per area of watershed varying from as low as 76 kg N km^{-2} yr^{-1} for the watersheds of northern Canada to 1,450 kg N km^{-2} yr^{-1} for the watersheds of the North Sea (Figure 5-12). As stated above, the export of nitrogen from nonpoint sources dominates over wastewater and other point sources for all regions (Howarth et al. 1996; Howarth 1998). The flux of nitrogen from a region per area of watershed—both the total flux and the flux from nonpoint sources—is weakly correlated with population density (Figure 5-13).

The International SCOPE Nitrogen Project constructed mass balances for reactive nitrogen under human control at the scale of large regions (Howarth et al. 1996). For imports, the analysis carried out by the International SCOPE Nitrogen Project considered application of nitrogen fertilizer, nitrogen fixation by agricultural crops, deposition from the atmosphere of oxidized forms of nitrogen (which are presumed to come primarily from fossil-fuel combustion in the temperate zone; Holland et al. 1999), and the import or export of nitrogen in food and animal feedstocks. Sewage was not considered a net source, since it is a recycling of nitrogen that was brought into a region for agricultural purposes or directly as nitrogen in food. Similarly, deposition of ammonium and organic nitrogen from the atmosphere were not considered as inputs, since these are largely a recycling of nitrogen volatilized into the atmosphere from agricultural sources within the same region (Howarth et al. 1996; Howarth 1998).

Surprisingly, the International SCOPE Nitrogen Project found that the export of nitrogen from the landscape to the coast in the temperate zones of North America and Europe is a linear function of the import of reactive nitrogen forms into the region by human activity (Figure 5-14; Howarth et al. 1996; Howarth 1998). On average for the temperate regions of North America and Europe, 20 percent of the nitrogen inputs under human control flow out of regions to coastal waters. The majority of the human-controlled nitrogen inputs are either denitrified or stored in the ecosystems in regions. Unfortunately, the nature of these sinks—includ-

A

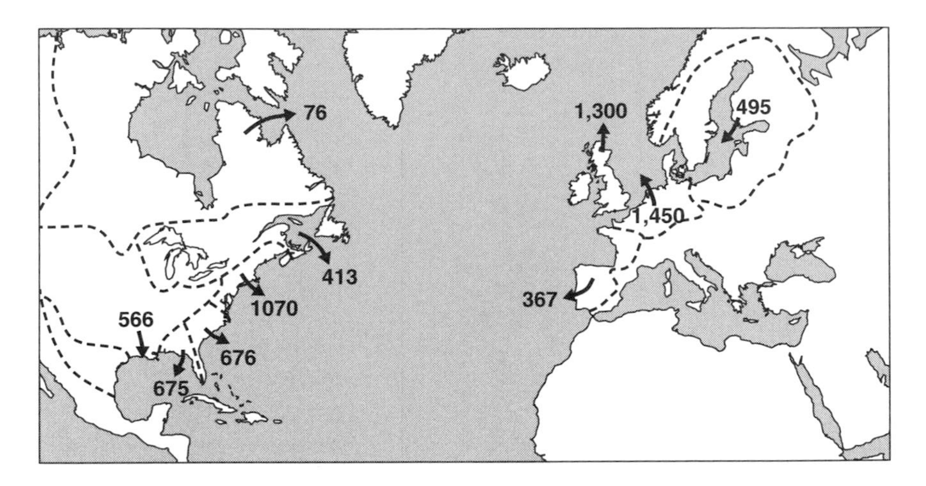

B

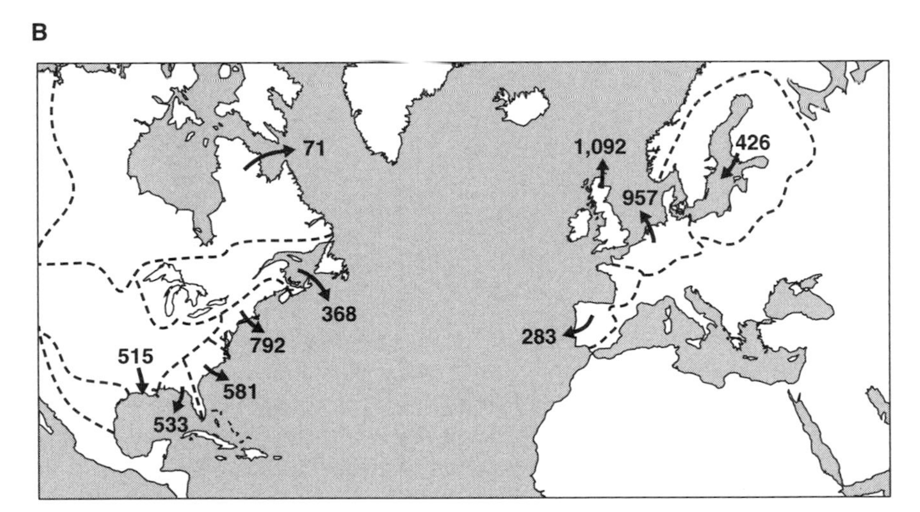

FIGURE 5-12 Regional export of total nitrogen to the North Atlantic coast per area of watershed (kg N km^{-2} yr^{-1}). (A) Total nitrogen fluxes in rivers and in sewage treatment plants. (B) Fluxes in rivers that only originate from nonpoint sources of nitrogen in the landscape (modified from Howarth 1998).

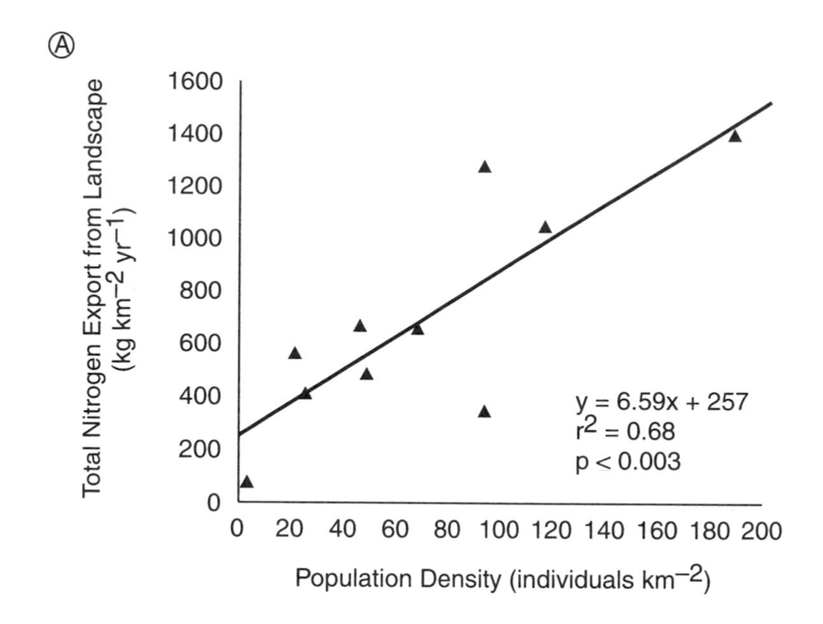

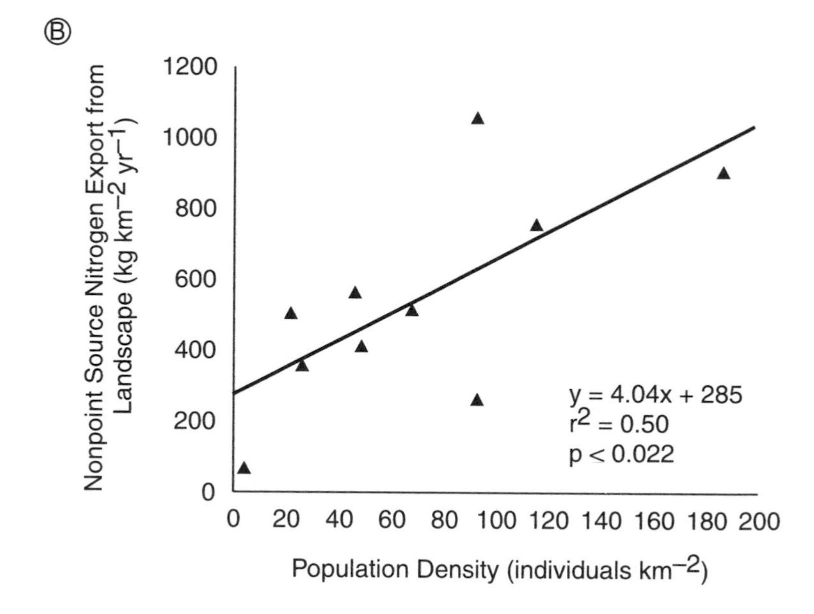

FIGURE 5-13 The relationship between population density and the export of nitrogen in rivers to the coast for temperate regions surrounding the North Atlantic Ocean. Each point represents one region. (A) The total nitrogen export from the region in rivers. (B) The flux of nitrogen from nonpoint sources in the region, independent of upstream sources (modified from Smith et al. 1997).

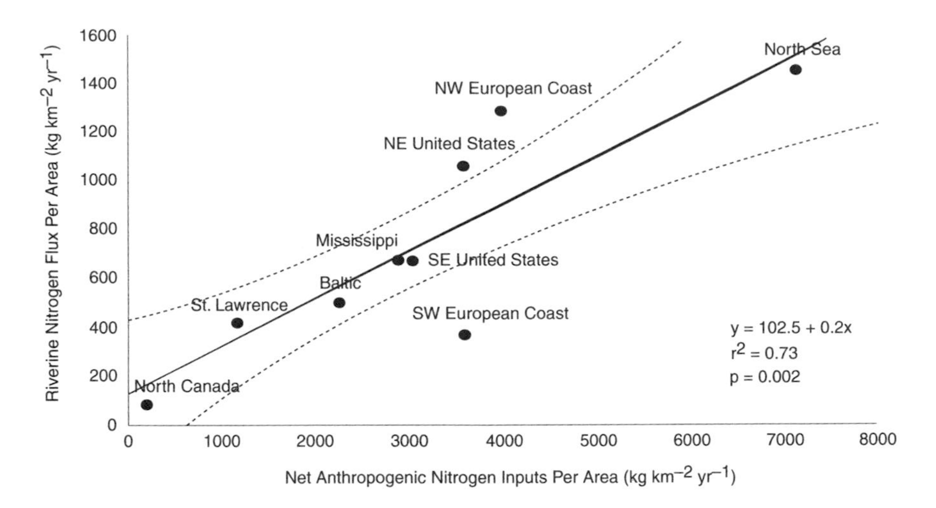

FIGURE 5-14 A comparison of human-controlled inputs of nitrogen to a region and nitrogen export from the region to the coast in rivers, for temperate regions surrounding the North Atlantic Ocean. Note that the export of nitrogen from a region is linearly related to the inputs of nitrogen to the region. The dashed lines refer to the 95 percent confidence limits around the regression line (solid line; modified from Howarth et al. 1996).

ing whether or not they will change with time—is poorly known (Howarth et al. 1996).

The International SCOPE Nitrogen Project further used regressions (Figure 5-15A-D) to suggest that deposition from fossil-fuel sources (NO_y deposition) per unit mass introduced into the landscape is a better predictor of nitrogen export to coastal waters ($r^2 = 0.81$) than is fertilizer application ($r^2 = 0.28$; Howarth 1998). Furthermore, a simple multiple regression that used both NO_y deposition and agricultural inputs (i.e., the sum of fertilizer, nitrogen fixation in agriculture, and net movements of nitrogen in foodstocks) was constructed to predict nitrogen export to the coast. The best overall fit was obtained by a curve where the NO_y deposition term was seven times greater than the agricultural input term (Howarth et al. 1996). This suggests that, per unit mass, nitrogen from fossil fuel sources may contribute more to the nitrogen flux in rivers to the coast than do agricultural sources. Of course, in many areas the total inputs of nitrogen as fertilizer are far greater than are the inputs from NO_y deposition. For example, in the Mississippi River basin the total inputs of nitrogen as fertilizer far exceed those from NO_y deposition; consequently, agri-

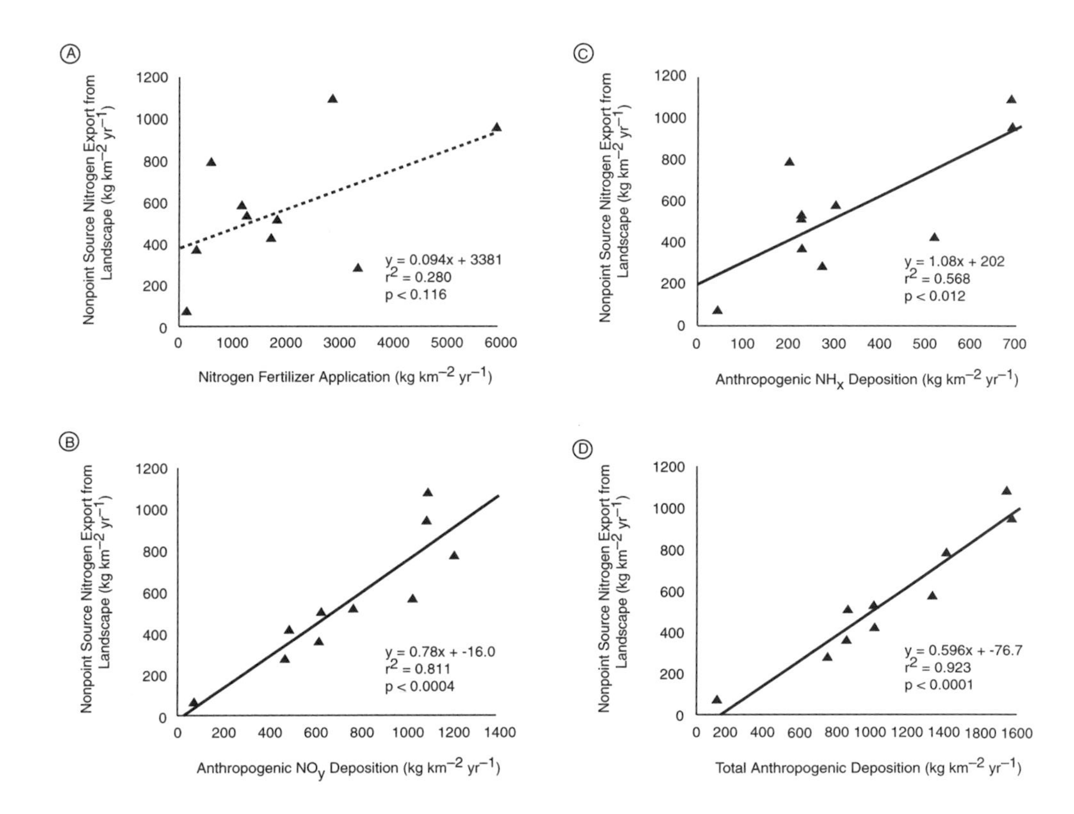

FIGURE 5-15 Analyses carried out during the International SCOPE Nitrogen Project suggest that nitrogen from deposition from fossil fuel sources (NO_y deposition) per unit mass introduced into the landscape is a better predictor of nitrogen export to coastal waters ($r^2 = 0.81$) than fertilizer application ($r^2 = 0.28$; modified from Howarth 1998).

culture is the greatest contributor to the nitrogen export from that basin (Goolsby et al. 1999).

The best regression fit for the export of nitrogen from nonpoint sources for the temperate regions of the North Atlantic Ocean results from using the sum of NO_y deposition and ammonium deposition to predict nitrogen export (Figure 5-15A-D; $r^2 = 0.92$; Howarth 1998). The ammonium deposition is strongly tied to livestock densities (Bouwman et al. 1997), which suggests that livestock wastes contribute disproportionately to the nitrogen pollution of surface waters by agriculture and, together with the fossil-fuel source, are often major factors in nitrogen export to the coastal oceans at the scale of large regions (Howarth 1998).

INSIGHTS FROM THE SPARROW MODEL APPLIED TO THE NATIONAL SCALE

Another useful large-scale approach to assessing sources of nitrogen and phosphorus in surface waters was taken by Smith et al. (1997). By applying the Spatially Referenced Regressions on Watersheds (SPARROW) model (Appendix D) to a set of data from 414 stations in the National Stream Quality Accounting Network, Smith et al. (1997) concluded that just over half of the streams and rivers in the United States probably have total phosphorus concentrations in excess of 0.1 mg l^{-1} (Figure 5-16). Furthermore, they concluded that livestock waste production is the single largest source of phosphorus contamination leading to elevated phosphorus concentrations nationally (Smith et al. 1997). Mean values for "land-water delivery factors" (the percent of the original source of phosphorus that actually reaches surface waters) were estimated as approximately 0.07 and 0.11, respectively, for phosphorus from fertilizer application and phosphorus from livestock wastes. That is, the analysis by Smith et al. (1997) suggested that per mass of phosphorus, phosphorus from livestock wastes was 50 percent more likely to be exported to surface waters. Note that these delivery factors are estimated as part of the model in determining the best fit between nutrient sources and concentrations.

With respect to nitrogen, Smith et al. (1997) concluded that much of the United States probably exports less than 500 kg N km^{-2} yr^{-1}, but that export is probably much higher in much of the Mississippi River basin and in the watersheds of the northeastern United States (Figure 5-17). For the areas of export over 1,000 kg N km^{-2} yr^{-1}, Smith et al. (1997) concluded that fertilizer was the largest source of nitrogen overall (48 percent), followed by atmospheric deposition (18 percent) and livestock wastes (15 percent). To some degree, this result is driven by the large area of the Mississippi River basin—this basin represents 41 percent of the area of the lower 48 states, and is a region where fertilizer application greatly exceeds

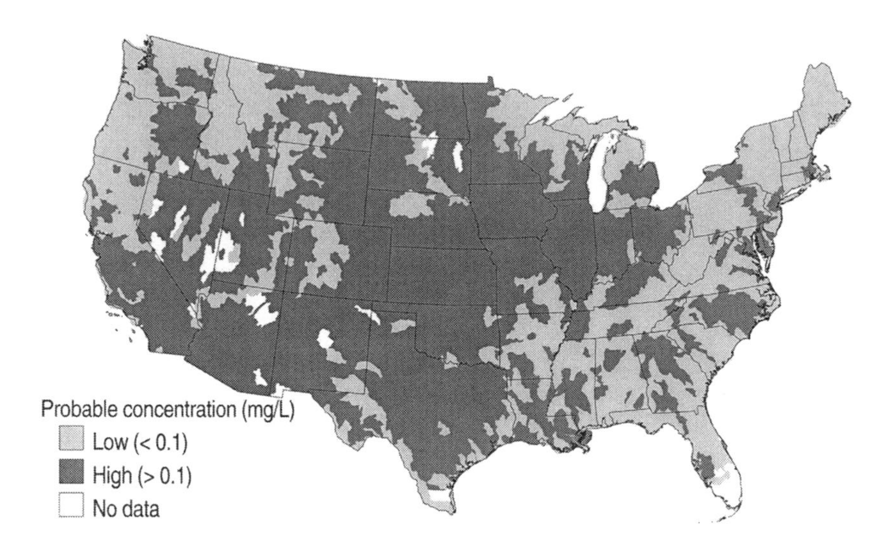

FIGURE 5-16 Classification of predicted total phosphorus concentrations in surface waters of the United States as estimated from the SPARROW model (Smith et al. 1997).

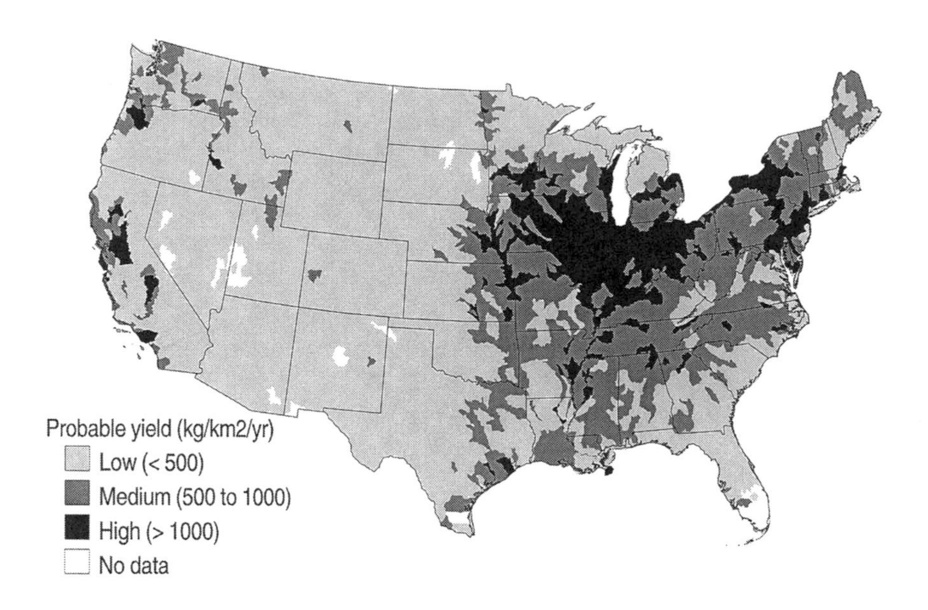

FIGURE 5-17 Predicted local total nitrogen yield in hydrologic cataloging units of the conterminous United States. Local yield refers to transport per unit area at the outflow of the unit due to nitrogen sources in the unit, independent of upstream sources (Smith et al. 1997).

NO_y deposition (Howarth et al. 1996). Conversely, in the northeastern United States, atmospheric deposition is the largest nonpoint source of nitrogen to surface waters (Howarth et al. 1996; Jaworski et al. 1997; Smith et al. 1997).

For the portions of the United States where total nitrogen export was over 1,000 kg N km^{-2} yr^{-1}, the SPARROW model estimated land-water delivery factors of 0.24 for livestock wastes, 0.32 for fertilizer application, and 1.62 for atmospheric deposition (Smith et al. 1997). Note that for both livestock waste and fertilizer, the delivery factors are greater for nitrogen than for phosphorus (by two- to four-fold). This is consistent with the known greater mobility of nitrogen in dissolved forms in surface and groundwater and in volatile forms in the atmosphere.

There are some biases in the land delivery factors for nitrogen fertilizer and for atmospheric deposition, as Smith et al. (1997) did not include in their analysis the nitrogen fixation by agricultural crops or dry deposition from the atmosphere. Nitrogen fixation by agricultural crops tends to be correlated with nitrogen fertilizer application in the United States, and both are sources of nitrogen to downstream ecosystems (Howarth et al. 1996). Howarth et al. (1996) also demonstrated that, on average, for the portions of the United States that export over 1,000 kg N km^{-2} yr^{-1}, nitrogen fertilizer makes up just over 60 percent of the sum of fertilizer application plus nitrogen fixation by agricultural crops. Adjusting the land-water delivery factor from Smith et al. (1997) to include nitrogen fixation as a source of nitrogen yields a new land-water delivery factor of 0.20 for the combined nitrogen from fertilizer and nitrogen fixation (a value comparable to that for nitrogen loss from livestock waste determined by Smith et al. 1997).

For atmospheric deposition, Smith et al. (1997) reported a land-water delivery coefficient of 1.62, suggesting that more nitrogen runs off the landscape from a depositional source than actually falls in deposition. This clearly cannot be so, and the most likely explanation for this high delivery factor is that the deposition estimates used for input were only for wet deposition of NO_y, and did not include NO_y dry deposition or wet or dry deposition of ammonium and organic nitrogen (personal communication, Smith 1999). On average for areas in the United States receiving fairly high levels of atmospheric deposition, wet NO_y deposition is approximately 25 percent of total atmospheric deposition (wet and dry of both reduced and oxidized forms). (Although there is a great deal of uncertainty associated with this estimate [Johnston and Lindberg 1992; Lovett and Lindberg 1993; Whelpdale et al. 1997; Holland et al. 1999; Valigura et al. 2000]). Using this value as a correction factor for the land-water delivery factor for nitrogen deposition of Smith et al. (1997), leads to a land delivery factor of approximately 0.40 in areas of high nitrogen export in the United States.

A comparison of these revised delivery factors of 0.40 for total atmospheric deposition of nitrogen (NO_y) and 0.20 for nitrogen fertilizer application plus nitrogen fixation by agricultural crops leads to the conclusion that nitrogen from depositional sources is about two-fold more mobile in the landscape than is nitrogen running off agricultural fields. This conclusion is consistent with that from the regional analysis of the International SCOPE Nitrogen Project discussed earlier, which also demonstrated the greater mobility of nitrogen from NO_y deposition (Howarth et al. 1996). Together, these results suggest that while the global mobilization of newly available nitrogen is greater through fertilizer production than through fossil-fuel combustion (Galloway et al. 1995; Vitousek et al. 1997), the nitrogen from fossil fuel sources may be disproportionately important to coastal eutrophication and other adverse impacts of nutrient over-enrichment.

Overall, the conclusions reached by Smith et al. (1997) from their SPARROW analysis agree remarkably well with the conclusions of the International SCOPE Nitrogen Project (Howarth et al. 1996; Howarth 1998), with one exception. Results from the Project show that livestock wastes are a more significant source of nitrogen to surface waters than predicted by Smith et al. (1997); the SPARROW analysis finds livestock wastes to be the major source of phosphorus, but a lesser source of nitrogen.

NUTRIENT BUDGETS FOR SPECIFIC ESTUARIES AND COASTAL WATERS

Knowledge of nutrient inputs to an estuary is essential for management of nutrient over-enrichment problems, and nutrient budgets have now been prepared for many estuaries. Several of these have recently been summarized by Valigura et al. (2000). Often, nutrient inputs are estimated as part of some larger scientific research project and are published in the peer-reviewed scientific literature. More frequently, the budgets are prepared as management tools and are either not published, or are published as government or consulting company reports (Valigura et al. 2000). Documentation of the data sources and approaches used is sometimes missing and is seldom fully adequate for independent review.

No standard methodologies exist for estimating nutrient inputs to estuaries, and many different approaches have been used. In some cases, nutrient budgets are based on export-coefficient models, where nutrient exports are estimated from literature values as a function of land-use types without independent verification of fluxes (Chapter 8). In other cases, budgets are based on empirically derived loading coefficients for the actual watershed. These approaches work well for determining the importance of point-source inputs such as wastewater treatment plants. However, without proper calibration, estimates for nutrient inputs from non-point sources can be misleading. Estimating the importance of atmospheric deposition as a source is particularly problematic when using export-coefficient models.

For example, export-coefficient models simply take empirical data, and apply it through series of relatively straightforward calculations to obtain an estimate of the total load. In the simplest form (which is often the form used), the approach uses published coefficients for various land use types in the watershed (developing these coefficients is not straightforward, thus often the coefficients were derived for regions other than that within which the watershed resides). In a simple hypothetical watershed, published coefficients might suggest that farmland exports X g N m^{-2} yr^{-1}, forests export Y g N m^{-2} yr^{-1}, and urban lands export Z g N m^{-2} $year^{-1}$. These values are multiplied by the area of each land type in the watershed to get the export for the watershed as a whole. Atmospheric deposition of nitrogen (NO_y) presents an immediate problem in that these models have historically not worried about whether the export coefficients used were derived for areas with high or low atmospheric deposition of nitrogen. Thus, atmospheric deposition has been ignored, and so the export from forests is generally treated as a background, natural flux. This erroneously implies that no amount of atmospheric deposition of nitrogen will increase the export of nitrogen from forests. Presumably, the approach could be improved so that forest export varied depending on deposition, but to date, no specific efforts to address this problem have been successfully completed.

Almost all nutrient budgets for estuaries rely on gauged stream discharge data where these are available. However, for many estuaries (including major ones such as Chesapeake Bay, Delaware Bay, and the Hudson River), significant portions of the watersheds are not gauged because of the difficulty in gauging tidal streams and rivers (Valigura et al. 2000). Where available, data on concentrations of total phosphorus and nitrogen are used in these budgets, but for many estuaries only inorganic dissolved nutrients are measured (Valigura et al. 2000). These problems add considerable error to the nutrient budgets.

Methodologies for determining the sources of nutrients and the magnitude of the load contributed by each are poorly developed at the scale of individual estuaries, and there is an urgent need for developing better approaches, particularly with regard to atmospheric deposition of nitrogen onto the landscape. The large-scale and regional analyses discussed above (the International SCOPE Nitrogen Project and the SPARROW analysis) provide a potential framework based on quantifying inputs to the watershed, but these analyses are relatively recent and have not yet been applied to the management of most estuaries. In an effort to determine the validity of using SPARROW-derived estimates for a given estuary, Valigura et al. (2000) conducted a preliminary comparison of SPARROW-derived estimates with independently derived estimates of nitrogen loading to 27 estuaries on the Atlantic and Gulf of Mexico coasts of the United States. Based on that comparison, Valigura et al. (2000) concluded that while SPARROW accurately predicted the mean loading to the estuaries as a group, it did a poor job of predicting the load to any one particular estuary (i.e., a linear regression of the SPARROW estimates and the locally derived estimates had a slope of 1 and an R^2 of 0.49). However, as with many such analyses involving locally derived information, the observed data from each estuary varies in quality and quantity and the methods used to calculate estimates varied as well. Thus, the locally derived estimates were not obtained from directly comparable data sets and most were not verified. Thus the poor match between SPARROW predictions and local estimates may lie with the quality of the individual estimates for the 27 estuaries. (Chapters 7 and 8 expand on the limitations imposed on understanding individual estuarine behavior by inconsistent observations.)

Perhaps the greatest uncertainty with estuary nitrogen budgets concerns the contribution of atmospheric deposition. In most classical estuarine studies, nitrogen inputs from the atmosphere were completely ignored. This has changed since Fisher and Oppenheimer (1991) pointed out the potential importance of atmospheric deposition as a source of nitrogen to Chesapeake Bay, and since Paerl (1985) showed the importance of atmospheric deposition as a nitrogen source to the coastal waters of North Carolina. However, even many nutrient budgets constructed during the last decade have no estimate for the input of nitrogen from atmospheric deposition. In many other estuaries, budgets estimate the importance only of direct deposition onto the surface waters of the estuary itself (and generally only wet deposition, not dry deposition), and do not estimate deposition onto the landscape with subsequent export to the estuary.

Available evidence (although constrained by limited monitoring) indicates that direct deposition onto the water surface alone (not including the contribution of nitrogen which falls on the landscape and is then exported to estuaries) contributes between 1 percent and 40 percent of the total nitrogen input to an estuary—depending in large part on the relative area of the estuary and its watershed (Nixon et al. 1996; Valigura et al. 2000). In estuaries where the ratio of the area of the estuary to the area of its watershed is greater than 0.2, direct atmospheric depositions usually make up 20 percent or more of the total nitrogen loading (Valigura et al. 2000). Where the ratio of the estuarine area to the area of its watershed is less than 0.1, atmospheric deposition directly onto the water surface generally makes up less than 10 percent of the total nitrogen input (Valigura et al. 2000).

For estuaries that have relatively large watersheds, the deposition of nitrogen from the atmosphere onto the landscape with subsequent runoff into the estuary is probably greater than the deposition of nitrogen directly onto the water surface. Unfortunately, the magnitude of this flux is poorly characterized for most estuaries. The deposition onto the landscape can be estimated for most watersheds, although the error associated with these estimates can be considerable due to inadequate monitoring and the difficulty with measuring dry deposition. The larger problem, however, is with determining what portion of the nitrogen deposition is retained in the landscape and what portion is exported to rivers and the coast. The two major approaches for making this determination are to use statistical models or to use process-based models on nitrogen retention in the watershed. In their application to estuaries, both approaches are quite recent and are relatively untested. There is an urgent need for further development and evaluation of these techniques; however, it appears that the statistical approaches have led to more reliable estimates, for reasons discussed below.

Both the SPARROW model and regressions comparing nitrogen flux in rivers to sources of nitrogen across landscapes (used by the International SCOPE Nitrogen Project) represent examples of statistical approaches that appear to provide reliable estimates of the portion of the nitrogen deposition retained in the landscape versus what is exported to rivers and coastal areas. Jaworski et al. (1997) used a similar approach in the northeastern United States, comparing atmospheric deposition and riverine flux for 17 watersheds with relatively little agricultural activity or sewage inputs. This led to the conclusion that approximately 40 percent of the nitrogen deposition is exported from the landscape (correcting their analysis by assuming that dry deposition is equal to wet deposition), a value remarkably similar to the results from applying the SPARROW model at the national scale. By applying this result to other watersheds in the northeast, including those with agricultural activity, Jaworski et al. (1997) estimated that between 36 percent and 80 percent of the total nitrogen flux in rivers was originally derived from atmospheric deposition onto the landscape. Note that the riverine nitrogen fluxes were estimated at U.S. Geological Survey (USGS) gauging stations above the tidal portions of these rivers, and generally excluded the large urban influences at the river mouths.

In another recent effort, a National Oceanic and Atmospheric Administration (NOAA)-sponsored project brought together researchers from around the United States to examine atmospheric deposition to coastal waters (Valigura et al. 2000). Valigura et al. (2000) summarized and compared the four different approaches included in the NOAA project, including a process-based model and an application of the statistical approach used by SPARROW. They report that, for 42 estuaries in the United States, atmospheric deposition onto the landscape contributed between 6 percent and 50 percent of the total nitrogen load to the receiving body. Jaworski et al (1997) and Valigura et al. (2000) give estimates in common for only one river/estuary—the Hudson-Raritan—and for this system, their estimates are similar to the statistical model results, but quite different from the process-based model estimates. Jaworski et al. (1997) estimate that 34 percent of the nitrogen flux in the Hudson comes from atmospheric deposition onto the landscape, after correction for the point source inputs from New York City (Hetling et al. 1996). In contrast, estimates from the process-based model indicated 9 percent of the nitrogen flux of the Hudson-Raritan total nitrogen load comes from nitrogen deposition onto the landscape. The statistical SPARROW model approach estimated the flux to the estuary from atmospheric deposition onto the watershed as 26 percent for this system.

Great uncertainty about the importance of atmospheric deposition as a nitrogen source to specific estuaries may exist. However, there is little doubt that the relative importance of fossil-fuel combustion versus agricultural activity in controlling atmospheric deposition of nitrogen to estuaries depends both on the nature and extent of farming activities in the watershed and on the nature and extent of fossil-fuel combustion in the airsheds upwind of the watershed. In estuaries fed by watersheds with little agricultural activity but significant loads of atmospheric pollution (such as the Connecticut and Merrimack rivers and most of the northeastern United States), atmospheric deposition of nitrogen from fossil-fuel combustion can account for up to 90 percent or more of the nitrogen contributed by nonpoint sources. On the other hand, for watersheds such as the Mississippi Basin where agricultural activity is high and atmospheric pollution from fossil-fuel combustion is relatively low (Figure 5-18), agricultural sources dominate the fluxes of nitrogen. Interestingly, the major hot-spots of agricultural activity that dominate the nitrogen fluxes for the Mississippi and Gulf of Mexico appear to be far from the Gulf in Iowa, Illinois, Indiana, Minnesota, and Ohio (Goolsby et al. 1999).

For many estuaries, both atmospheric deposition of nitrogen derived from fossil-fuel combustion (NO_y) and nitrogen from agricultural sources are likely to be major contributors. For example, the model used by managers to estimate nitrogen inputs to Chesapeake Bay predicts that

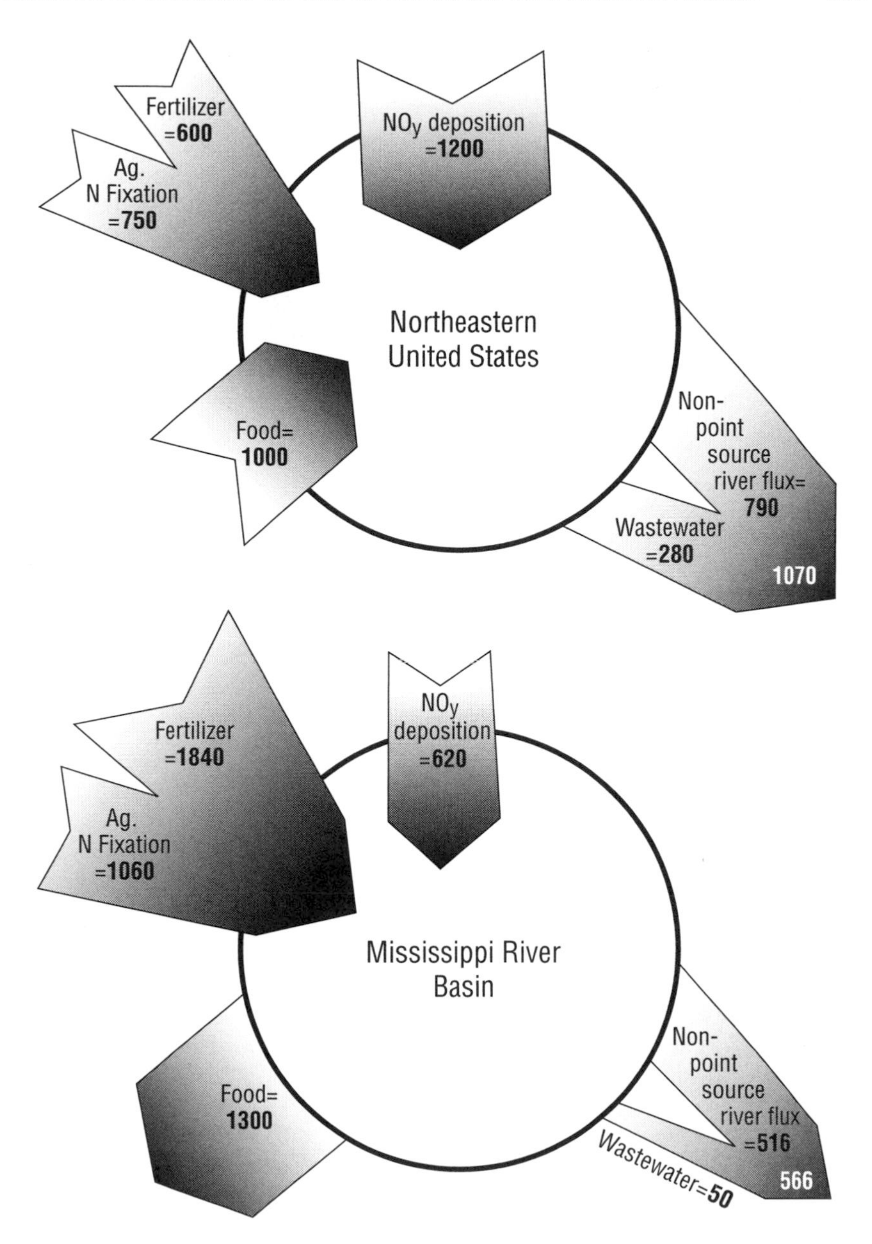

FIGURE 5-18 A comparison of human-controlled inputs of nitrogen and nitrogen losses (kg N km^{-2} yr^{-1}) as food exports and in riverine exports between the northeastern United States and the Mississippi River basin. Note that, on average, nitrogen is exported in foods and feedstocks from the Mississippi basin and imported to the northeastern United States.

agriculture contributes 59 percent of the nonpoint source inputs, NO_y deposition onto the landscape is slightly less important (Magnien et al. 1995). The comparative analysis of Jaworski et al. (1997), on the other hand, suggests that atmospheric deposition is the dominant source of nitrogen from nonpoint sources in the major tributaries of Chesapeake Bay. Further study and analysis is necessary to determine whether Jaworski et al. (1997) have overestimated the importance of atmospheric deposition or whether Magnien et al. (1995) have underestimated it.

However, the process-based model of nitrogen retention used by Magnien et al. (1995) has not been independently verified and is subject to large uncertainties (Boesch et al. 2000). Small changes in the assumed ability of forests to retain or export nitrogen from atmospheric deposition can lead to large changes in the relative importance of NO_y deposition to the bay. As discussed above, there is great variation among forests in their ability to retain nitrogen from atmospheric deposition, and regional and large-scale analysis of nitrogen fluxes for the United States indicate a greater mobility of nitrogen from deposition (less retention) than is often found in small-scale watershed studies. Further, the model of Magnien et al. (1995) does not include some of the latest findings on nitrogen export from land, such as the large export of nitrogen in dissolved organic forms that was noted above (Campbell et al. 2000).

A recent report from the Environmental Protection Agency (EPA) estimates that between 10 percent and 40 percent of the total nitrogen input to estuaries comes from atmospheric deposition, including deposition directly onto the water surface and onto the watershed (EPA 1999c). However, it must be stressed that very few of the individual studies upon which this conclusion is based had adequate methodologies for determining the input of nitrogen from atmospheric deposition, particularly the indirect input through atmospheric deposition onto the landscape with subsequent runoff into the estuary. Many of these studies have probably underestimated the importance of this pathway, and it seems likely that atmospheric deposition is a greater input to estuaries than suggested by the 1999 EPA report.

OCEANIC WATERS AS A NUTRIENT SOURCE TO ESTUARIES AND COASTAL WATERS

In addition to receiving nutrient inputs from land and from atmospheric deposition, estuaries can receive nutrients across their boundary with the ocean. This term is often ignored, but can be substantial. For example, Nixon et al. (1995) estimate that for total nutrient inputs to Narragansett Bay, 15 percent of the nitrogen and 40 percent of the phosphorus inputs come from offshore, oceanic sources; despite this, the net flux of both nitrogen and phosphorus for Narragansett Bay is an export of these nutrients from the estuary to offshore waters (Nixon et al. 1996). On the other hand, Chesapeake Bay is a net importer of phosphorus from offshore ocean waters, although it too is a net exporter of nitrogen (Boynton et al. 1995; Nixon et al. 1996). The physical circulation pattern of an estuary is a major determinant in the importance of nutrient import to the estuary from offshore sources. Partially mixed estuaries (such as Chesapeake Bay) and fully mixed estuaries (such as Narragansett Bay) often import nutrients from offshore, whereas salt-wedge estuaries (such as the southwest pass of the Mississippi River and Oslo Fjiord) and hypersaline estuaries (such as portions of Shark Bay, Australia) do not (Howarth et al. 1995).

Offshore waters on the continental shelf can themselves receive nutrients from several sources, including deep ocean water, river and sewage inputs from land, and direct deposition from the atmosphere (Nixon et al. 1996; Prospero et al. 1996; Howarth 1998). The relative importance of these sources varies among the coastal waters of the United States, in part because of differences in ocean circulation patterns (particularly advection of water from the deep ocean—water that is extremely high in nutrients—onto the continental shelf). For most of the continental shelf area of the United States, this advection of water is the dominant nutrient input. However, input from the Mississippi River is the dominant source for the Gulf of Mexico. Human activity has tended to greatly increase inputs of nitrogen from rivers and atmospheric deposition, but has had no impact on the advection of water from the deep ocean onto the continental shelf. Consequently, human activity has almost tripled nitrogen input to the Gulf of Mexico, but has increased nitrogen inputs to the waters on the continental shelf of the northeastern United States by only 28 percent (Table 5-1). Of course, much of this input in the northeastern United States is concentrated in the plumes of a few rivers, such as that of the Hudson River, and these waters may therefore be experiencing eutrophication (Howarth 1998).

Rate of Change of Nutrient Inputs to the Coast

Historical data on fluxes of total nitrogen in rivers are rare, but data for trends in nitrate concentrations are available for many rivers going back to the early 1900s. Since human activity preferentially mobilizes nitrate over other forms of nitrogen in rivers (Howarth et al. 1996), these historical nitrate data are valuable in tracking the effects of humans on nitrogen fluxes to the coast. For the Mississippi River, the nitrate flux to the Gulf of Mexico is now some three-fold larger than 30 years ago, and most of this increase occurred between 1970 and 1983 (Figure 5-19;

TABLE 5-1

North Atlantic Continental Shelves	Nitrogen Sources (Tg yr-1)			Increase Due to Humans (%)
	Rivers and Estuaries	Direct Atmospheric Deposition	Deep Ocean	
North Canada rivers	0.16 (0.16)	0.10 (0.03)	0.77	7
St. Lawrence basin	0.34 (0.11)	0.13 (0.01)	1.26	25
Northeast coast of the United States	0.27 (0.03)	0.21 (0.01)	1.54	28
Southeast coast of the United States	0.13 (0.03)	0.06 (0.01)	1.36	11
Gulf of Mexico	2.10 (0.50)	0.28 (0.03)	0.14	275
North Sea and Northwest Europe	0.97 (0.14)	0.64 (0.02)	1.32	98
Southwest European coast	0.11 (0.04)	0.03 (0.001)	0.20	40

TABLE 5-1 Sources of nitrogen to the continental shelves of the temperate zone portions of the North Atlantic Ocean. Flux from rivers and estuaries is the direct input of rivers that discharge onto the continental shelf, minus nitrogen consumed in estuaries. Atmospheric deposition estimates are those directly onto the waters of the continental shelf and do not include deposition onto the landscape (which is part of the flux from rivers and estuaries). The flux from the deep ocean represents the advection of nitrate-rich deep Atlantic water onto the continental shelf. Data for modern values are means reported by Nixon et al. (1996). Pristine values as outlined by Nixon et al. (1996) for their treatment of modern estimates, but with data for pristine river fluxes from Howarth et al. (1996) and for pristine values of deposition from Prospero et al. (1996). "Increase due to humans" is the percentage comparison of total modern inputs compared to pristine inputs. Fluxes from the deep ocean are assumed not to have been affected by human activities (modified from Howarth 1998).

Goolsby et al. 1999). Similarly, nitrate fluxes in many rivers in the northeastern United States have increased two- to three-fold or more since 1960, with much of this increase occurring between 1965 and 1980 (Figure 5-20; Jaworski et al. 1997). Interestingly, most of the increase in nitrate in the Mississippi River was due to increased use of nitrogen fertilizer (Goolsby et al. 1999), whereas most of the increase in nitrate in the northeastern rivers was due to increased nitrogen deposition from the atmosphere onto the landscape, with the nitrogen originating from fossil-fuel combustion (Jaworski et al. 1997). The increase in nitrate flux in the

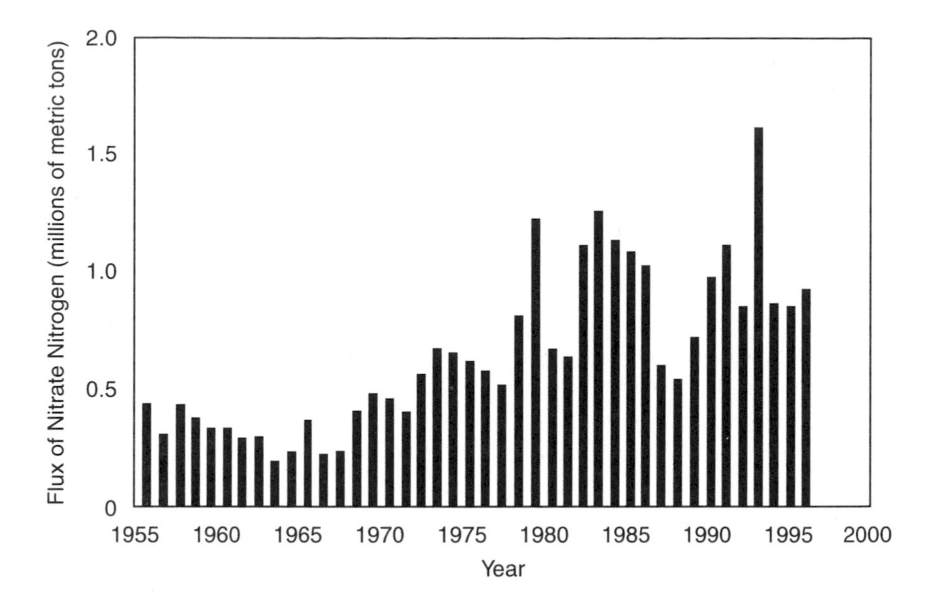

FIGURE 5-19 Bar chart showing the annual flux of nitrogen as nitrate (NO_3) from the Mississippi River basin to the Gulf of Mexico, indicating significant increases beginning in the late 1970s (modified from Goolsby et al. 1999).

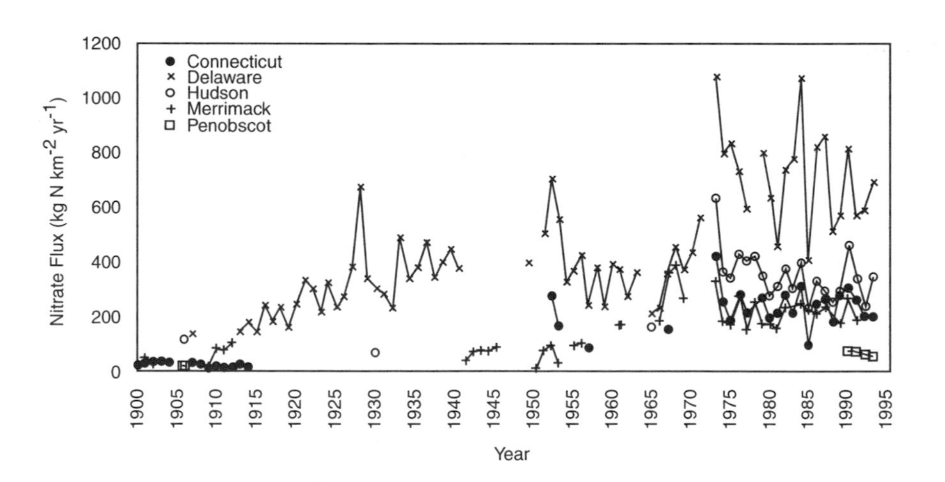

FIGURE 5-20 Flux of nitrate nitrogen from five major rivers in the northeastern United States from the early 1900s to 1994 (modified from Jaworski et al. 1997).

northeastern rivers during the 1960s and 1970s, and its stabilization since then, closely parallels the trend in human inputs of nitrogen to the landscape during that time (Jaworski et al. 1997).

In contrast to nitrogen, phosphorus fluxes to estuaries have often changed little over the past several decades. For the Mississippi River, data on total phosphorus flux are only available since the early 1970s, but there has been no statistically significant change since then (Goolsby et al. 1999). Smith et al. (1987) used data from 300 river locations throughout the United States to compare water quality trends from 1974 to 1981. Many rivers showed no trend during that time; rivers that had a trend in total phosphorus flux were equally divided between those that showed an increase and those that showed a decrease. Where total phosphorus fluxes increased, it was generally attributable to increased use of phosphorus fertilizer in the watershed. Decreases in total phosphorus fluxes were generally a result of point source reductions (Smith et al. 1987). Smith et al. (1987) also analyzed the national river data for trends in nitrate flux from 1974 to 1981. For nitrate, most rivers showed a marked increase in flux during that time, particularly for rivers in the eastern United States. This increased nitrate flux was attributed both to agricultural activity and to nitrogen deposition (Smith et al. 1987).

IMPLICATIONS FOR ACHIEVING SOURCE REDUCTIONS

Human activity has an enormous impact on the cycling of nutrients and especially on the movement of such nutrients as nitrogen and phosphorus into estuaries and other coastal waters. Although much effort has been made in the United States to improve control of point sources of pollution, nonpoint sources as urban runoff, agricultural runoff (particularly from animal feeding operations), and atmospheric deposition are generally of greater concern in terms of impact on nutrient enrichment and eutrophication of coastal waters. While sewage inputs dominate in some estuaries, nonpoint sources dominate nationally. Insufficient effort has been expended on controlling nonpoint sources of nitrogen and phosphorus, and there are few comprehensive plans for managing nutrient enrichment of the nation's coastal waters, particularly from nonpoint sources. Efforts to manage nonpoint and point sources of nitrogen and phosphorus are needed to reduce adverse impacts of nutrient over-enrichment in the nation's rivers, lakes, and coastal waters.

There is evidence that both atmospheric deposition of nitrogen from fossil-fuel combustion and agricultural sources of nitrogen contribute nitrogen to coastal waters. The relative importance of these varies among estuaries, but recent evidence indicates that the amount of nitrogen from deposition has been historically underestimated as an input to many estuaries, particularly by the indirect pathway of nitrogen deposited onto the landscape and then exported to the estuary. Recent evidence also indicates that per unit input to the landscape, nitrogen from fossil-fuel combustion is more important than nitrogen from fertilizer and, in turn, contributes disproportionately in the input of nitrogen to coastal waters.

Much uncertainly remains regarding the fluxes of nitrogen from the atmosphere to the landscape and to estuaries, and this is a critically important research priority. Although understanding some details regarding the atmospheric transport and fate of biologically available nitrogen will require additional research, the significant role atmospheric deposition of nitrogen plays in nutrient over-enrichment in some regions is clear. Addressing this component of the problem will require coordinated efforts over many states, clearly dictating a federal role in the effort. The regional nature of the atmospheric component of nitrogen loading argues that nutrient management should be a significant component of efforts to reduce air pollution and should be a key consideration during re-authorization of the Clean Air Act.

In general, sources of nutrients to estuaries have been poorly characterized, and in some cases sources have been mistakenly characterized because some land-use export-coefficient models used for characterization are inadequately verified. There are currently no easy-to-use and reliable methods for the manager of an estuary to determine the sources of nutrients flowing into that estuary. As will be discussed in Chapter 8, enhanced and coordinated monitoring efforts will be a key component of any local, regional, or national effort to reduce the impacts of nutrient over-enrichment.

Some critical questions related to understanding the sources of nutrients most affecting eutrophication and other impacts of nutrient over-enrichment remain unanswered. For instance, nitrogen deposition and fate in urban and suburban areas is poorly known, and wet nitrogen deposition in coastal areas is poorly understood. There is only a limited understanding of dry deposition in any environment, and understanding this in coastal areas and over water is challenging. Research efforts to expand understanding of atmospheric deposition of nitrogen should be expanded.

Changes in agricultural production systems are concentrating large amounts of nutrients in localized areas, thereby increasing the risk of nutrient leakage to the environment. Most of this concentration is associated with animal feedlots and with the long-distance transport of feedstocks. Changes in farm practices are driven by economics, and this concentration and long-range transport provide economic advantages to the producers; the larger costs, such as the external cost of nutrient exports to estuaries, remain unaddressed. As is discussed further in Chapter 9, a balanced and cost-effective nutrient management strategy will require an understanding of both the relative importance of various sources of nutrients, and the economic costs associated with reducing the loads attributable to each.

6

What Determines Susceptibility to Nutrient Over-Enrichment?

KEY POINTS IN CHAPTER 6

To plan effective strategies for managing coastal nutrient over-enrichment, managers need to understand how different types of estuaries respond to nutrient inputs. This chapter reviews the wide variety of processes controlling the susceptibility of coastal systems to nutrient enrichment and discusses existing approaches to estuarine classification that may be useful in assessing susceptibility. It finds:

- a widely accepted estuarine classification scheme is a prerequisite for a systematic approach to extending lessons learned and successful management options from one estuary to others;
- such a classification scheme should allow categorization of relatively poorly known systems on the basis of a minimum suite of measurements;
- quantitative classifications that provide insights into the relative importance of the different factors controlling estuarine dynamics have the most potential for predictive analysis; and
- a high priority should be the development of a national framework of "index sites" within which there would be an integration and coordination of environmental monitoring and research, with the goal of developing a predictive understanding of the response of coastal systems to both nutrient enrichment and nutrient reduction.

Coastal zone managers strive to accommodate human actions while minimizing negative impacts on coastal ecosystems. Successful management requires considerable information at a variety of levels, including an understanding of systems in their natural, pristine condition as well as how natural systems respond to human activities (Karr and Chu 1997). Coastal waters are particularly complex because different types of estuaries, embayments, and shelf systems differ in their responses to nutrient enrichment. This means that varying levels of nutrient input can cause very different responses in different systems.

This chapter examines classification schemes that could enhance attempts to understand, predict, and manage eutrophication and other impacts of nutrient over-enrichment in the nation's coastal waters. It reviews existing and developing estuarine classifications. While the emphasis is on estuaries, where impacts of nutrient loading are most acute, the approach is equally appropriate for coastal systems in general, including those in the nearshore region of the continental shelf.

The diverse physical settings (defined by a number of parameters including geology, soil type, climatic setting, and topology) of estuaries and their watersheds give rise to different types of estuarine systems. While there are numerous similarities between all estuaries, there are also some basic differences. For example, both a drowned river valley estuary, such as Chesapeake Bay, and a bar-built estuary, such as Plum Island Sound in northeastern Massachusetts, have temporal and spatial patterns of salinity that reflect seasonal variations in freshwater discharge. However, while Chesapeake Bay is a deep-water, plankton-dominated system where waters have a long residence time, Plum Island Sound is a shallow, emergent, marsh-dominated system where waters have a short residence time. The expected quantitative values for indicators of ecological health or for susceptibility to nutrient over-enrichment are not the same for these two systems, even though many of the same biological or ecological attributes may work as indicators in these disparate situations. Knowledge of the physical setting and the undisturbed ecosystem condition must underpin any monitoring and management effort to restore a coastal system impaired by nutrient over-enrichment.

MAJOR FACTORS INFLUENCING ESTUARINE SUSCEPTIBILITY TO NUTRIENT OVER-ENRICHMENT

Certain key characteristics appear to be of primary importance in determining estuarine response to nutrient enrichment. These factors range from biotic factors to physical setting to hydrodynamic regime. Twelve of the most important factors are:

1. **Physiographic setting.** Characterization of the physiographic setting could include a geomorphic descriptor of an estuary (e.g., inverted continental shelf estuary like the Mississippi River plume, coastal embayment, and drowned river valley), a descriptor of the

major biological community(ies) (e.g., mangrove swamp, emergent marsh macrophyte, rocky intertidal, coral reef, and planktonic system), and a description of the biogeographic province as used by Hayden and Dolan (1976), Briggs (1974), and Gosner (1971). Physiographic setting largely determines the primary production base.

2. **Primary production base.** The term primary production base refers to various primary producers that have unique temperature, substrate, light, and nutrient requirements and thus respond differently to nutrient loading. Susceptibility will vary across estuaries with different primary production bases. Examples of major types of primary producer communities include: emergent marshes and swamps, attached intertidal algae, benthic microalgae, drifting macroalgae, seagrasses, phytoplankton, and coral.

3. **Nutrient load.** Nutrient load is the total amount of various nutrients contributed by the upstream landscape and atmosphere (Figure 6-1A&B). Coastal systems are among the most heavily loaded ecosystems on Earth. Even small nutrient losses per unit area of the terrestrial landscape become immense when scaled over the entire expanse of terrestrial watersheds.

4. **Dilution.** Dilution of watershed-derived nutrients occurs due to a variety of mixing processes upon entry into an estuary. It makes a difference whether a given nutrient load is distributed over 1 km^2 versus 1,000 km^2 or mixed into 10^6 versus 10^{10} m^3. Dilution is one of the dominant factors used to predict lake eutrophication, and Nixon (1992) showed a strong relationship between areal nutrient load, which partially accounts for dilution, and primary production in a wide variety of estuaries. Areal nutrient load is the magnitude of the nutrient load (e.g., kg yr^{-1}) scaled to the surface area of the receiving estuary (e.g., m^2, thus kg m^{-2} yr^{-1}). More recently the National Oceanic and Atmospheric Administration (NOAA) has incorporated estimates of dilution into their susceptibility classifications, using dissolved concentration potential and estuarine export potential.

5. **Water residence time, TR, and flushing.** Steady state conditions in a waterbody are affected by the fluxes into and out of the system. Residence time of water in an estuary or part of an estuary is an important temporal scale for relating physical phenomena to

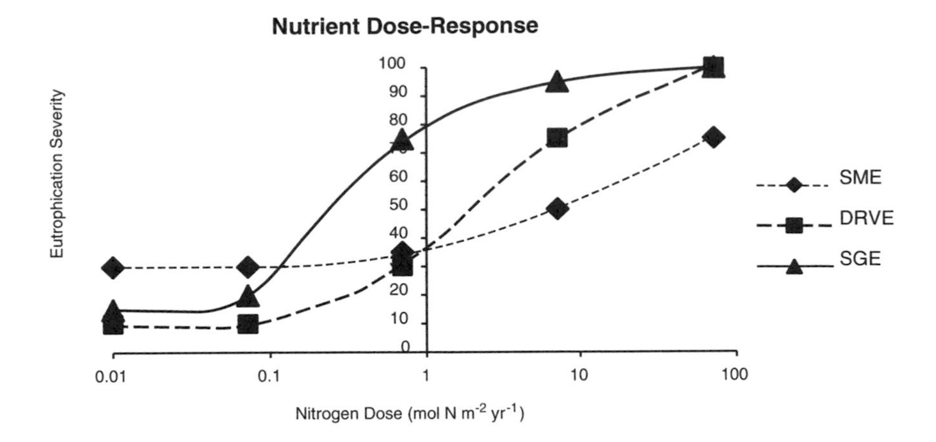

FIGURE 6-1 As nutrient loading is increased over the range of globally observed levels, it is hypothesized that different types of estuaries differ in their susceptibility to eutrophication. (A) This figure shows hypothetical Dose-Response Curves for three major types of coastal systems: Salt Marsh Dominated Estuary (SME), Plankton Dominated Drowned River Valley Estuary (DRVE), Seagrass Dominated Estuary (SGE). At the lowest levels of nitrogen loading (0.01 to 0.01 moles N m^{-2} yr^{-1}), each of the systems is likely to be oligotrophic (low level of productivity). Salt marsh estuaries are naturally more productive than seagrass dominated and plankton dominated drowned river valley estuaries. Seagrass dominated estuaries are likely to be the most sensitive to nutrient enrichment, shown by the rapid rise in eutrophication severity as nitrogen loading is increased from 0.1 to 1.0 moles N m^{-2} yr^{-1}. Salt marsh estuaries are expected to be the least sensitive to nutrient enrichment, which is illustrated by the slow rise in eutrophication severity only after nitrogen loading exceeds 1.0 moles N m^{-2} yr^{-1} (unpublished figure by C. Hopkinson).

ecological processes related to nutrient loading[1] (Malone 1977; Cloern et al. 1983; Vallino and Hopkinson 1998; Howarth et al. 2000) (Box 6-1). For example, phytoplankton blooms can occur only when the plankton turnover time is shorter than the water

[1] Many estuaries can be described as hypersaline (Chapter 5). Thus, some consideration was given to including salinity as one discriminator of estuarine class. However, because both dilution and water residence time play a role in determining salinity, it was felt that adding salinity as a specific discriminator of estuarine class would be redundant.

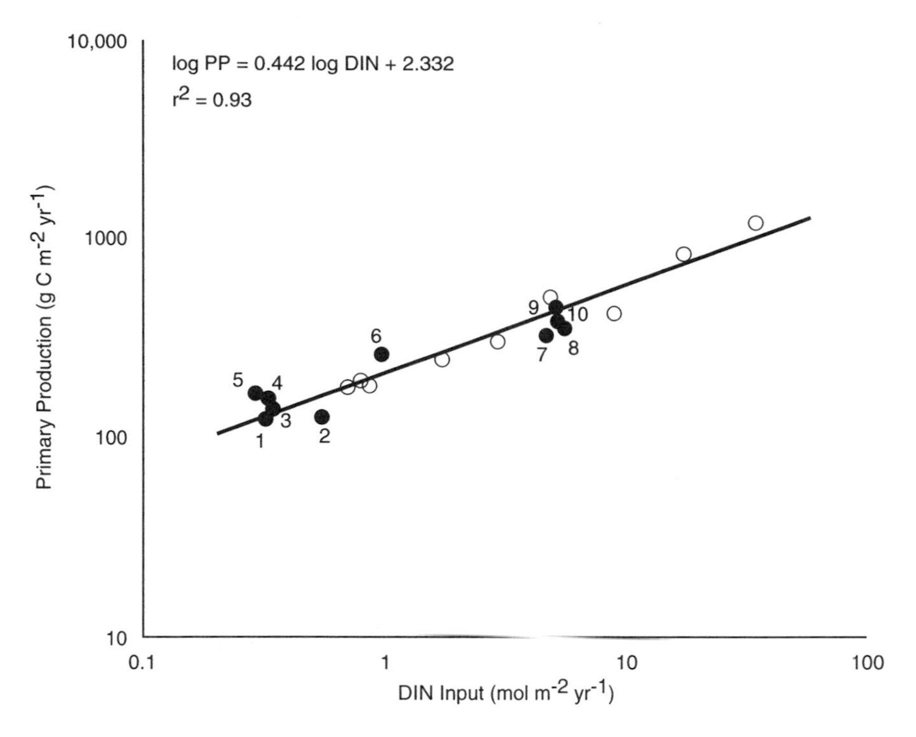

(B) Primary production by phytoplankton (^{14}C uptake) as a function of the estimated rate of input of dissolved inorganic nitrogen (DIN) per unit area in a variety of marine ecosystems. The open circles are for large (13 m^3, 5 m deep) well-mixed mesocosm tanks at the Marine Ecosystem Research Laboratory (MERL) during a multi-year fertilization experiment (Nixon et al. 1986; Nixon 1992). Natural systems (solid circles) are: (1) Scotian shelf, (2) Sargasso Sea, (3) North Sea, (4) Baltic Sea, (5) North Central Pacific, (6) Tomales Bay, California, (7) continental shelf off New York, (8) outer continental shelf off southeastern U.S., (9) Peru upwelling, (10) Georges Bank (modified from Nixon et al. 1996).

residence time. If both water residence time and phytoplankton turnover time are one day, there is no chance of a bloom; algae are flushed from the system as fast as they multiply. Alternatively, if the residence time is seven days and phytoplankton turnover time is one day, phytoplankton can double seven times prior to being exported and an initial algal population of 5 µg chl-a l^{-1} can become a 640 µg l^{-1} bloom, given no other losses. There are other ecological processes whose time scales also can be compared to residence time to determine their potential influence on eutrophication. The

BOX 6-1
The Effect of Residence Time:
The Hudson River Estuary as a Case Study

Estuaries vary greatly in their susceptibility to eutrophication (Bricker et al. 1999). The Hudson River estuary receives extremely high inputs of nutrients, both from wastewater treatment plants in New York City and environs and from non-point sources in the watershed. However, several studies in the estuary during the 1970s showed fairly low rates of primary production (Malone 1977; Sirois and Fredrick 1978). The low production, despite high nutrients, resulted from short water residence times and perhaps light limitation from relatively deep mixing of the water column.

During many summers in the 1990s, the freshwater discharge into the estuary was less than seen during the 1970s (Howarth et al. 2000). This increased the water residence time, increased the stratification in the estuary, and also led to greater water clarity due to less input of sediment and/or lessened resuspension of bottom sediments (Figure 6-2). Thus, not only did phytoplankton have longer to grow before being advected out of the estuary, but light limitation was lessened, increasing growth rates. The result is the estuary became much more productive. During many times in the 1990s, rates of production were high enough to classify the estuary as highly eutrophic (Howarth et al. 2000). Thus, climatic variation can make an estuary more or less susceptible to eutrophication. Future climate warming in the northeastern United States is likely to result in lessened freshwater discharge (Moore et al. 1997), aggravating eutrophication in the Hudson and similar estuaries with short water residence times (Howarth et al. 2000).

FIGURE 6-2 Relationship between freshwater discharge and (A) Gross Primary Productivity (GPP), (B) water residence time, (C) stratification, (D) and light penetration during 25 cruises conducted during the spring, summer, and fall of 1994, 1995, and 1997 in the Hudson River estuary. Squares represent times when tidal amplitude was less than 1.15 m; triangles represent greater tides. The dashed line in (A) indicates the approximate value for GPP above which an estuary is considered to be eutrophic. Note that high rates of GPP only occur when freshwater discharge is less than 200 $m^3 s^{-1}$, and are more likely when tidal amplitudes are low. Freshwater discharge data are from the USGS's monitoring station at Green Island, New York (USGS 1999a). Discharge at Green Island constitutes approximately 67 percent of the total estimated freshwater input to the Hudson estuary and is well correlated with these total inputs (Howarth et al. 2000; used with permission from Springer-Verlag).

Figure 6-2 on next page

BOX 6-1 Continued

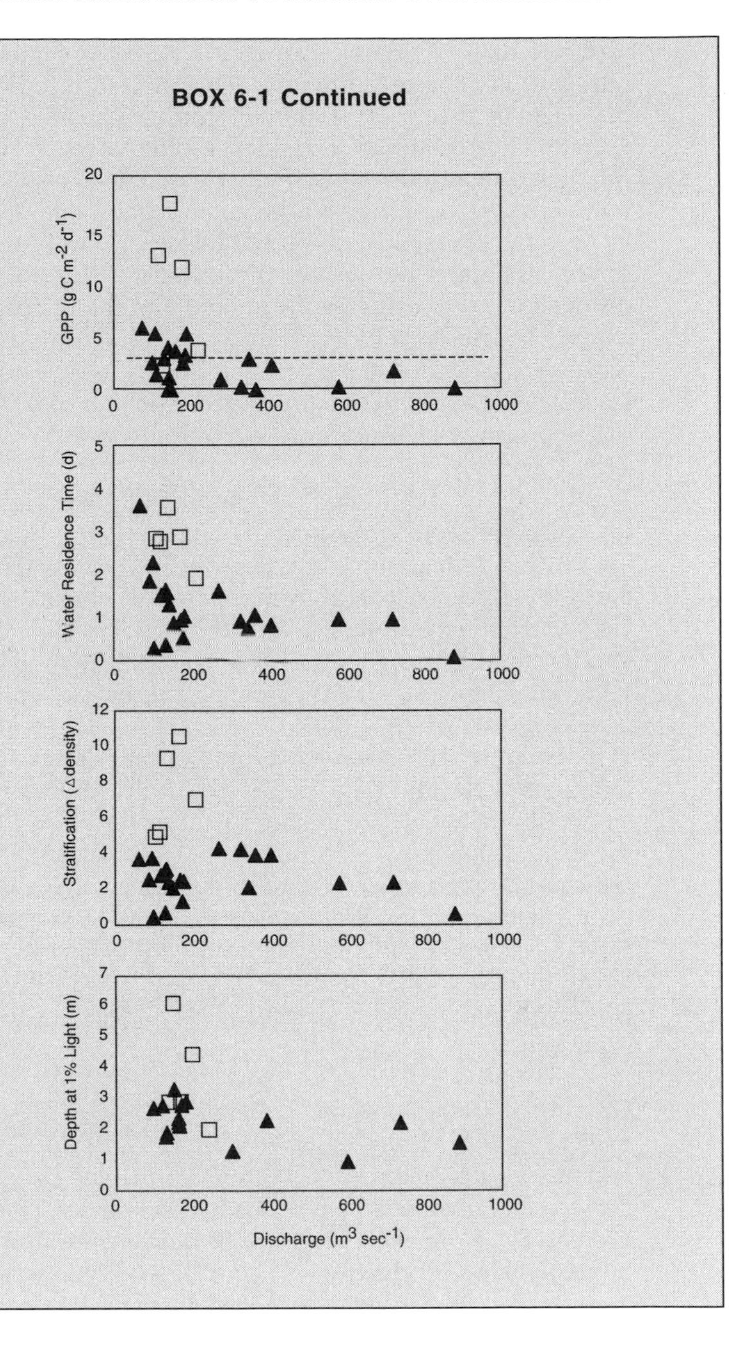

turnover time of organic nitrogen (i.e., conversion to inorganic nitrogen) in comparison to residence time can indicate whether this might be an important source of nitrogen fueling phytoplankton growth. For example, if residence time is seven days, organic nitrogen compounds with a lability (i.e., able to be decomposed or remineralized) or turnover time greater than 14 days are not likely to be remineralized to the inorganic form, and be available to phytoplankton while within the estuary. The fraction of total nitrogen input to estuaries from land and the atmosphere that is exported varies as a function of water residence time in the system, as is the fraction of input that is denitrified in estuaries (Nixon et al. 1996). NOAA has incorporated estimates of flushing in their development of an estuarine susceptibility index.

6. **Stratification.** Stratification is an important physical process affecting eutrophication. Stratification can maintain phytoplankton in the nutrient rich, photic zone (Malone 1977; Howarth et al. 2000) and isolate deeper waters from reaeration. Most hydrodynamic classifications include a measure of stratification intensity (Hansen and Rattray 1966). NOAA considers stratification to be an important component of their developing estuarine susceptibility index.

7. **Hypsography.** Hypsography describes the relative areal extent of land surface elevation, and might be a useful indicator of estuarine susceptibility to nutrient enhanced eutrophication. Knowledge of the relationship between estuarine area and elevation/depth will indicate the percentage of area potentially colonizable by emergent marsh, intertidal flats, submerged aquatic vegetation, phytoplankton, macroalgae, etc. Overlaid with measures of water turbidity and stratification, it might be possible to illustrate the spatial extent of sites potentially susceptible to a variety of eutrophication symptoms.

8. **Grazing of phytoplankton.** Grazing by benthic filter feeders acts to clear particles from the water column, and can limit the accumulation of algal biomass (Cloern 1982). Alpine and Cloern (1992) showed that filter feeding benthos in San Francisco Bay effectively decreased the estuarine response to nutrient loading (in terms of phytoplankton production). There is some conjecture as to the importance of what were once vast filter feeding oyster populations in Chesapeake Bay and whether these acted to decrease the intensity of phytoplankton blooms in the past (Newell 1988). Zooplankton grazing can exert a strong influence on phytoplankton

blooms and eutrophication symptoms in lakes, but this phenomenon remains relatively unexplored in coastal systems (Ingrid et al. 1996). Likewise, the feeding activity of top predators, which can "cascade" down to influence zooplankton sizes, abundances, and grazing pressure on phytoplantkon in estuaries (Carpenter and Kitchell 1993), remains poorly understood.

9. **Suspended materials load and light extinction.** Suspended load and light are two important factors related to nutrient over-enrichment (Box 6-2). Light is a primary factor controlling primary production. Researchers have shown that light can play a critical role in determining the response of estuarine systems to nutrient loading (e.g., Cloern 1987, 1991, 1996, 1999). In northern San Francisco Bay, high turbidity from watershed sediment erosion reduces light levels to such an extent that primary production is light-limited year round. A new conceptual model of coastal eutrophication (Cloern 1999) suggests that it is the interaction of nutrient loading and other stressors/factors that determines estuarine response.

10. **Denitrification.** Denitrification is the process whereby nitrate is converted to gaseous nitrogen and N_2O, and thereby made biologically unavailable. Denitrification provides a sink for nitrogen in estuarine systems; it essentially counteracts allochthonous nutrient inputs to estuaries and thereby can reduce eutrophication responses. Denitrification has been shown to be proportional to the rate of organic nitrogen remineralization in sediments (Seitzinger 1988), which is coupled with the magnitude of primary production that is oxidized by the benthos (Nixon 1981; Seitzinger and Giblin 1996). The relationship between denitrification and eutrophication is not simply linear. There are potential indirect effects of eutrophication that limit denitrification. For example, bottom water anoxia limits nitrification and hence denitrification in sediments and bottom waters. High sulfide concentrations, which are also associated with anoxic conditions, inhibit nitrification as well (Joye and Hollibaugh 1995). Knowledge of the magnitude of denitrification can help predict the eutrophication response of an estuary because nitrogen that is denitrified is largely unavailable to support primary production.

11. **Spatial and temporal distribution of nutrient inputs.** Distribution of nutrient inputs varies along the expanse of an estuary (Vallino and Hopkinson 1998). The potential effect of nutrient

BOX 6-2
An Index of Susceptibility for Estuarine Phytoplankton Communities

The quantity of nutrients into an estuarine system is not the only factor affecting susceptibility. Many estuaries maintain low algal biomass and low primary production under nutrient-rich conditions, such as North San Francisco Bay. How is it that some highly nutrient loaded systems do not show symptoms of eutrophication?

Through research aimed at understanding the mechanisms controlling bloom dynamics in the San Francisco Bay system, Dr. James Cloern and colleagues at the U.S. Geological Survey (USGS) in Menlo Park, California, have developed an index of the sensitivity of a particular estuarine system to changes in nutrient concentration (Cloern 1999). The index is based on a model of phytoplankton population growth, where growth is the product of the carbon assimilation rate and the ratio of chlorophyll to carbon. The model includes functional responses of population growth to photosynthetic efficiency, light availability, temperature, photo-adaptation and nutrient availability. Because light energy can be a major resource that can and often does limit algal growth and production in estuarine ecosystems, a phytoplankton resource limitation plot, with light resource plotted against nutrient resource, indicates whether phytoplankton growth rate is more sensitive to changes in light or nutrients. Where the ratio of growth-rate sensitivity to light and nutrients, R, is greater than one ($R>1$) growth rate is strongly limited by light availability; ratios less than 1 ($R<1$) indicate strong nutrient limitation (Figure 6-3A&B). The boundary line between light and nutrient limitation shifts with changes in the physiological state of the phytoplankton. Thus it is possible to use measures of the light and nutrient resources in an estuary to make judgments about the relative strength of light and nutrient limitation over time and space.

The light-nutrient limitation index is appealing for its simplicity. Using this approach, assessments of the sensitivity of an estuary to nutrient loading can be done using measurements of turbidity, light, and nutrient concentrations. The approach has been validated in applications to numerous estuaries in North America and Europe that demonstrate a range of temporal and spatial variations in the magnitude of nutrient and light limitation. Assessment results could be used to guide management strategies for individual systems. For instance, management strategies that emphasize nutrient reductions would be high priority in estuaries classified as nutrient sensitive; other strategies would be more important in estuaries classified as nutrient insensitive.

The light-nutrient sensitivity index is but one tool available to assess and control the eutrophication threat to coastal ecosystems (Cloern 1999). Other important factors include the effects of residence time and phytoplankton grazing. Furthermore, the index gives no information about harmful algae, macroalgae, or seagrass epiphytes. Because of its simplicity, robustness, and ease of application, however, even with its limitations this index makes a significant contribution to developing a classification of estuarine sensitivity to nutrient enrichment that can be used to help manage nutrient over-enrichment problems.

continued

BOX 6-2 Continued

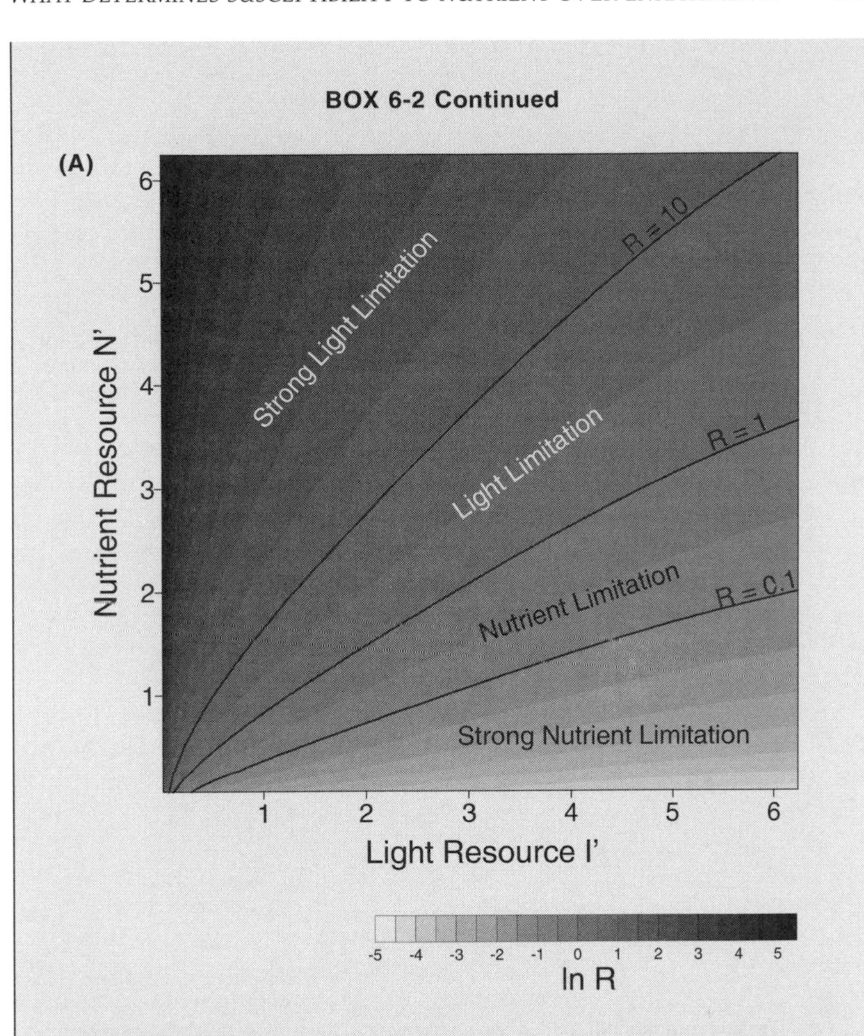

FIGURE 6-3 Illustration of a classification that determines the relative importance of light and nutrients in controlling estuarine trophic state (Cloern 1999). Phytoplankton light and nutrient resource limitation can be calculated as the ratio (R) of growth-rate sensitivity to light and nutrients. (A) Large values of R (greater than 10) are resource combinations where growth rate is strongly limited by light availability: small values of R (less than 0.1) are regions of strong nutrient limitation. The line R = 1 defines the combinations of I' and N' for which growth rate is equally limited by light and nutrient resources. This figure was produced from interpolation of calculated values of R (used with permission from Kluwer Academic Publishers).

continued

BOX 6-2 Continued

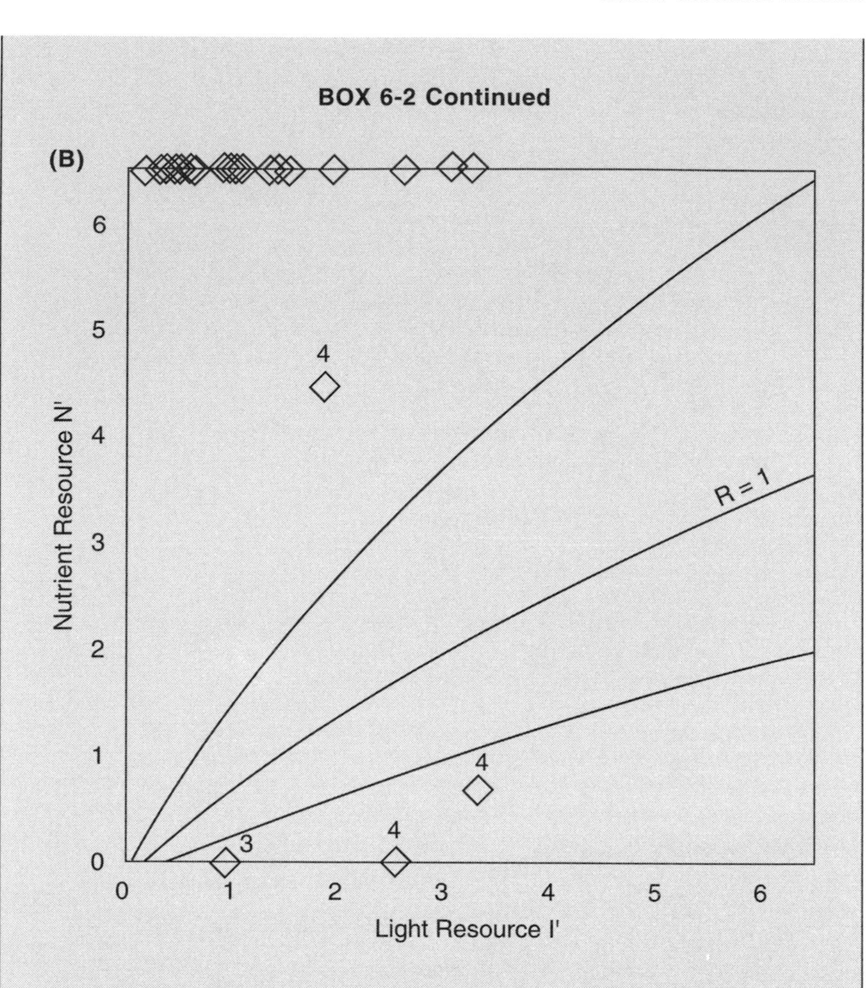

(B) The light and nutrient resources for phytoplankton growth in South San Francisco Bay are shown for measurements made between January 1992 and November 1993 (data from Wienke et al. 1993; Caffrey et al. 1994; Hager 1994). The graph illustrates that South San Francisco Bay is strongly light limited (high nutrient resource and low light resource) during all times the year except March and April (indicated by numbers 3 and 4, where nutrient resource is lower than light resource). From a resource manager's perspective, it would not be cost effective to reduce nutrient loading to the Bay, as eutrophication or other impacts associated with nutrient over-enrichment have not impaired water quality. This is largely the result of the fact that nutrients only control phytoplankton growth during two months of the year (modified from Cloern 1999).

inputs will also vary depending on the location of the input. Seasonal variation in agricultural activity (such as fertilizer application, fossil fuel combustion, or precipitation) results in changes in nutrient loads that a recovery body may see during the year. Furthermore, variation in the load contributed by one of many tributaries may also vary, resulting in seasonal and geographic variability.

12. **Allochthonous organic matter inputs.** Organic matter contributes directly to eutrophication. The relative magnitude of inorganic versus organic nitrogen load influences the balance between autotrophic and heterotrophic metabolism (Hopkinson and Vallino 1995). The relative magnitude of dissolved versus particulate organic matter loads influences residence time of inputs, as particles are preferentially trapped by processes operating in the estuarine turbidity maximum and by gravity. The carbon:nitrogen stoichiometry of organic matter remineralized by the benthos and denitrification further influence the balance between autotrophic and heterotrophic processes in estuaries. Algal blooms are an example of an autotrophic process and net oxygen uptake is an example of a heterotrophic process.

Recognizing that there are basic differences in estuarine susceptibility to nutrient over-enrichment, the development of a conceptual framework or classification scheme for organizing how to think about susceptibility will enable scientists and managers to better understand and predict the effects of human activities on estuarine and coastal ecosystems, and thus to more effectively manage human activities. Variations in the 12 factors discussed above result in different responses. Hence, systematic variation in these 12 factors can result in systematic responses. Thus, coupling an understanding of how these factors vary from estuary to estuary of known load-response behavior can lead to a predictive framework or classification scheme.

It is not adequate to understand eutrophication and other processes in a few, well-studied coastal systems. Useful understanding will require a systematic means of extending the results from one coastal system to others that have not been studied extensively. There are too many estuaries in North America to carry out comprehensive ecosystem studies of all those undergoing nutrient enrichment, and scientists and managers do not presently have a broad enough understanding of estuarine and coastal processes to choose representative systems for detailed analysis. Development of a scheme for classification of estuarine and coastal systems is a prerequisite to understanding and reducing the effects of nutrient over-enrichment. A classification system would provide a language with which to describe the similarities and differences between systems. By enabling researchers to apply more effective and rigorous hypothesis testing, such a classification scheme could provide greater guidance for focused and effective research.

COASTAL CLASSIFICATION

Scientists and resource managers have used classification systems for decades to organize information about ecological systems. Yet the classification of estuarine and coastal systems remains a difficult topic because they exhibit such dynamic changes in time and space. As discussed earlier, each estuary or coastal system possesses a set of characteristics (e.g., morphology, river flow, tidal range, circulation, productivity, etc.) that are controlled, to a large degree, by local geology and climate. A classification scheme should have the ability to:

- encompass broad spatial and temporal scales,
- integrate structural and functional characteristics under different disturbance regimes,
- convey information about mechanisms controlling estuarine or coastal features, and
- accomplish its goal at low cost with a high level of uniform understanding among resource managers.

A useful classification scheme should allow classification of relatively unknown systems on the basis of a minimum suite of measurements (Jay et al. 1999) obtainable from climate records, maps, remote sensing, or ocean monitoring.

To envision how a useful scheme might be constructed, an examination of existing schemes and their value for understanding estuarine response to nutrient loading is warranted. In reviewing existing estuarine classification systems, it is important to bear in mind that only in the past decade or so, however, have classification systems been developed to increase our understanding and ability to predict the effects of enhanced nutrient delivery to coastal ecosystems.

Most classifications of estuaries are based upon physical parameters and geomorphic characteristics. Three basic types of estuarine classification include geomorphic, hydrodynamic, and habitat enumeration approaches.

GEOMORPHIC CLASSIFICATION

Pritchard (1952, 1967) and Dyer (1973) used a geomorphic approach to classify estuaries. From a geomorphological standpoint, Pritchard identified four primary subdivisions of estuaries: 1) drowned river valleys (e.g., Chesapeake Bay), 2) fjord-type (e.g., Penobscot Bay), 3) bar-built (e.g., Laguna Madre), and 4) estuaries produced by tectonic processes (e.g., San Francisco Bay). This approach has intuitive appeal, but it generally lacks a quantitative framework allowing further elaboration. Thus, this approach is of limited usefulness in understanding, predicting, and managing estuary response to nutrient loading.

HYDRODYNAMIC CLASSIFICATION

The hydrodynamic approach to estuarine classification focuses on the interaction in narrow estuaries of tidal currents and river flow. Tidal currents provide energy for mixing while river flow is a source of stratification or buoyancy. Stommel and Farmer (1952) divided estuaries into 4 categories based on stratification: 1) well mixed, 2) partially mixed, 3) fjord-like, and 4) salt wedge. This simple classification was made quantitative by defining a stratification number G/J, where G is energy dissipation over a defined channel length and J is the rate of gain of potential energy of water moving through the estuary over the same length (Ippen and Harlemann 1961; Prandle 1986). By incorporating Richardson number and critical depth criteria, Fischer (1976), Simpson and Hunter (1974), and Nunes Vaz and Lennon (1991) have made variations on this classification.

The one-parameter (e.g., G/J) classifications capture one important aspect of estuarine circulation but they have no direct relation to the various estuarine types (Jay et al. 1999). Accordingly, two-parameter classifications were developed that describe the interaction of geomorphology, fresh water, and tides. The most widely known two-parameter scheme (Figure 6-4; Hansen and Rattray 1966) employs two parameters to classify estuarine circulation: 1) a stratification parameter, $\partial S/S_0$ and 2) a circulation parameter, U_s/U_F. The first parameter describes stratification as the ratio of the top-to-bottom salinity difference to mean salinity over the section. The circulation parameter is a ratio of the net surface current to the mean freshwater velocity through the section. Numerous clarifications, modifications, and additions have been made to this approach, most focused on providing a closer connection between the density field and tidal processes (Fischer 1976; Officer 1976; Oey 1984).

Hansen and Rattray (1966) identified seven types of estuaries, basically following the conventional usage of Stommel and Farmer, but further differentiating physically significant differences of regime. Type $1_{(a+b)}$ estuaries have net flow seaward at all depths and upstream salt transfer controlled by diffusion (1_a–well-mixed with slight salinity stratification, 1_b–strong stratification). Type $2_{(a+b)}$ estuaries have net flow reversal at depth with both advection and diffusion contributing to upstream salt flux (2_a–slight stratification, 2_b–strong stratification). Type $3_{(a+b)}$ is distinguished from type 2 primarily by the dominance of advection in accounting for salt flux. (Type 3_b estuaries are so deep that the salinity gradient and circulation do not extend to the bottom) (i.e., fjords). In Type 4 estuaries (salt wedge), the stratification is still greater and the flow grades from a thick upper layer flowing over a thin lower layer to a shallow surface layer flowing with little influence over a deep lower layer. The Hansen and Rattray (1966) circulation-stratification diagram illustrates that a range of circulation patterns and estuarine morphologies is possible for a certain degree of stratification (Box 6-3 and Figure 6-4).

HABITAT CLASSIFICATION

There is a long history of classifying environments on the basis of their plant community composition or sediment characteristics. In the early 1900s there were several attempts to classify wetlands, especially peatlands of Europe and North America, on the basis of the combined chemical and physical conditions of the wetland along with the vegetative community description. These early classifications served as models for more inclusive classifications developed by the U.S. Fish and Wildlife Service in the 1950s and 1970s.

In the 1950s, the U.S. Fish and Wildlife Service developed a classification scheme to inventory the distribution, extent, and quality of remaining wetlands in relation to their value as wildlife habitat. Twenty types of wetlands were described, including "coastal saline areas." Four overall categories were arranged by life forms of vegetation in order of increasing water depth or frequency of inundation. The scheme was elegantly simple, with salinity being the sole chemical criteria. The classification became known as the Circular 39 Classification (Shaw and Fredine 1956).

In 1979, the National Wetlands Inventory classification was adopted as the primary wetland classification scheme for U.S. wetlands. Developed by the U.S. Fish and Wildlife Service, the classification scheme included deepwater habitats and wetlands (Cowardin et al. 1979). The classification and following inventory was intended to describe ecological taxa, to arrange them in a system useful for resource managers, to provide units for mapping, and to provide uniformity of concepts and terms. The classification has a hierarchical approach that uses systems, subsystems, classes, subclasses, dominance types, and special modifiers to more pre-

BOX 6-3
Hansen and Rattray Classification Scheme

Hansen and Rattray (1966) developed a classification scheme with relatively simple parameters and good predictive ability to deal with salt transport mechanisms. The scheme allows calculation of the relative importance of diffusive (tidal) salt transport relative to total landward salt flux needed to maintain the salt balance. This kind of predictive ability is important for understanding relationships between nutrient loading and eutrophication processes.

Since the original work was done, others have improved the parameterization of tidal processes. Jay and Smith (1988) directly represented the forcing of residual circulation. Friederichs and Madsen (1992) suggested a modification to address the influence of tidal flats. Hearn (1998) considered the effect of surface heating and evaporation on narrow Mediterranean estuaries. Because of the types of forcing factors chosen as parameters, classification schemes based on Hansen and Rattray are restricted to narrow embayments, fjords, and river-estuaries where these factors predominate. Systems dominated by wind forcing, such as broad, shallow embayments, and plumes, are unsuitable because they are too wide to have substantial river flow per unit width or they have no lateral boundaries.

The Hansen and Rattray (1966) approach has been criticized because it fails to incorporate the inherent variability within individual estuaries and in the same estuary at different times. As new hydrodynamic classification schemes are developed, they should incorporate variability. Examples of such variability include spring-neap variability of stratification and variability associated with freshwater inputs. One measure of variability might include the ratio of cross-estuary to along-estuary salinity gradients (Geyer et al. 1999). By addressing estuarine variability as part of estuarine classification, such systems will better articulate the connections between estuarine structure and estuarine processes.

cisely define wetlands and deepwater habitats. A stated goal is to eventually inventory the wetlands across the United States at a scale of 1:24,000.

The National Wetlands Inventory classification includes four levels, three of which include coastal habitats: marine, estuarine, and riverine. Subsystems that are of interest to coastal scientists and managers include subtidal and intertidal regions. The lowest hierarchical level, the modifier level, more precisely describes the water regime, salinity, pH, and soil.

A few habitat-based classification schemes have included forcing function criteria. Odum et al. (1974) developed a classification and functional description of coastal ecosystems that included major forcing functions and stresses that influence the distribution of systems. A hierarchical approach was used as opposed to a quantitative approach, and thus the forcing functions and stresses were not parameterized.

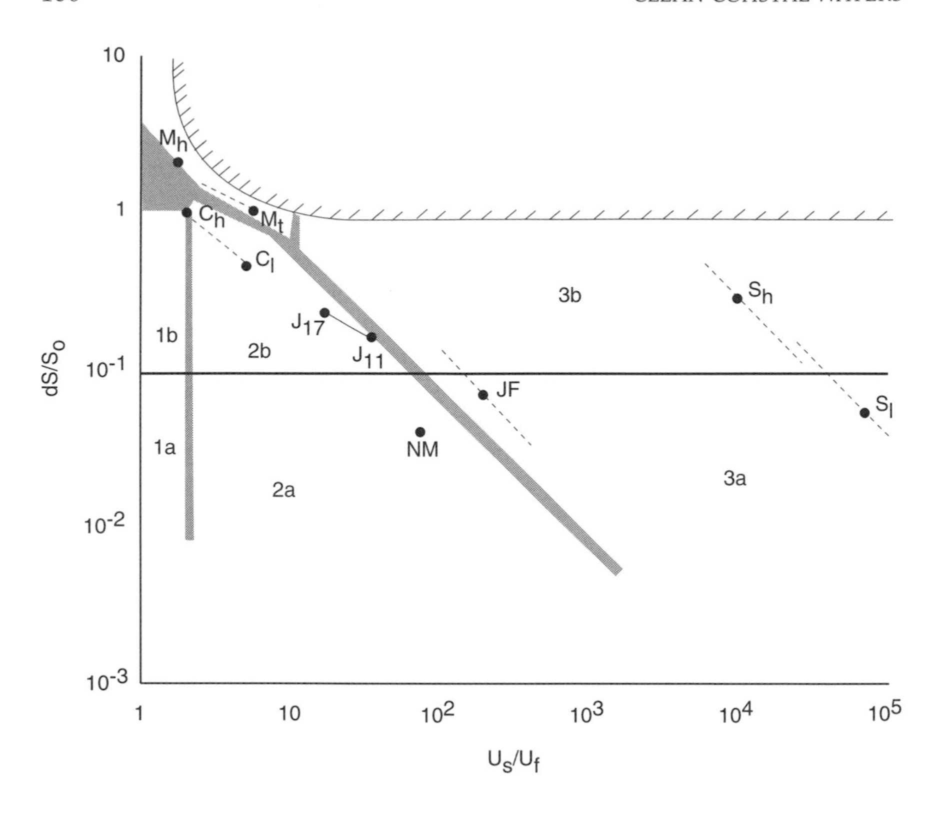

FIGURE 6-4 An example of a physically-based estuarine classification. Hansen and Rattray developed a classification for differentiating estuaries on the basis of factors controlling circulation. The first factor describes the degree of estuarine stratification, which is related to the difference in surface and bottom water salinity ($\partial S/S_0$), and the second factor describes the relative importance of freshwater flow on net flow (U_s/U_f). This figure shows how several North American estuaries differ according to these criteria. There is a gradient of increasing stratification from Type 1 to Type 4 estuaries. The Columbia River (C_h and C_l) is a type 1 or 2 estuary depending on time of year and freshwater runoff; the James (J_{17} and J_{11}) and Narrows of the Mersey (NM) estuaries are Type 2 estuaries; the Strait of Juan de Fuca (JF) and Silver Bay (S_h and S_l) systems are Type 3; and the Mississippi River (M_h and M_l) is a Type 4 estuary. Subscripts $_h$ and $_l$ refer to high and low river discharge conditions, respectively. Recent research on estuarine susceptibility to eutrophication suggests that stratification is a major determinant in estuarine response to nutrient loading. The Hansen and Rattray classification enables stratification to be quantified and thus may be useful in developing an eutrophication classification (modified from Hansen and Rattray 1966).

The objective of most habitat type classification schemes is to impose boundaries on natural ecosystems for the purposes of inventory, evaluation, and management. This approach works well for aiding in the inventory and evaluation of wetlands as well as estuarine habitats, but it generally lacks a logical or quantitative framework. The use of the approach as a tool to assess susceptibility to nutrient loading is unproven.

HYBRID CLASSIFICATION

Jay et al. (1999) borrowed from a highly effective hierarchical geomorphic classification scheme with a proven predictive ability for fluvial systems (Montgomery and Buffington 1993) to develop a geomorphic estuarine classification with a hierarchical structure. Several recent fluvial classification systems are based on a hierarchical ranking of linkages between the geologic and climatic settings, the stream habitat features, and the biota (Hawkins et al. 1993). The geomorphic and climatic processes that shape the abiotic and biotic features of streams provide a conceptual and practical foundation for understanding the structure and processes of fluvial systems. Furthermore, an understanding of process allows streams to be viewed in a larger spatial and temporal perspective, and to infer the direction and magnitude of potential changes due to natural and human disturbances. A stream classification system based on patterns and processes and how they are expressed at different temporal and spatial scales can aid successful management (Rosgen 1994).

The Jay et al. (1999) classification for estuaries provides a means to identify environments found in various types of estuarine subsystems, relates estuarine types to dominant sediment transport processes, and allows a prediction as to whether sediment transport is limited by transport capacity or sediment supply for coarse and fine sediments. Linkage with hydrodynamic classification schemes is through the non-dimensional hydrodynamic parameters associated with each sediment transport forcing mode. Six transport processes are parameterized: 1) net motion of river flow (Q_R), 2) oscillatory tidal flow (Q_T), 3) internal circulation or buoyancy forcing (Q_I), 4) atmospherically forced circulation (Q_A), 5) transport and resuspension by wind waves and swell (Q_W), and 6) transport by sea ice (Q_H). Each forcing process has a representative time scale and descriptive non-dimension hydrodynamic parameters responsible for distinctive modes of sediment transport. Extension of this hierarchical geomorphic classification to address issues of ecological importance such as eutrophication has not been attempted, but the inclusion of particle trapping and residence time processes suggests that this may be a profitable avenue for further investigation.

THE NATIONAL OCEANIC AND ATMOSPHERIC ADMINISTRATION'S NATIONAL OCEAN SERVICE CLASSIFICATION SCHEMES

NOAA's National Ocean Service have been working to develop methods to gauge the susceptibility of estuaries to nutrient over-enrichment for over a decade. The first index developed, the "dissolved concentration potential" (DCP), integrated nutrient loads with an estimate of estuarine dilution and flushing. The dilution parameter is proportional to estuarine volume and the flushing parameter is calculated with the Ketchum (1951) fractional freshwater method, which is derived from the replacement of the freshwater component of the total system volume by river flow.

$$DCP = \frac{Q_f}{V_f} \times \frac{1}{V_t} \times \overline{N}$$

where Q_f is freshwater discharge and V_f is freshwater volume in the estuary, V_t is total estuarine volume and $\overline{N}$ is mean nutrient load for all estuaries.

The DCP provides an estimate of average nutrient concentration throughout an estuary assuming there is no biological processing. Systems with a high DCP tend to concentrate nutrient inputs, while systems with a low DCP strongly dilute or flush nutrients. NOAA categorizes estuaries as having low, medium, and high susceptibility to nutrient loading on the basis of DCP concentration, less than 0.1 mg l^{-1}, 0.1 to 1.0 mg l^{-1}, and greater than 1.0 mg l^{-1}, respectively. It would be interesting to determine the degree to which a measure of eutrophication severity such as phytoplankton production or standing crop correlates with measures of nutrient load and DCP. This index has not been quantitatively compared to the Nixon model (Nixon 1992, 1997; Nixon et al. 1996), which is based solely on areal nutrient loading, in its ability to predict eutrophication.

DCP provides a quantitative measure of estuarine susceptibility to nutrient loading and is based on physical criteria including estuarine volume, volume of freshwater, and freshwater inputs. An underlying assumption is that the system is vertically homogenous (i.e., does not account for stratification). It could be applied at various times of the year to assess how susceptibility varies temporally. DCP relies exclusively on freshwater input as the mechanism for flushing. Flushing would be underestimated in systems where tides or winds are primary mechanisms controlling mixing, such as in many lagoonal systems. Spatial gradients in flushing cannot be accommodated with the DCP approach.

The "estuarine export potential" (EXP) is a second generation classification developed by NOAA to predict estuarine response to nutrient loading. EXP defines the relative capacity of estuaries to dilute and flush dissolved nutrient loads. It addresses several deficiencies of the DCP index by incorporating aspects of stratification and tidal range. In its present configuration, EXP is not a quantitative index; rather it categorizes systems into low, medium, and high flushing and dilution potentials. The EXP index estimates dilution potential from measures of estuarine volume and the presence of stratification, and it estimates flushing potential from measures of tidal range and the ratio of river runoff to estuarine volume. The method uses a decision-rule process and a combination of qualitative and quantitative measures. This approach has intuitive appeal (as does the geomorphic classification scheme), but it generally lacks a formal quantitative framework allowing further elaboration. While the relative simplicity of the approach can be seen as a strength (i.e., data from a few estuaries can be applied to a large number of estuaries), the semi-quantitative parameterization limits the predictive ability.

NOAA staff, in conjunction with leading estuarine scientists, have applied the EXP scheme to the 138 estuaries included in the National Estuarine Eutrophication Assessment (Box 6-4). Overall, EXP was found to be useful in developing an assessment of eutrophication susceptibility, being in the "ballpark" for about 85 percent of the estuaries. Problems in prediction were focused mainly in a few Maine estuaries, small estuaries in southern California, and the Puget Sound estuaries. The next logical step would be to quantify the extent to which the ability to predict estuarine susceptibility has improved with further development of the EXP index. Analyses of the relationship between areal nutrient loading, EXP, and trophic state would illustrate whether predictive ability has improved relative to the first approximation presented by Nixon (1992).

Results produced using the EXP approach are currently being reviewed by experts, while the National Ocean Service pursues more rigorous coupling of EXP with National Estuarine Eutrophication Assessment results. At the same time, the National Ocean Service is refining the approach to help increase both spatial and temporal resolution. This will enable the National Ocean Service to evaluate the relative susceptibility of various regions within estuaries during a range of runoff conditions (Bricker et al. 1999).

EXP refinements under consideration include the addition of factors thought to influence estuarine susceptibility, including temperature (to address biotic differences between biogeographic provinces), the importance of wind mixing, inlet configuration, estuarine plume exchange with nearshore oceanic water, and the ratio of shoreline length to estuarine surface area (which is believed to correlate with the importance of intertidal wetlands). NOAA also plans to explore spatial and temporal vari-

BOX 6-4
Determining which Estuaries are Naturally More Susceptible to Nutrient-Related Impacts: A Possible Approach

Estuaries can be classified based on physical transport processes that, in part, determine their susceptibility to nutrient-related water quality conditions. An index to quantify the transport processes, EXP, was developed using physical and hydrologic data, assembled by NOAA's National Ocean Service for 138 estuaries in the conterminous United States.

As a first approximation, the EXP index classifies an estuary's susceptibility to nutrient-related water quality concerns using two key physical factors: the dilution capacity of the water column and its flushing/retention time (as discussed earlier in this chapter, other factors can play an important role in determining susceptibility to eutrophication). Dilution capacity is determined by the volume of water available to dilute nutrient supplies. In vertically homogenous estuaries, the dilution volume is equal to the estuary volume. In contrast, for vertically stratified systems the dilution volume is limited to the upper layer of the water column. Flushing is the time required for freshwater inflow and tidal prism volume (modified by a re-entrainment coefficient) to replace the estuary volume. The index represents the average annual and system-wide conditions, providing an order-of-magnitude separation for the 138 coastal systems studied. Figure 6-5 provides some examples of results using this approach to classification.

The results indicate that there are substantial differences among the 138 estuaries. Dilution volume ranges over five orders of magnitude and flushing time ranges just under five orders of magnitude. Systems with relatively large volumes and short flushing times, such as large river systems (e.g., Columbia and Mississippi Rivers) are less susceptible to eutrophication due to nutrient loading. Systems with moderate volumes and long flushing times, such as Chesapeake Bay, are more susceptible to eutrophication.

One way to apply the susceptibility concept is to couple EXP with nutrient load estimates from each estuarine watershed (Figure 6-6). This provides a predicted nitrogen concentration in the water column that suggests, in a comparative sense, the potential for nutrient-related water quality symptoms. For example, higher nutrient concentrations imply the potential for more extreme expressions of nutrient-related symptoms.

Coupling EXP with nutrient load estimates also has the potential to suggest how responsive the system may be to additional nutrient loads or nutrient abatement strategies. For example, estuaries in the upper left portion of Figure 6-6 would have to add or reduce comparatively more nutrients to affect water column concentrations than estuaries in the lower right. Likewise, this work may begin to describe how changes to an estuary's physical environment could potentially alter its susceptibility to nutrient-related conditions. For example, the dilution or flushing components of EXP could be affected by alterations in freshwater inflow (e.g., diversions, impoundments, or consumptive loss) or tidal exchange (e.g., inlet modification, channel dredging).

continued

BOX 6-4 Continued

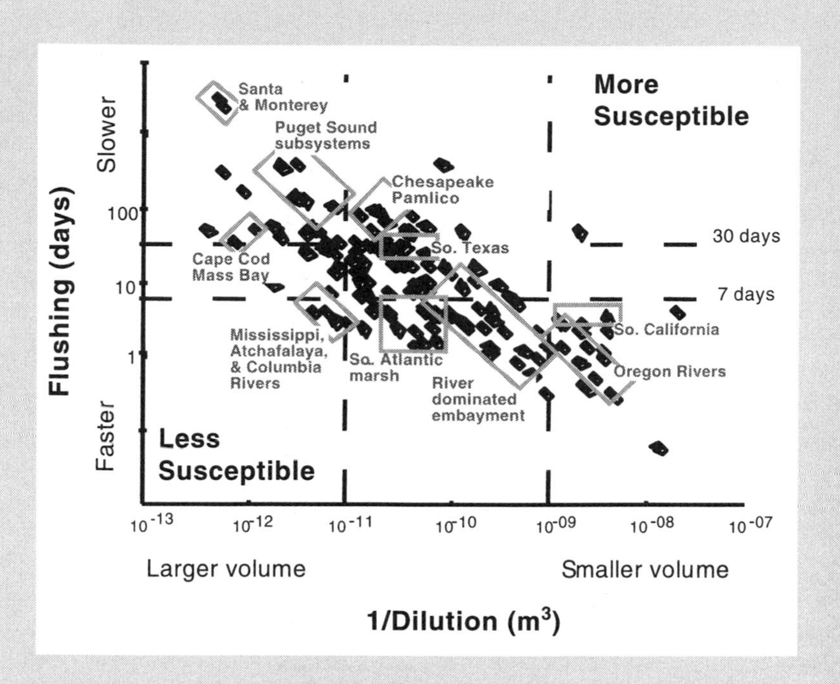

FIGURE 6-5 Coastal systems can be classified according to their dilution and mixing capacities. Here, NOAA has classified 138 coastal systems of the United States according to dilution (volume of estuarine water above the pycnocline) and flushing (based on time to replace estuarine volume by freshwater inflow or tidal prism volume). Coastal systems falling in the lower left region of the graph are those with extremely large dilution volumes and short flushing times. We would expect these systems to be the least susceptible to nutrient-enhanced eutrophication. Systems in the upper right region of the graph have the smallest dilution volumes and longest flushing times. We would expect these systems to be very susceptible to eutrophication.

continued

BOX 6-4 Continued

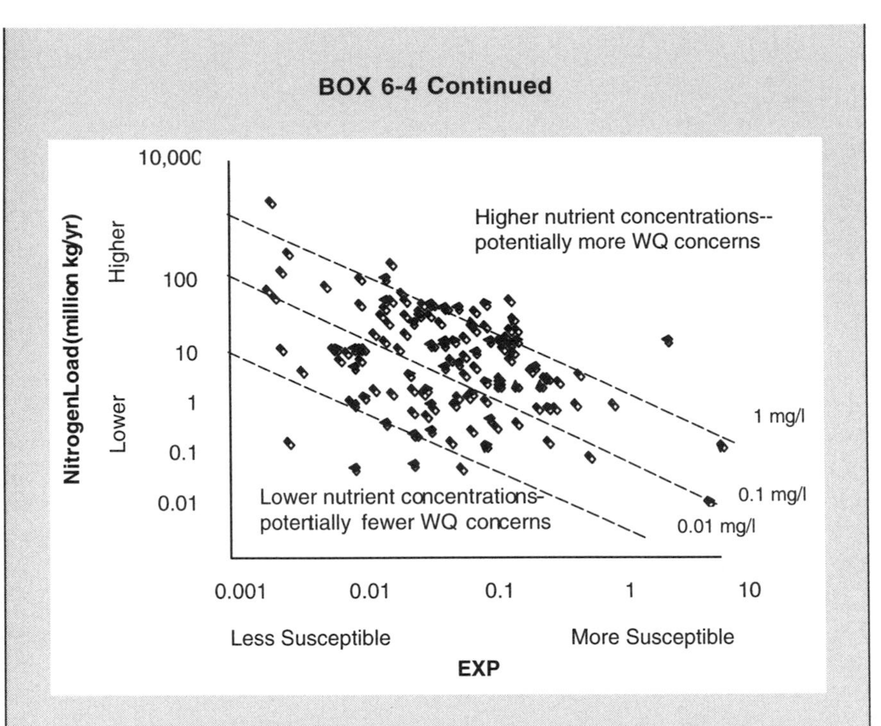

FIGURE 6-6 By coupling EXP (a measure of estuarine export potential) with an estimate of nitrogen load to each of 138 U.S. coastal systems (predicted from the USGS SPARROW model), it is possible to predict the average nutrient concentration in each system and hence its potential trophic state. Systems with a low nutrient load and low EXP (in the lower right region of the graph) are predicted to have the lowest nutrient concentrations. In contrast, those in the upper right should have the highest nutrient concentrations. NOAA is now in the process of comparing predicted nutrient concentration with measured trophic state as presented in the National Estuarine Eutrophication Assessment (Bricker et al. 1999). WQ stands for water quality. Lines of concentration indicate combinations of EXP and nutrient loading resulting in equal nitrogen concentrations.

ability of EXP in single systems. These efforts, when coupled with efforts to model nutrient loading to estuaries and more rigorous quantification of tidal and stratification parameters should improve the capability to predict estuarine susceptibility to nutrient enrichment (Bricker et al. 1999).

Most recently, NOAA has been addressing some of the deficiencies of the DCP index to incorporate measures of tidal flushing and stratification. Tidal flushing is addressed by incorporating a measure of the tidal prism into the overall flushing calculation, and stratification is addressed by calculating dilution (for stratified systems) on the basis of the freshwater fraction rather than total estuarine volume. Using updated information on nitrogen loading, calculated using USGS's Spatially Referenced Regressions on Watersheds (SPARROW) model and estuarine volumes, some patterns have been revealed in plots of revised DCP versus nitrogen load for many of the estuaries included in NOAA's National Estuary Eutrophication Assessment survey (Bricker et al. 1999; Box 6-4).

In this work, geographically and geomorphologically similar systems tended to cluster. For instance, estuaries in Maine fell out as a cluster as did large rivers, mid-Atlantic lagoons, and south-Atlantic marsh dominated estuaries. It is not obvious why these groupings occur, but this behavior may reflect similarities in stratification and circulation in local regions or it may reflect basic differences in physiography or primary production base (e.g., salt marsh dominated lagoons versus plankton dominated drowned river valleys). Initial comparisons to trophic state showed considerable deviations from predictions. For example, three clusters predicted to range from low to high susceptibility have been observed to be moderate to highly eutrophic. Again, an explanation for these patterns is not obvious and perhaps suggests further basic differences between various types of estuaries beyond that captured by measures of nutrient loading, dilution, and flushing. Perhaps a different set of rules governs the behavior of salt marsh and phytoplankton dominated estuaries. The DCP revisions summarized here represent preliminary results. They have not been thoroughly reviewed, but were presented to the committee to illustrate some of the directions NOAA's National Ocean Service is pursuing to increase understanding of susceptibility to nutrient loading.

NEXT STEPS

NOAA's DCP and EXP classification schemes are unique in that they were developed for the sole purpose of eutrophication understanding and prediction. However, neither of these schemes (including the revised DCP), have been tested rigorously to determine their ability to predict estuarine susceptibility to enhanced nutrient loading. Although physically based classification schemes have been useful in describing aspects of estuarine circulation, they have not been used systematically to understand responses to nutrient loading. For example, no classification scheme has been developed that encompasses the myriad of factors thought to be important eutrophication controls. While a statistical approach might be useful for identifying the suite of factors that best explain variability in eutrophication or other adverse impacts from nutrient over-enrichment across estuaries, such a relationship might not elucidate the actual mechanisms controlling eutrophication and thus we lose information perhaps critical to managing and reversing eutrophication trajectories. Improved classifications need to be developed that can be generalized to a broader range of features and processes relevant to estuarine ecosystems, especially those affecting the susceptibility of various estuaries to nutrient over-enrichment.

A Proposal to Select and Use Coastal Index Sites

There are too many estuaries in the United States for the nation to conduct comprehensive ecosystem studies of all those affected by nutrient enrichment. Although the federal government conducts monitoring activities at more than 15,000 sites nationwide (Pryor et al. 1998), these efforts are not sufficiently coordinated to provide a predictive understanding of the causes and effects of nutrient enrichment. Because it is generally understood that different types of coastal systems differ in their response to nutrient enrichment, it would be extremely useful for managers to have a framework of dose-response curves for each of the major types of coastal systems. With such a tool, coastal managers could predict the effects of both increased and reduced nutrient inputs. Thus, a system is needed to classify coastal systems into a number of major types that are likely to respond similarly to nutrient enrichment. Resource managers could then apply these unique dose-response curves to their estuary's particular conditions.

When considering questions about how to improve the integration of environmental monitoring and research across the nation's many existing networks and programs, the Environmental Monitoring Team of the Committee on Environment and National Resources (a committee of the Executive Office's National Science and Technology Council; Pryor et al. 1998) produced a three-tiered conceptual framework describing how federal environmental monitoring activities can fit together. The first tier includes inventories and remote sensing; the second includes national and regional surveys; the third tier includes intensive monitoring and research sites, or "index sites." The goal was to integrate activities across tiers and thus provide the understanding that will enable sound evaluation of the status, trends, and future of the environment. (This approach also is similar to one proposed by NOAA, the Environmental Protection Agency [EPA], and USGS in their draft coastal research and monitoring strategy.)

The Committee on the Causes and Management of Coastal Eutrophication recommends adoption of this three-tiered framework as a way of better integrating monitoring and research in support of improved management of coastal ecosystems. Of primary importance, the committee suggests that monitoring and research be conducted at a sufficiently high spatial and temporal resolution at Tier III "index sites" to develop predictive, cause-effect or dose-response models for the nation's major types of coastal systems.

The establishment of a national framework of index sites where monitoring and research are closely integrated would lead to the development of a predictive understanding of coastal system responses to anthropogenic activities, especially nutrient enrichment. At index sites, intensive monitoring and research activities would lead to a broad understanding of how and why estuaries respond as they do to nutrient inputs. Research should not be restricted to increased nutrient loading scenarios, but should also examine responses to nutrient loading reductions. We expect a different set of dose-response curves for nutrient reductions that would incorporate time lags, hystereses, and non-linear responses of biological systems. Predictive models would be developed at index sites. Index sites should be established for each of the major types of estuaries; we expect unique dose-response curves for each estuarine type. Examples of possible Tier III index sites would be the coastal long-term ecological research sites. Coastal long-term ecological research sites are funded by the National Science Foundation (NSF) at Plum Island Ecosystem in northern Massachusetts, Baltimore Ecosystem Study, Virginia Coast Reserve, Santa Barbara Channel, Everglades, and Georgia Coastal. Each estuarine type should possess a unique dose-response curve that relates primarily to variations in the major factors controlling estuarine susceptibility to nutrient over-enrichment (i.e., loading, dilution, and flushing).

How best to identify the major types of coastal systems is elusive. Based on our analysis, this committee believes that a combination of physiographic province and primary production base could serve as the key criteria for selecting the major types of coastal systems. Following this thinking, and based on physiographic characteristics, the twelve major types of coastal systems are:

1. open continental shelf (e.g., Georgia Bight, Monterey Bay);
2. coastal embayment (e.g., Massachusetts Bay, Buzzards Bay, Long Island Sound);
3. river plume (inverted) estuary (e.g., Mississippi River plume);
4. coastal plain or drowned river valley estuary (e.g., Chesapeake Bay, Hudson River, Charleston Harbor);
5. coastal plain salt marsh estuary (e.g., Plum Island Sound, North Inlet, Duplin River);
6. lagoon (e.g., Padre Island, Pamlico Sound);
7. fjord estuary (e.g., Penobscot Bay);
8. coral reef system (e.g., Kaneohe Bay);
9. tectonically caused estuary (e.g., San Francisco Bay, Tomales Bay);
10. large river, non-drowned river estuary (e.g., Columbia River);

11. seagrass dominated estuary (e.g., Tampa Bay); and
12. rocky intertidal, macroalgae dominated estuary (e.g., Casco Bay).

How to identify research teams qualified to conduct process-oriented research at the index sites is equally elusive. To make major advances in understanding coastal systems and in predicting the effects of increased nutrient loading will require interdisciplinary research coordinated among investigators working within the index sites. Research should emphasize major ecological questions that stress linkages between terrestrial and coastal ecosystems. The research should seek to understand the causes of major ecological and environmental changes, including eutrophication and how populations, communities, and ecosystems of the coastal systems respond to these changes. Research at index sites should include experimental studies across a range of appropriate spatial and temporal scales. Comparative approaches encompassing parallel studies in different coastal systems are likely to provide important insight in how systems respond to nutrient enrichment. There should be close coupling between experimental, descriptive, and comparative research, with simulation modeling used to guide the research and to facilitate comparison with research in other systems. Finally, for the research to be of public value, there is a need for the detailed, process-based models to be abstracted to "simple" dose-response curves that can be easily applied by coastal resource managers at the local level.

Index site research teams should be selected on a peer-reviewed competitive basis, similar to that employed by NSF for the selection of recent coastal long-term ecological research sites. There are probably only 12 to 24 research groups around the country qualified to conduct this type of research, including academic and federal groups. For index site research to be successful, the highest selection priority should be on the originality and quality of the research proposal and research group, followed by representation of the major estuarine types, by research site characteristics and suitability for conducting eutrophication research, and finally by geographic spread.

Development of predictive, mechanistic models requires the integration of process-oriented research with comparative studies of estuaries (Geyer et al. 1999). It is not adequate to understand eutrophication processes in only the few, well-studied index estuaries; however, a systematic means of extending the results from one estuary to others that have not been extensively studied is also required. Understanding of processes resulting from the detailed studies in index estuaries can be tested and broadened through comparisons conducted in other estuaries of similar "type" but which represent the range of physical, hydrological, and biological characteristics. Within each "type," responses are expected to vary according to the major factors that control the response to nutrient addition such as dilution, freshwater input, flushing due to gravitational, tidal and wind driven circulation, stratification, water clarity/turbidity, denitrification, and biological control. Tier II coastal systems should be used as sites where comparative research can be conducted. Tier II systems might include estuaries such as those in NOAA's National Estuarine Research Reserves and EPA's National Estuary Programs. At these sites, research is conducted and data collected at much lower temporal and spatial resolution. Integration within a national program, however, would ensure collection of data necessary for testing the predictive models developed at index sites. Using information from existing programs can be cost-effective, but this is not always the ideal approach because many of these programs do not administer their own monitoring but instead rely on state agencies. Also, research projects can be short term, whereas a long-term perspective is of critical importance.

The result of monitoring, research, and modeling conducted within a three-tiered national framework would be a series of dose-response curves tailored to each major type of coastal system (Figure 6-5). Specific dose-response curves tailored to individual estuaries on the basis of their unique characteristics could then be applied by local and state resource managers hoping to control or reverse eutrophication trends.

ADDITIONAL QUESTIONS

An understanding of the response of coastal waters to nutrient loading is developing slowly. Following the scientific lead of freshwater ecologists modeling the trophic state of lakes, marine ecologists are developing a predictive understanding of some of the key parameters controlling estuarine response to nutrient loading. There are numerous models that relate eutrophication or primary production to single variables, such as the filter feeding benthos and light availability (Alpine and Cloern 1992; Cloern 1999). Nixon (1992) described a strong relationship between rate of estuarine phytoplankton production and areal rate of nutrient loading. However, the extreme variation in response to any level of loading clearly demonstrates the importance of other factors that determine differences between estuaries. The next level of understanding may well result from incorporation of additional factors into models. Several groups around the world are taking the next steps of incorporating measures of circulation, stratification, mixing, dilution, and turbidity into their eutrophication models. NOAA is updating its DCP and EXP measures of estuarine susceptibility with current data and comparing predictions to their national dataset of estuarine trophic state. By incorporating new measures for estuarine susceptibility, predictions of estuarine response to

BOX 6-5
Eutrophication Reversal in Tampa Bay

Tampa Bay, Florida, is a seagrass dominated estuary that has also experienced nitrogen source reductions and concomitant reversals in eutrophication. Impacts to Tampa Bay from increasing population and industrial development resulted in high algal biomass and large seagrass reductions during the 1960s and 1970s. By 1982, seagrass coverage was only 72 percent of earlier estimates. However, in 1996 seagrass coverage had increased by 25 percent since 1982 (Ries 1993; Johansson and Greening 2000).

Nutrient loading reduction strategies were initiated in 1980, and since then, nitrogen inputs from sewage treatment plants have been reduced by 50 percent. As indicated in Figure 6-7, the pattern of eutrophication has been reversed in a time consistent with nutrient load reductions. Chlorophyll *a* concentrations began to decrease within three to five years of nutrient reductions and are now fluctuating close to targets set by resource managers. Seagrasses have taken longer to recover, lagging nutrient reduction by about eight years, but since 1988 coverage has been increasing about 200 hectares annually. With a management target of 15,378 hectares, recovery will take another 25 years to complete if recovery continues at present rates. The causes of lags are uncertain, but are thought to include continued release of nitrogen from internal nitrogen stores accumulated during earlier years of accumulation, and in the case of seagrass recovery, continued high epiphytic growth and high water column turbidity due to sediment resuspension in areas devoid of seagrass. It will be important to monitor Tampa Bay into the future to continue learning how the system responds, and how increased urban runoff from development, the growing role of atmospheric deposition from fossil fuel combustion, and complicating factors like dredging affect it.

FIGURE 6-7. Eutrophication reversal in Tampa Bay. As a result of large reductions in nitrogen loading, eutrophication is only slowly being reversed. (A) Recovery targets have been reached for chlorophyll *a* concentrations in Old Tampa Bay (modified from Greening 1999; Johansson and Greening 2000), (B) but are only slowly being approached for seagrass coverage. Nutrient reductions began in 1980 (modified from Greening 1999).

Figure 6-7 on next page

BOX 6-5 Continued

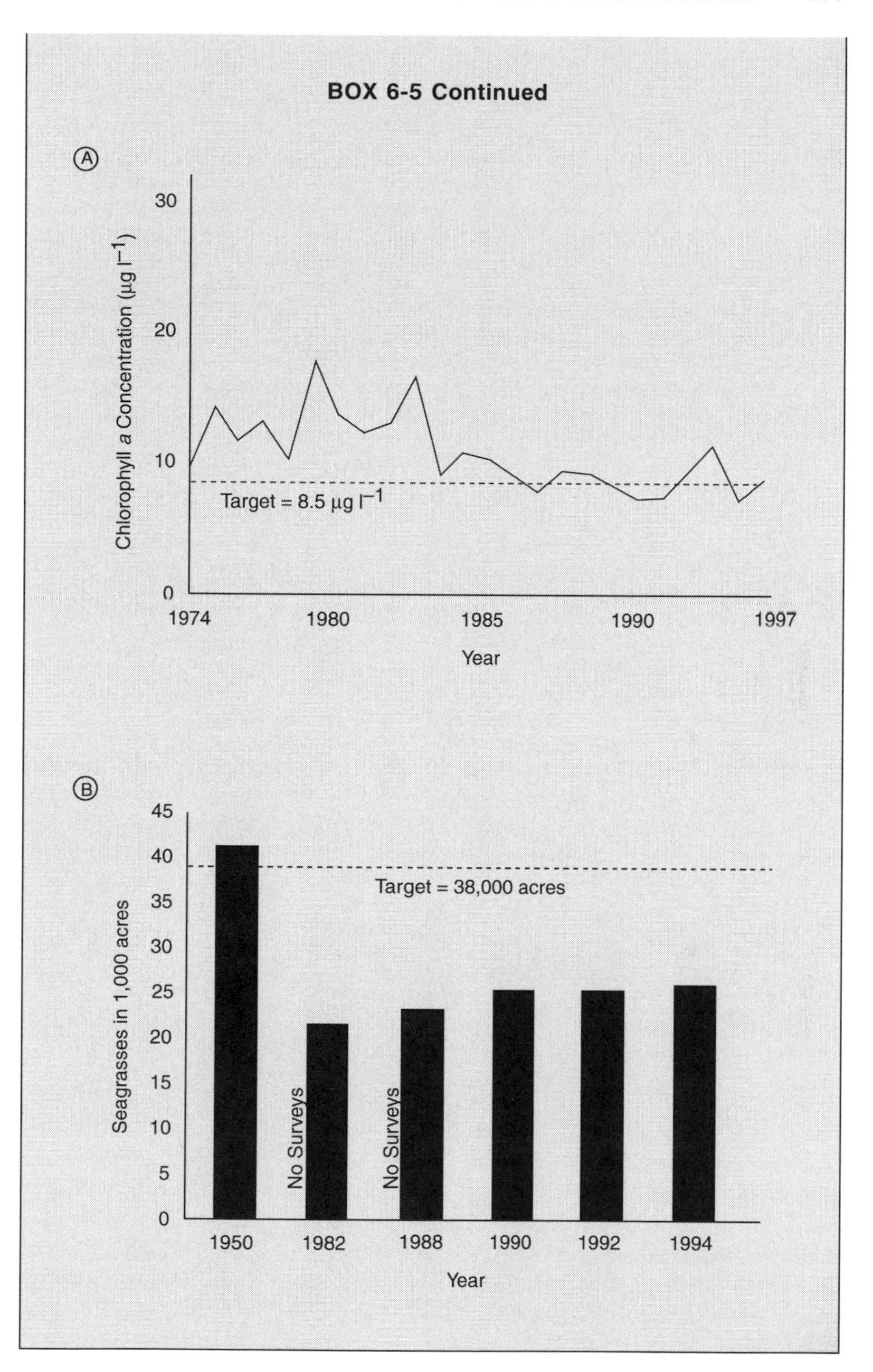

nutrient loading based on areal loading alone may be improved (Nixon 1992).

Scientists and managers are equally concerned about reversing the eutrophication trend observed in many of the nation's estuaries. Whether the same classification schemes being developed to predict effects of increasing nutrients will work equally well in predicting reversals, if nutrient inputs are reduced, remains to be determined. As some studies have indicated, there are non-linearities and thresholds in eutrophication response to increased nutrient loading. There are undoubtedly additional non-linearities and thresholds governing the response of estuarine systems to reductions in nutrient loading. While loading, dilution, mixing, flushing, circulation, and stratification may be the key parameters governing the initial response of a system to nutrient increases, internal stores of organic matter and internal nutrient cycling processes might be important factors governing eutrophication reversal. There are few locations where nitrogen loading reductions have occurred and even fewer where resultant changes in estuarine trophic status have been chronicled.

The Himmerfjärden, an estuary of the Swedish central Baltic coast, is one example where eutrophication has been reversed following reductions in nitrogen loading. Present nitrogen loads are less than 10 percent of pre-treatment input levels. Primary production, chlorophyll concentrations, and water transparency have all improved since treatment began. In this system, the reversal showed little lag in response following source reductions, presumably reflecting the rocky nature of the coastal zone and small internal stores of organic nutrients (Granéli et al. 1990; Elmgren and Larsson 1997).

Tampa Bay, Florida, is another coastal system where local managers are reversing eutrophication. In this seagrass-dominated estuary, however, significant nutrient reductions have not resulted in eutrophication reversals to the extent observed in the Himmerfjärden. Time lags in recovery are hypothesized to result from large internal sources of detrital nitrogen accumulated over years of earlier eutrophic conditions. Recovery is likely to be slow until these historic stores of nitrogen are reprocessed and either flushed from the system or denitrified. Box 6-5 describes the Tampa Bay reversal in greater detail.

Part III

Understanding Abatement Strategies

7

The Role of Monitoring and Modeling

KEY POINTS IN CHAPTER 7

This chapter reviews monitoring and modeling and how each can best be used to increase understanding of coastal nutrient over-enrichment and develop management approaches. It finds:

- There is still great need for better technical information on status and trends in the marine environment to guide management and regulatory decisions, verify the efficacy of existing programs, and help shape national policy.
- Effective marine environmental monitoring programs must have clearly defined goals and objectives; a technical design based on an understanding of system linkages and processes; testable questions and hypotheses; peer review; methods that employ statistically valid observations and predictive models; and the means to translate data into information products tailored to the needs of their users, including decisionmakers and the public.
- There is no simple formula to ensure a successful monitoring program. Adequate resources—time, funding, and expertise—must be committed to the initial planning. The program should address all sources of variability and uncertainty, as well as cause and effect relationships. A successful monitoring program requires input from everyone who will use the data—scientists, managers, decisionmakers, and the public.
- Calibrated process models of estuarine water quality tend to be more useful forecasting (extrapolation) tools than simpler formulations, because they tend to include a greater representation of the physics, chemistry, and biology of the physical system being simulated.
- When model results are presented to managers, they should be accompanied by estimates of confidence levels.

- Agencies should develop standards for storing and manipulating hydrologic, hydraulic, water quality, and atmospheric deposition time series. This will make it easier to link models that may not have been developed for similar purposes.
- Managers are often concerned with the effects of nutrient loading on commercial and recreational fisheries and other higher trophic levels. These linkages are not always clear, and the use of modeling to understand cause and effect relationships is in its infancy. The lack of knowledge about the connections among nutrient loadings, phytoplankton community response, and higher trophic levels makes modeling difficult. New models are needed that use comparative ecosystem approaches to better understand key processes and their controls in estuaries.

In 1990, a major report on marine environmental monitoring (NRC 1990) concluded that: "There is a growing need for better technical information on the condition and changes in the condition of the marine environment to guide management and regulatory decisions, verify the efficacy of existing programs, and help shape national policy on marine environmental protection." The situation has not improved dramatically in the decade since this statement was published.

Environmental monitoring involves the observation or measurement of an ecosystem variable to understand the nature of the system and changes over time. Monitoring can have other important uses beyond mere observation. For instance, compliance monitoring can trigger enforcement action. In research, monitoring is used to detect interrelationships between variables and scales of variability to improve understanding of complex processes. The data acquired during monitoring can be used to specify parameters needed to create useful models and to help calibrate, verify, and evaluate models.[1] When planning a monitoring program, important decisions must be made before the first observation is made, including what to measure, where to measure, when, how long, at what frequency, and which techniques to use. How these decisions are made often reflects important underlying and frequently unstated assumptions concerning how the ecosystem functions.

Monitoring can play an important role in understanding and mitigating nutrient over-enrichment problems by helping pinpoint the nature and extent of problems. Because nutrient over-enrichment often results in local problems, the management responses, including monitoring programs, is typically local. This local emphasis influences the scales of

[1] Calibration consists of the tuning of the model to a set of field data, preferably data that were not used in the model construction. Verification is the statistical comparison of the model output to additional data collected under different forcing and boundary conditions. Evaluation involves the comparison of model output with data collected after implementation of an environmental control program.

measurements; resources available to the program; decisions about what, how, and when to monitor; and the comparability of monitoring results among programs.

One of the biggest challenges to effective monitoring is deciding how to allocate scarce resources. If the goal is to map a coastal characteristic with a given accuracy, then statistical techniques (e.g., Bretherton et al. 1976) provide methodologies to estimate the expected error associated with any given array of sensors. If, however, particular areas must be protected, for example a swimming beach or a fish farm, then monitoring efforts must be more focused.

Unfortunately, the spatial coherence scales of eutrophication and related processes are often very small in comparison to the body of water in which they occur, with the result that what constitutes a significant variation from normal can be difficult to determine. Also, the distribution of affected areas within a given system can be patchy. When resources to support monitoring are limited, decisions concerning where to monitor may favor economically or politically sensitive regions.

Typical monitoring programs are built around fixed devices and sampling schemes. To design an appropriate sampling scheme, an estimate of the important scales of variability must be made. Sampling does not need to take place at all of these scales, but, if a particular scale is not sampled, its effects must be averaged out of the record by the design of the measurement device. Otherwise, the resulting record would appear to have significant variability at scales where, in fact, it does not. Thus, through careful design, a program can conserve resources and sample only the important scales.

For example, both semi-diurnal and diurnal tidal variations often affect an estuary. Nonetheless, these scales may not be the dominant scales at which eutrophication or other adverse processes take place. By averaging over a tidal cycle, the important parameters may be sampled at a lower repetition rate and still retain all the important information. Additional savings may be obtained if that sampling need not occur throughout the year. In many locations, cold temperatures, reduced metabolic rates, reduced discharge, and increased wind stirring (during certain seasons) eliminate the potential development of hypoxic conditions. If monitoring such conditions is the program goal, the monitoring may be restricted or discontinued during these seasons.

Modeling and monitoring share a close interdependence. Modeling synthesizes the results of observational programs. As such, models provide important assistance for the development of monitoring arrays. Monitoring data, however, are necessary for the calibration, verification, and post-auditing (or evaluation) of models. They also provide the initial conditions, boundary conditions, and forcing functions for these models.

Finally, they provide data for assimilation when the models are used in a predictive mode. In these cases, real data are blended with model output to keep the model from diverging too far from reality.

At least two kinds of data are necessary to run models accurately. For water quality models of receiving basins, the first category includes necessary model input parameters, such as inflows, input loads, wind vectors, hypsographic data, and tides. For watershed models, key data includes topography, precipitation, and land use characteristics. The second data category contains measured values that correspond to model output (e.g., flows, velocities, concentrations, ambient loads) for purposes of calibration, verification, and post-auditing.

An iterative process of modeling, verification through careful statistical comparison of model output with observations (Willmott et al. 1985), and model modification is necessary (e.g., Herring et al. 1999) to obtain results in which managers can have confidence. Useful models require close interaction among model developers, field scientists who monitor and describe the real world, and theoreticians who explain the observations. Once quantitative measures of a model's ability to calculate the state of the system on certain space and time scales are specified, managers can determine whether the observed level of reliability is acceptable.

It can be argued that no model is truly able to predict, that is, to provide perfect estimates of future conditions. The term "predict" is used in this chapter to mean "forecast" or "estimate" for future or hypothetical conditions. The accuracy of such predictions will vary depending on the degree of integration of those who monitor with those who model. Prediction is the ultimate management use of models. While one can argue about the relative predictive skill of existing models, it is clear that prediction is an important goal justifying their development.

The detail and complexity of a model is often reflected in the amount of data required to initialize and run the model. Many mathematically simple models require extensive and expensive monitoring programs to provide data before they can produce accurate results. Thus, the level of model sophistication does not necessarily indicate savings in the resources that must be devoted to monitoring in order to produce accurate hindcasts or predictions.

Finally, mention should be made of the use of data assimilation. Numerical models have a tendency for their computation results to drift away from reality as they are run for longer and longer periods of time. One method used to correct this problem is to assimilate field data as they become available. If one observes a discrepancy between observations and model output, the model state is pulled back toward the observed state of the system being modeled. There exist many numerical techniques for achieving this goal. Meteorologists have used this approach for many years, and physical oceanographers and biological oceanographers are beginning to incorporate it into their models.

It must be remembered, however, that models are not a substitute for measurements. A properly calibrated and verified model can be useful for producing estimates of future conditions and guiding management, but field measurements, when available, are always superior to model computations.

INTRODUCTION TO MONITORING

Monitoring provides long-term data sets that can be used to verify or disprove existing theories developed from shorter, more focused data sets. Monitoring characterizes the scales of variability, in both space and time, thus allowing modification of sampling schemes to maximize the use of available resources. In particular, monitoring allows determination of long-term climatic scales of change, which can be mistaken for trends in shorter records.

Exploratory data analysis suggests that carefully manipulated data sets from monitoring programs, along with a fair share of serendipity, may result in new insights into functional relationships among variables of an ecosystem (Tukey 1977). While this is clearly an avenue of productive future research, the number of examples of such insight remains small.

Focused monitoring programs are generally established in response to, rather than in anticipation of, a problem. This means that baseline information can be missing from a monitored region. Once established, monitoring programs are useful for identifying events, but unless maintained for long periods, their utility for determining the existence of a trend is far less. In a similar sense, they are also useful for monitoring the effectiveness of remediation activities, if maintained for sufficiently long periods (i.e., periods longer than the natural scales of variability of the system).

Long-term monitoring programs are necessary to isolate subtle changes in the environment. Only through data gathering programs that are sufficiently interdisciplinary in their design it will be possible to develop and test hypotheses concerning the processes and impacts of eutrophication.

As pointed out by the 1990 National Research Council report *Managing Troubled Waters: The Role of Marine Environmental Monitoring*, monitoring is generally carried out to gather information about regulatory and permit compliance, model verification, or trends in important environmental or water quality parameters. These data can play an important role in: 1) defining the severity and extent of problems, 2) supporting integrated decisionmaking when coupled with research and predictive modeling, and 3) guiding the setting of priorities for management programs. Because establishing and maintaining targeted monitoring programs is expensive and complex, greater use of environmental data collected for a variety of purposes is gaining appeal. For example, data collected through the National Pollutant Discharge Elimination System permitting process (the process of permitting point source pollutant discharges in compliance with the Clean Water Act) has demonstrated great utility for developing and evaluating the effectiveness of regional stormwater management plans and for characterizing local stormwater discharges in diverse settings (Brush et al. 1994; Cooke et al. 1994). These efforts to use data derived for regulatory purposes can provide valuable insights into the impact of land use on the concentration of a variety of constituents and thus have implications for developing loading estimates and other watershed management applications. Efforts should be made to encourage greater accessibility to similar permitting data, including associated metadata, and compliance with accepted collection and analysis protocols.

Ever widening use of electronic storage and management of data sets and the greater accessibility provided by the internet hold great potential for reducing the cost of environmental monitoring by obtaining full value from data already being collected. Such a shift in philosophy, while already under way, would be facilitated if the basic guidelines and philosophies espoused herein are more fully integrated with established or contemplated regulatory monitoring plans. The committee believes that monitoring data are frequently not accessible to all who could benefit from their use (Chapter 2). Data management and the development of informational synthesis products should, therefore, be a major part of all monitoring programs—federal, state, and local. These data and syntheses should be available quickly to all users who could benefit from them at a reasonable cost. The internet offers a relatively simple, widely accessible route for distribution.

ELEMENTS OF AN EFFECTIVE MONITORING PROGRAM

Effective marine environmental monitoring programs must have the following features: clearly defined goals and objectives; a technical design that is based on an understanding of system linkages and processes; testable questions and hypotheses; peer review; methods that employ statistically valid observations and predictive models; and the means to translate data into information products tailored to the needs of their users, including decisionmakers and the public (NRC 1990).

Monitoring programs are costly undertakings and need to be carefully planned with specific goals in mind. They often are established in response to strong public pressure, leading to situations in which program managers are expected to perform good science in a situation driven not by scientifically justifiable design but by political expediency. Legal mandates may cause duplication of effort, leave gaps in the required data records, and monitor the wrong system measures. Under the best of conditions, only a limited number of measures can be monitored for a sustained period of time. If resources are inappropriately used, the situation worsens.

A related problem arises with poor sampling design. If the wrong questions are being asked, undersampling may result in not being able to sort out a weak anthropogenic signal from the natural variations in the environment. Alternatively, oversampling may result in wasted resources. Problems arise with sampling spatial scale, as well as temporal scale. Monitoring for regulatory compliance is often inappropriate for determining regional and national trends.

There is no simple formula that will ensure a successful monitoring program, but much has been written on the topic over the years. In planning a monitoring system, there must be implicit decisions about how monitoring information will be used to make decisions (Box 7-1). It is imperative that all stakeholders—public, managers, policymakers, and scientists—be involved in the plan's development, understand the implications of the various options, and agree on what results can be expected at what times in the course of the program. It is important that everyone involved harbor realistic expectations. Natural systems are complex and highly variable in time and space. Risk-free decisionmaking is an impossible goal (NRC 1990).

Current monitoring programs generally do not provide integrated data across multiple natural resources at the different temporal and spatial scales needed to develop sound management policies (CENR 1997). A number of issues must be addressed in order to enhance the probability of success of future monitoring programs (NRC 1990; CENR 1997; Nowlin 1999).

First, it is imperative that adequate resources—time, funding, and expertise—be committed to the initial planning of a monitoring program if the probability of success is to be maximized. These resources must be used to design a program that incorporates all sources of variability and uncertainty, as well as the best scientific understanding of cause and effect relationships. Objectives and information needs must be defined before the program design decisions can be made rationally. Successful monitoring programs strive to gather the long time series data needed for trend detection, but at the same time must be flexible to allow reallocation of resources during the program. They need to use adaptive strategies for

BOX 7-1
A National Coastal Monitoring Program:
A Danish Example

Serious signs of environmental degradation, including much publicized episodes of oxygen deficiency in the Kattegat during the 1980s, led the Danish government to create the *Action Plan Against Pollution of the Danish Aquatic Environment with Nutrients* in 1987. The *Action Plan* called for total discharges of nitrogen and phosphorus from agriculture, individual industrial outfalls, and municipal sewage works to be reduced by 50 percent and 80 percent, respectively. Because of great uncertainty about the sources of discharges as well as about the effectiveness of intervention measures, three related programs were initiated: 1) a nationwide monitoring program, 2) a marine research program, and 3) a wastewater research program. Using $1.8 million, universities, consulting firms, the Danish environmental protection agencies, and local governments created a highly effective, joint effort that has resulted in a very broad and detailed understanding of coastal nutrient over-enrichment.

The Danish Nationwide Monitoring Program was undertaken to: 1) describe the quality of the aquatic environment; 2) determine where, how, and why environmental changes occur; 3) assess the effectiveness of environmental programs; and 4) determine compliance with water quality objectives. Fundamental requirements of the program were to describe geographical variation and short- and long-term temporal variation so that impacts could be identified and defined with an acceptable degree of certainty. The scope of the monitoring program was extensive and included descriptions of oxygen concentrations, marine sediments, benthic fauna, benthic vegetation, zooplankton, phytoplankton, nutrient concentration and loading, atmospheric inputs, and hydrographic conditions. Monitoring of all estuaries, bays, and coastal waters was undertaken by county governments, while open marine waters were monitored by the National Environmental Research Institute. The monitoring program incorporated data from several other national forestry, fishery, and meteorological institutes. Study results were stored in a centralized, systematic database designed to provide ready access to all potential end-users, from government officials to the public. Data and metadata are compiled and presented in a goal-oriented manner (Figure 7-1).

A final component of the *Action Plan* was to evaluate the results of the Nationwide Monitoring Program after its completion, and then refine to the program to better meet national nutrient pollution reduction goals. The evaluation resulted in several suggestions for modifying components of the monitoring program. One important conclusion was to better couple modeling and monitoring efforts. Choice of monitoring sites often can be guided by modeling, both with respect to data requirements for model development and with respect to sites identified by the model as being sensitive to nutrient over-enrichment. Another conclusion was that

continued

BOX 7-1 Continued

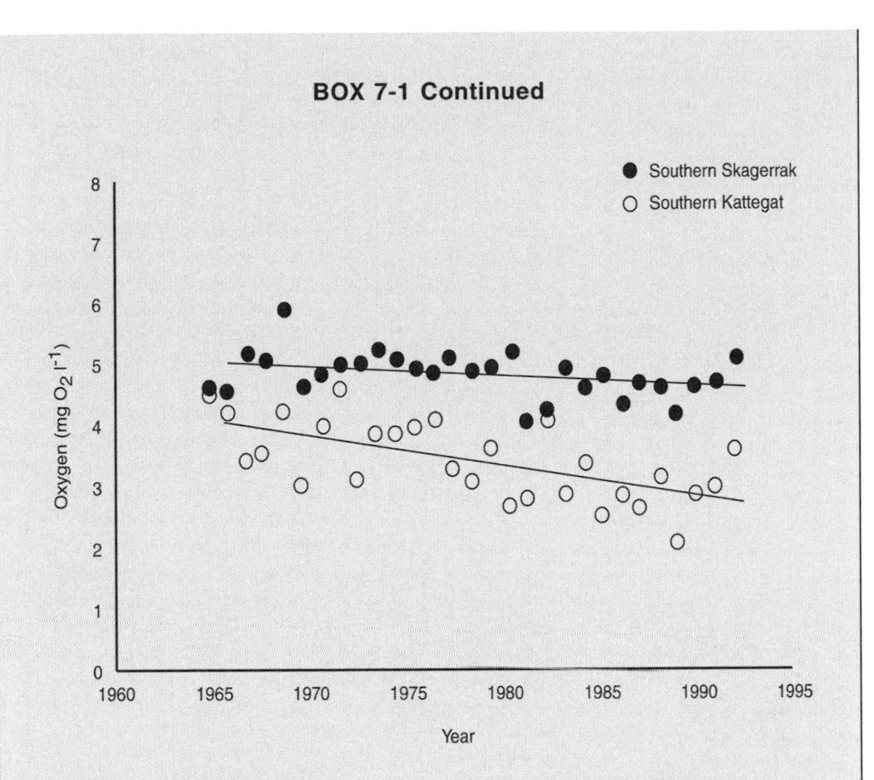

FIGURE 7-1 The long-term record of mean oxygen concentration under the pycnocline in the southern Skagerrak and southern Kattegat in Denmark during late summer over the period 1965-1993. Such long-term records have been collected and interpreted under the Danish Nationwide Monitoring Program and have proven to be invaluable sources of information for defining spatial and temporal scales of variability in environmental conditions and also for examining relations with variables suspected to be contributing to coastal eutrophication (modified from Christensen 1998).

biological, chemical, and physical monitoring data should be coupled to obtain a better understanding of interrelationships. It was also suggested that monitoring efforts be reduced in areas that experience significant natural variation because it is difficult to demonstrate cause and effect in such places.

incorporation of future scientific and technological advances. Finally, expectations and goals must be carefully defined, clearly stated, and agreed to by all involved. These actions require input from everyone who will use the data—scientists, managers, policymakers, and the public. As is often the case, resource limitations may play a role in the monitoring program strategy. Effective priority setting must be based on a full understanding of the impact such decisions will have on the reliability of key model output or other information derived from the monitoring program.

Retrospective analyses of historical data and preliminary research are invaluable in steering the design of sampling strategies. Planners must clearly define what will be considered a meaningful change in the parameters selected for measurement, in light of expected levels of natural variability. At this point, based on existing understanding of how the ecosystem functions, selection of parameters to measure can be made. These decisions will necessarily be influenced by existing technology, signal to noise ratios in the environment, and the existence of surrogate variables.

If the motivation for the monitoring program includes management and regulatory issues, the decision of what to monitor will depend on the final desired state of the system and preconceived ideas of how the system functions. Monitoring strategies for nutrients are often more complicated than the strategies commonly applied to other pollutants (e.g., toxins or carcinogens). For example, if a particular carcinogen is being regulated in order to maintain its concentration below a given level in an estuary, then this chemical constituent's concentration is the variable targeted for measurement. Similarly, if an ambient nutrient concentration in a stream is to be used to help calculate the nutrient load that stream is contributing to a downstream receiving body, then such a concentration may be targeted for measurement. Unfortunately, this similarity does not hold for monitoring nutrient concentrations in the actual coastal receiving body, because the effects of nutrients on an ecosystem are complex and not necessarily measureable. For example, if an estuary is nitrogen limited, primary productivity may be stimulated by nitrogen loading. This increased primary productivity will then remove nitrogen from the water column at a high rate and tie it up in organic matter. Thus the ambient nitrogen concentration in the water column of the estuary may never rise significantly, or remain elevated long enough to be observed, even as eutrophication takes place. Thus, the ambient concentration of a nutrient in receiving waters rarely reflects the degree to which the body has been affected by nutrient over-enrichment.

Monitoring is often targeted to study more characteristics than simply the eutrophication of a particular coastal waterbody. Often, monitoring programs focus broadly on health of the ecosystem. The decision frequently is made to monitor a variable that is believed to integrate the effects of numerous processes (Naiman et al. 1998). Such variables could be the areal extent of seagrass beds, the number of commercially harvestable oysters, or the number of waterfowl nesting in a region. The choice of which variables to monitor makes a statement, both about the values that the manager and advisors share and about their inherent beliefs concerning how the system functions. If eutrophication is a concern and the extent of seagrass beds is the observable monitored, there is an inherent belief that this characteristic parameter varies with the degree of nutrient enrichment of the system. The economic and ecological impacts of coastal eutrophication are often not demonstrable from the available scientific data (Chapter 4). Therefore, monitoring should include biological, physical, and chemical properties on all relevant time and space scales. Monitoring should be on scales appropriate to capture all variability and linkages between variables. Often, this will mean that biological and chemical measurements will need to be made on finer scales than is presently the norm.

The duration of any monitoring program is particularly important. Since the purpose of monitoring involves, among other things, the detection of trends, the length of monitoring must be sufficiently long to allow separation of naturally occurring trends from anthropogenically induced changes. Unfortunately, the political will to maintain long-term funding for monitoring programs is often lacking because such programs rarely (and were never intended to) produce major breakthroughs in understanding. The U.S. Geological Survey (USGS) stream monitoring program has provided excellent data on stream flow and nutrient content for many years. These long data sets allow monitoring of changes in runoff characteristics on decadal time scales and development of statistical models of discharge and load. But the gradual reduction of this network over recent years, primarily because of budget pressures, has had dramatic effects, reducing our capability to estimate flow. The data collected by USGS are invaluable, and continuation of this monitoring is essential. However, USGS monitoring network was not designed specifically to assess inputs to coastal regions. The committee concludes that there are major missing pieces in the resultant data set that are needed to support the management of healthy coastal ecosystems; for instance, monitoring sites "below the fall line" (the transition point between lowland and upland portions of rivers, marked by waterfalls and other rocky stretches that limit navigability) are few and far between. Since many of our older, eastern cities arose at these transition points, the network is failing to cover areas down river containing significant population centers. An important aspect of any discussion of national monitoring should be expansion of the USGS monitoring program so that it better assesses nutrient inputs to estuaries and tracks how these change over time.

The distinction between monitoring and research can be vague. Few programs have been able to combine the need for multiple use measurements effectively, but careful evaluation of multiple tier measurements (NRC 1990; CENR 1997) will help alleviate these problems. A careful program of quality control must be established at the outset, and a data management structure to convert data into knowledge must be designed. This will include analysis, interpretation, and modeling and must be carefully attuned to the myriad sources of data and uses of the resultant knowledge. Finally, rapid dissemination of data and knowledge products to all users is essential.

The final step in design of any successful monitoring program should be establishment of a process of independent review and, if necessary, protocol modification. Such reviews are necessary for monitoring programs to take advantage of new measurement and analysis techniques, and they help determine if programs are effectively answering the questions for which they were designed.

Much of the preceding discussion can be applied equally to national, regional, and local monitoring programs. Although the national scale of nutrient over-enrichment of coastal waters in recent years has been documented (NOAA 1999a), the United States still lacks a coherent and consistent strategy for monitoring the effects of nutrient over-enrichment in coastal waters. This seriously constrains the ability to assess the effects and costs of eutrophication regionally or nationally. The National Oceanic and Atmospheric Administration (NOAA) National Estuarine Eutrophication Assessment (Bricker et al. 1999) is an admirable effort, but it is limited by the inconsistency of data collection. To avoid similar limitations in the future, a national monitoring strategy for the United States should be implemented by all relevant agencies. Such a national monitoring strategy should recognize regional differences and should be based on a classification scheme that reflects an understanding of the similarities and differences among estuaries, as discussed in Chapter 6. The development of such a classification scheme is a major research challenge.

DEVELOPING QUANTITATIVE MEASURES OF ESTUARINE CONDITIONS

The primary quantitative indicators may include variations in algal composition, elevated concentration of chlorophyll *a*, and an increase in the extinction coefficient for a given waterbody (which is adequately reflected by the depth of light penetration adequate to support photosynthesis). Because algal composition changes are typically determined with expensive techniques such as microscopy or pigment analyses, this indicator is less useful for large-scale screening analyses. Secondary indicators of eutrophication include changes in the dissolved oxygen regime of the system; changes in the areal extent of seagrass beds; and incidence of harmful algal blooms or nuisance blooms and changes in their frequency, duration, and areal extent.

An assessment of the condition of the nation's estuarine and coastal waters needs to be based on a consistent, broad-scale monitoring. This requires a reduced suite of measures that are easy to obtain, easy to calibrate, and that do not require massive commitment of resources to cover the broadest geographic region. For mesoeutrophic (or less impacted) systems, annual measures in July or August when symptoms could be expected to be most severe, should suffice to characterize worst-case conditions. For eutrophic or hypereutrophic systems more frequent measurements are needed, at least twice per year. Light conditions, chlorophyll *a*, gross primary production and respiration from diurnal patterns in dissolved oxygen content, and mean dissolved oxygen content likely could all be obtained through month-long near-surface and near-bottom deployment of off-the-shelf sensors and digital recording packages.

Although federal, state, and local agencies conduct monitoring activities at thousands of sites nationwide, the efforts are not sufficiently coordinated to provide a comprehensive assessment of the trophic status of U.S. coastal ecosystems (see "The State of the Nation's Ecosystems"; http://www.heinzctr.org/). In response to this shortfall in our assessment capability, the Environmental Monitoring Team, established by the Committee on Environment and Natural Resources, has recommended a three-tiered conceptual framework for integrating federal environmental monitoring activities. The first tier includes inventories and remote sensing; the second tier includes national and regional surveys; the third tier includes intensive monitoring, research, and modeling. Integration across tiers should provide the understanding that will enable evaluation of the status, trends, and future of the state of the environment. This three-tiered framework is a sound approach for gathering and integrating information, and can be applied to monitoring coastal systems.

A broadscale assessment of the trophic status of the nation's estuaries should be based on inventories and remote sensing at Tier I sites. Tier I sites should be representative of the 12 major types of coastal systems described in Chapter 6, the four major estuarine circulation types, and the seven biogeographic provinces. It would also be useful to sample both from systems believed to be generally healthy and ones believed to have experienced nutrient related problems.

No such consistent, coherent monitoring program now exists. One program, the Environmental Protection Agency's (EPA) Environmental Monitoring and Assessment Program, is at a similar scale, but with broader (non-coastal) goals and sampling protocols. Local monitoring of these variables occurs, for example, at National Estuary Program sites, but the programs are subject to funding vagaries. The Committee on Environment and Natural Resources framework document recommends monitoring some 20 to 25 sites to identify national trends, far less than necessary to adequately represent the diversity of estuarine types, circulation types, and biogeographic provinces necessary to fully characterize the status of coastal waters. Rather, up to 25 sites will be required as Tier III, or index sites, to conduct the temporally and spatially intensive, monitoring and process-based research necessary to develop predictive models able to determine cause and effect relationships.

Monitoring to increase basic scientific understanding of the ecosystem and for validating process-based predictive models should be carried out at an intermediate number of Tier II sites. At these sites, basic data should be collected to determine the food web structure, including secondary producers, primary production structure, nutrient dynamics, hydrodynamics, and details about external nutrient loads. Estuaries within NOAA's National Estuarine Research Reserve System (NERRS) and/or EPA's National Estuary Program (NEP) might be appropriate Tier II sites. However, the current suite of sites under these two programs was not selected with nutrient over-enrichment or the variability of estuarine types in mind. If nutrients are the major interest driving study of the Tier II sites, many new sites will need to be added, and some of the current NEERS and NEP sites could probably be dropped.

This monitoring scheme is not without problems. In systems dominated by submerged aquatic vegetation, benthic macrophytes, or coral reefs, the proposed measures may be inadequate. To document nutrient impacts, it will be necessary to look for trends in parameters such as areal extent of submerged aquatic vegetation or changes in algal growth on corals and epiphytic growth on seagrasses. These measures tend to require intensive efforts, however, and thus are contrary to the idea of using simple, cheap measures wherever possible. When placing recording devices in estuaries, one should be in the surface layer and one in the deep layer, but the location of other stations remains an open question. Decisions should be based on local knowledge of sensitive areas of the estuary.

DEVELOPING QUANTITATIVE MEASURES OF WATERSHED CONDITIONS

For a number of reasons, the development of watershed monitoring programs has proceeded for many years relatively independently of receiving water monitoring programs. The relative temporal stability of terrestrial distributions, when compared to estuarine and coastal distributions, has resulted in the development of map-based studies. While the development of quantitative measures of watershed condition has proceeded rapidly, empirical studies that test for significant relationships between watershed metrics and ecological condition (e.g., presence or abundance of species, water quality) are still few in number (Johnston et al. 1990). There is a clear need to identify the most important watershed metrics to monitor. In addition, it is essential to be aware of the assumptions and constraints that are implicit in the metrics, a problem that is not unique to watershed studies. For example, the selection of the land cover categories to be used in the analysis partially determines the results and the spatial scale of the data—both the total extent of the area and the grid cell size—and thus can strongly influence the numerical results (Turner et al. 1989a, b).

In agriculture-dominated watersheds, monitoring and management practices take on their own unique character. The long-term monitoring of various soil and water indicators is essential to document the status and trends of nutrient sources and coastal impacts. Monitoring is also essential to document any changes in nutrient inputs, system response to best management practice implementation, and impact of any land management changes. Effective monitoring strategies, however, must be spatially extensive as well as sufficiently frequent to detect real and statistically valid changes. Unfortunately, the outcomes and benefits of most monitoring programs will not be manifested for several years before yielding useful information. Monitoring programs are costly, labor intensive, and in most cases will need to be in place for several years. Overcoming these challenges, through documentation and education, is critical to the continued support of existing monitoring programs.

The long history of watershed monitoring in the United States provides examples of programs that have experienced varying degrees of success. These programs can be grouped into those that focus on nutrient sources and those that concentrate on coastal water impacts. It must be emphasized that the choice of both soil and water quality indicators will vary from situation to situation. Soil and water quality monitors will have to decide what they need to measure and how often (Sparrow et al. 2000).

Environmental concerns have forced many state and federal agencies to consider adopting standard soil phosphorus fertility tests as indicators of the potential for phosphorus release from soil and its transport in runoff. Environmental threshold levels range from two (Michigan) to four (Texas) times agronomic thresholds. In most cases, agencies proposing these thresholds plan to adopt a single threshold value for all regions under their jurisdiction. However, threshold soil phosphorus levels are too limited to be the sole criterion to guide manure management and phosphorus applications. For example, adjacent fields having similar soil test phosphorus levels but differing susceptibilities to surface runoff and erosion due to contrasting topography and management should not have similar soil phosphorus thresholds or management recommendations (Sharpley 1995; Pote et al. 1996). Therefore, environmental threshold soil phosphorus levels will have little value unless they are used with estimates of site-specific potential for surface runoff and erosion.

The intent of these soil phosphorus thresholds was to limit the land application of phosphorus, particularly in manures, biosolids, and other by-products. In all cases, the legislation was repealed because it directly related these thresholds to water quality degradation in a technically indefensible way. New legislation in various stages of development and adoption (i.e., Arkansas, Maryland, and Texas) will state that standards or threshold values will be based on the best science available and on soil-water relationships being developed (Lander et al. 1997; Simpson 1998). This course is also followed in the joint EPA-U.S. Department of Agriculture (USDA) strategy for sustainable nutrient management for animal feeding operations (USDA and EPA 1998b). This draft strategy proposes a variety of voluntary and regulatory approaches, whereby all animal feeding operations develop and implement comprehensive nutrient management plans by 2008. These plans deal with manure handling and storage, application of manure to the land, record keeping, feed management, integration with other conservation measures, and other options for manure use. The draft strategy is out for public comment, and will be revised and in place by the end of 2001 for poultry and swine operations and by 2002 for cattle and dairy facilities. This leaves scientists only two to three years to develop "the best science available" that includes technically defensible thresholds or indicators. Irrespective of how these thresholds are developed, a plan to monitor these indicators will be needed to establish baseline data and to document any changes in status as a result of land use changes and best management practice implementation.

In the United States, states are required to set their own water quality criteria, but so far only 22 states have quantitative standards and only Florida has adopted the federal EPA levels (Parry 1998). These standards include designated uses, water quality criteria to protect these uses, and an anti-degradation policy. Where water quality standards are not attained, even after best management practices have been implemented, response actions are defined through the total maximum daily load process of the 1998 Clean Water Act (EPA 1998b). Rather than just addressing constituent concentrations in stream and rivers, the total maximum daily load (TMDL) process considers system discharge and thereby the total constituent load, as well as the designated use and potential impact on the receiving waterbody. This approach offers tremendous advantages over other approaches and can be expected to become widely implemented. Again, spatially and temporally extensive monitoring programs will be needed to assess compliance and remediation impacts.

It would be useful to combine the two approaches—source monitoring and impact monitoring—to benefit from the best aspects of each. One approach to a soil and water quality index would be to integrate soil fertility measures and land management with a site's potential to transport nutrients to water bodies in surface and subsurface runoff. This approach is being advocated by researchers and an increasing number of advisory personnel to address nutrient management and the risk of nutrient transport at multi-field or watershed scales (Lander et al. 1997; Maryland General Assembly 1998; USDA and EPA 1998b). In cooperation with research scientists, USDA's Natural Resource Conservation Service has developed simple nutrient indexes as screening tools for use by field staff, watershed planners, and farmers to rank the vulnerability of fields as sources of nitrogen and phosphorus loss (Sharpley et al. 1998). The indices account for and rank transport and source factors that control nitrogen and phosphorus loss in subsurface and surface runoff and identify sites where the risk of nutrient movement is expected to be higher than others. These indexes have been incorporated into state and national nutrient management planning strategies that address the impacts of animal feeding operations on water quality to help identify agricultural areas or practices that have the greatest potential to impair water resources (Simpson 1998; USDA and EPA 1998b).

Inherent differences exist in the geography and biology of agricultural regions, and the same applies to forests, wetlands, and water. Since water bodies reflect the lands they drain (Hunsacker and Levine 1995), an ecoregional framework that describes similar patterns of naturally occurring biotic assemblages, such as land-surface form, soil, potential natural vegetation, and land use was proposed by Omernik (1987) and later refined by EPA (1996). The ecoregion concept provides a geographic framework for efficient management of aquatic ecosystems and their components (Hughes 1985; Hughes et al. 1986; Hughes and Larsen 1988). For example, studies in Ohio (Larsen et al. 1986), Arkansas (Rohm et al. 1987), and Oregon (Hughes et al. 1987; Whittier et al. 1988) have shown that distributional patterns of fish communities approximate ecoregional boundaries as defined a priori by Omernik (1987). This, in turn, implies that similar water quality standards, criteria, and monitoring strategies are likely to be valid in a given ecoregion (EPA 1996).

CONTROLLING COSTS

As with all extensive monitoring programs, one of the main challenges is controlling the costs. Cost concerns apply not only to the farmer wanting to characterize his own property but also to the policymaker seeking to draw conclusions at a watershed, regional, or national scale.

Whatever the technology used to assess soil and water quality, a universal requirement will be locating the places where measurement occurs so that data can be integrated for assessment at larger scales. When working at small scales, such as on a particular farm developing a runoff management plan, great accuracy is not essential. But understanding cumulative inputs and larger scales require more accuracy and consistency. Meeting this requirement is being made increasingly easy by the falling prices of Global Positioning System technology, which provides quick, precise location information.

One strategy to reduce monitoring costs is to recruit unpaid volunteers to do part of the work (Box 7-2). Such volunteers are often obtained by building relationships with local groups who care about the issues or locality (e.g., small environmental advocacy groups, school groups, or senior citizen groups). Working with these stakeholder groups can have important benefits. In the Chesapeake Bay area, for instance, stakeholder alliances have developed among state, federal, and local groups to work together to identify critical problems, focus resources, include watershed goals in planning, and implement effective strategies to safeguard soil and water resources (Chesapeake Bay Program 1995, 1998).

INTRODUCTION TO MODELING

One system is said to model another when the observable parameters in the first system vary in the same fashion as the observable parameters in the second. Models, therefore, may take many forms. They may be empirically derived statistical relationships plotted on a graph. They may be systems that have little in common with the system being modeled other than similar variations of, and relationships between, observables; a classic example from physics is the use of water flow through pipes to model the flow of electrons through an electrical circuit. Such system models are often called analogs. Models may be scaled approximations to the real system, such as the Marine Ecosystems Research Laboratory mesocosms in Narragansett Bay (Frithsen et al. 1985). Or they may be numerical models run on computers that are based either on first principles or empirical relationships. Each type of model has benefits and drawbacks.

Statistical models are empirical and are derived from observations. The relationships described must have a basis in our understanding of processes, if we are to have faith in the predictive capabilities of the model (Kinsman 1957). Furthermore, extrapolation from empirical data is known

BOX 7-2
Using Volunteer Observers in Monitoring Programs

Monitoring the many parameters important for understanding and managing watersheds is an intensive undertaking. Although many of these parameters require significant expertise and technology to obtain valid observations, some useful observations can be collected through techniques that can be learned by non-experts. Concerned and educated citizens represent a potentially large, and, until recently, virtually untapped reservoir of human resources that could be brought into action.

Involving citizens has many advantages beyond the obvious "free labor." For instance, public involvement can instill a sense of stewardship for the local environment. These same citizens then become a resource to elected officials and decisionmakers by providing first-hand observations and informed opinions about what the public wants. Citizen participation can increase the range and impact of existing monitoring efforts. All agencies have limited budgets and staffs, and resources are generally directed first toward the most severe problems, while less degraded areas—equally in need of monitoring to keep watch that problems do not occur or worsen—are ignored by necessity. But volunteer efforts can address these other sites, providing valuable baseline and trend information on watershed conditions. Such information can give early warning when problems start to develop in new areas, so such problems can be addressed before they become severe.

Local participants can be a critical resource for agency staff because they can bring an historical perspective and often know the landscape intimately. At times, public involvement in monitoring can help ensure data continuity, if volunteer efforts continue unabated when staff turnover within public agencies occurs during the course of a long-term monitoring activity.

Obviously, there are some potential disadvantages to public involvement in monitoring. For instance, it can be a challenge to maintain volunteer interest over long periods of time, which can lead to data gaps, and turnover in volunteer labor can lead to problems in data collection consistency (Ralph et al. 1994). Citizens participating as volunteers in monitoring activities must, of course, be trained, which requires resources and organization from the relevant agency. Even with basic training, volunteers are best used to take water samples and perform routine scientific tasks, and generally do not work with sophisticated equipment or participate in the collection of biological samples. However, there are a number of important monitoring activities well within the capabilities of average citizens, using basic equipment and not requiring specialized skills. These include:

Photographs. Historical and time-series photographs of sites in a watershed can provide important information to managers. People's family photo albums often contain important images of past conditions, uses, and resources.

Water samples. Local volunteers can easily be trained to take periodic water samples, and volunteers can be linked in a network to provide wide coverage of a watershed. This is especially effective for monitoring easily observable parameters

continued

BOX 7-2 Continued

such as suspended sediment. This kind of activity needs to be coordinated by an appropriate regulatory agency, which would supply the sample bottles, provide training, process samples, and maintain a database of results.

Habitat measurements. Stream morphology is an integrative measurement of overall watershed condition, and pools are particularly sensitive to change. Volunteers can help make simple habitat measurements, such as counting the number of large pools or other features, and tracking such areas to monitor for change, particularly after large storm events.

Riparian area surveys. Volunteer help is especially valuable for one-time surveys to establish baseline conditions, because they can cover large areas. Because riparian areas are critical to watershed health, such surveys can be very valuable. Volunteers can be used to survey marked plots in riparian areas to identify and count trees and other species, and even return periodically to note changes in species composition and growth and mortality.

Volunteer participation in monitoring programs is not a replacement for professional expertise in all instances, but the value of getting local citizens involved may make the effort worth considering.

to be uncertain. Thus, these models are most judiciously used in the range of observational situations used to derive the model. Analog models are most useful for explaining processes in general ways to people lacking technical expertise, rather than in understanding or predicting information about the system being modeled, since they involve neither observations nor fundamental principles associated with the system's behavior. Scaled models are useful for both prediction and understanding. One must, of course, always remain cognizant that system function may be scale dependent. Thus, problems similar to those encountered with the extrapolation of statistical models exist when one extrapolates the results of scaled experiments to full-sized natural systems. Numerical models are most useful when they are based on first principles. The ability to describe system functions in terms of mathematical equations often gives the impression that the underlying principles are fully understood, as might be the situation in basic physics. Unfortunately, empirical coefficients introduced into these equations often hide the degree of uncertainty concerning these principles.

A subset of numerical and statistical models often encountered in watershed modeling uses empirical relationships, such as a runoff coefficient, coupled with event mean concentrations (i.e., a flow-weighted average concentration) to estimate constituent loads. Alternatively, export coefficients (e.g., kg ha yr^{-1}), might be employed. These models may or may not be linked to a geographical information system for land use data and are often implemented on spreadsheets, and therefore could loosely be called spreadsheet models. These models are especially sensitive to empirical coefficients that do not always correspond to known parameter ranges, and should be checked carefully to ensure that they reflect a reasonable physical reality.

Although each modeling approach involves a unique set of problems, some are more suited than others to a particular situation. Understanding involves the development of heuristic and conceptual models, followed by carefully developed analytical and/or numerical models. These may not need to involve the total complexity of the system being modeled, if interest is focused on a particular sub-process. Management decisions may not allow the luxury of complete knowledge of the system. Results from analogous situations elsewhere, statistical (correlation) models, and other simplified models may provide sufficient predictive skill even though they do not incorporate full understanding of the processes involved.

Because they yield clear numerical results with which to gauge progress, models have a strong appeal to policymakers and managers, an appeal that can sometimes bring false confidence and misconception (Boesch et al. 2000). Complex numerical models are gaining greater acceptance by managers. It has been said that while all models are wrong, some models are useful. It is the talent of the proficient and successful modeler to understand for what problems the model is useful and to be able to explain its limitations. Numerical models must be stable, consistent, convergent, and of known accuracy (Messinger and Arakawa 1976). Further credibility of model results can be achieved through a careful process of calibration, verification, and periodic post-auditing.

Skill assessment is a term used to describe the estimation of the improvement in predictability of future states of the system through use of the model, as opposed to some simpler scheme such as persistence. A whole field of study has grown up to facilitate this effort, focused primarily on numerical models (Lynch and Davies 1995). General guidance in evaluation of the goodness-of-fit of hydrologic and water quality models that produce time series of hydrographs and quality parameters is provided by Legates and McCabe (1999). Such efforts are a necessary but time-consuming and costly undertaking.

Assembling the types of data necessary for running and calibrating a model is typically expensive and time consuming. A sensitivity analysis can provide guidance for this effort. When first encountering a model, a user establishes a realistic set of input parameters and systematically varies the input parameters to examine the effect on output values, in the manner suggested in Figure 7-2. This often indicates the relative importance of model input parameters and indicates which parameters should receive the most management attention. In addition, sensitivity analysis may indicate a need for measurements corresponding to model output where they may be missing. The overall goal is to learn as much about the models and from the models as possible before investing large amounts of time and money in data collection and processing.

In spite of the expense of data collection, modeling exercises without field data for calibration and verification, or simply for a reality check, usually have minimal credibility. In the absence of data for the watershed or receiving waterbody being studied, it should be demonstrated that the model satisfactorily represents the same physical processes on a similar watershed or waterbody. At worst, model output should be compared to

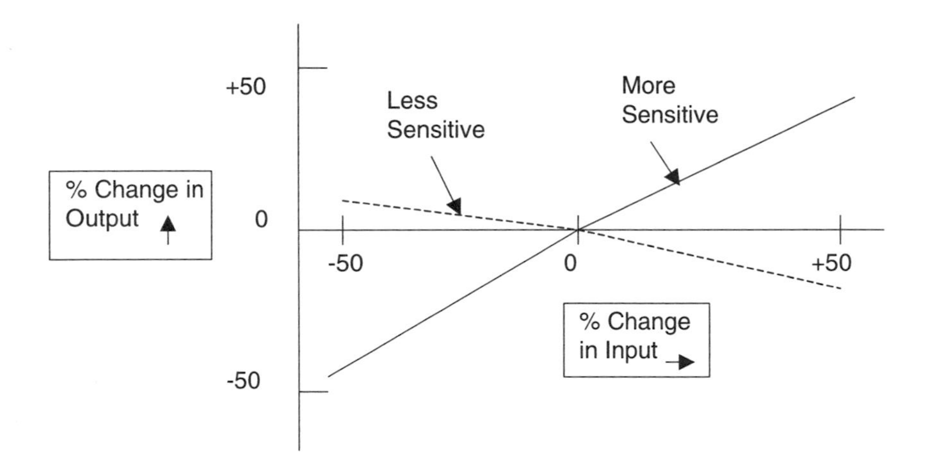

FIGURE 7-2 Example of sensitivity analysis: the relative change in an output variable from a model due to the relative change in an input variable to a model. The analysis is performed about a reference condition. An output variable (e.g., watershed runoff) is more "sensitive" to an input variable (e.g., watershed area) if the slope is steep, whereas a small slope illustrates relative insensitivity of an output variable (e.g., watershed runoff) to a change in an input variable (e.g., evaporation). Negative slope simply means an inverse relationship between an output and input variable; the magnitude of the slope is still the criterion that dictates whether an input variable is "sensitive" or not (unpublished figure by W. Huber).

theoretical solutions for simple problems. The modeler should have confidence that the processes are being represented in a way that assures realistic model estimates.

Numerical models solve equations, often differential equations, on a computer. The ability of these models to hindcast observations accurately is, too often, only qualitatively assessed. Certain notable exceptions, however, should be indicated. Radach and Lenhart (1995) and Varela et al. (1995) compare two years of North Sea nutrient concentration and phytoplankton data, with the annual cycle of the same parameters derived from the European Regional Seas Ecosystem Model (Baretta et al. 1995). Since the data were, in general, not used to tune the model parameters, the comparison represents a verification test of the model. Part of the Chesapeake Bay Program included a significant modeling effort. The model was calibrated using three years of data (Cerco and Cole 1993). Statistics that group data in large space-time bins for comparison indicate that this model generally reproduces calibration data to better than a 50 percent relative error and better than 20 percent for nitrogen and dissolved oxygen (Figure 7-3). The calibrated model was verified against a 30-year record with similar results for dissolved oxygen and chlorophyll (Cerco and Cole 1995). Other three-dimensional eutrophication models have been used successfully, with only moderate variations of the tunable free parameters (Hydroqual, Inc. 1995).

DiToro et al. (2000) calibrated a eutrophication model to a series of mesocosm experiments (Frithsen et al. 1985; Box 7-3). Since the mesocosms are well-mixed systems, there is no need for a hydrodynamic variable and no errors need be attributed to problems associated with this variable. The probability distributions of the variables are, in general, well reproduced near the means, but outliers are poorly predicted. Furthermore, there are significant phase errors in the model predictions of certain variables. This implies that comparison of means over long periods appear to be better than comparisons over shorter periods.

It should be stressed that, even when statistical or at least graphical comparison between observations and model output is made, the results are limited primarily to calibration. The cogent argument is often made that all available data should be used to calibrate a model to provide the best estimates of free parameters. Such an approach is clearly defensible if the model is to be used solely to forecast situations within the range of variability of parameter space sampled during calibration. It is far less clear that the unverified model will perform well when the boundary conditions, forcing functions, and loadings vary significantly from those used during the calibration phase. While such models are expected to be based on more fundamental assumptions than simple correlation, the tunable parameter values defined may still hide uncertainty concerning

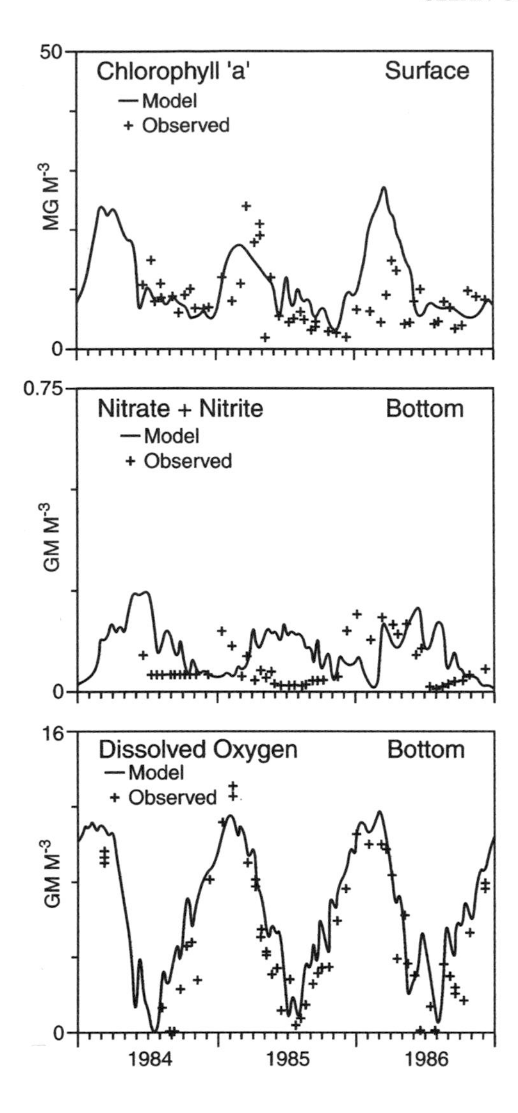

FIGURE 7-3 Comparison of modeled and observed surface chlorophyll *a*, bottom nitrate+nitrite, and bottom dissolved oxygen for the years 1984-1986 at a main stem station in Chesapeake Bay between the mouths of the Patuxent and Potomac Rivers. The model results are from the CE-QUAL-ICM model. Solid, continuous lines show the model output; dots with lines represent observations (mean and observed variability). Note the difference in phase between modeled and observed bottom nitrate+nitrite, which occurs due to the inability of the model to capture large observed negative nitrate fluxes accurately. The bottom dissolved oxygen concentrations are well reproduced by the model (modified from Cerco and Cole 1993).

BOX 7-3
Comparing Water Quality Model and Mesocosm Experimental Data

A series of mesocosm experiments were run at the University of Rhode Island Marine Ecosystems Research Laboratory in the early 1980s. Mesocosms, large (13.1 m^3) continuously stirred reactors, were filled with water from Narragansett Bay and a layer of bay sediment placed at the bottom of the tanks. Water was replaced four times daily at a mean rate of 480 liters per day. NH_3, PO_4, and SiO_2 were added in the molar ratios appropriate to sewage entering the upper bay. These nutrients were added at 1, 2, 4, 8, 16, and 32 times the areal average of nutrient addition to the bay. The six manipulated mesocosms and three control mesocosms were monitored for slightly more than two years.

A modified version of a model called the Water Quality Analysis Simulation Program (Appendix D) was calibrated to the ensemble of experimental configurations. Thus, one set of calibration coefficients was used to model both natural and highly eutrophic conditions. For the set of conditions illustrated in Figure 7-4, the model reproduced observed variations of dissolved oxygen very well. Chlorophyll *a* concentrations tended to be over-predicted, although the magnitudes of the peaks were generally captured. Nitrate and silica were not well reproduced because of problems with the sediment flux model for these two parameters. Even in these cases, however, the magnitude and shape of the annual variation was well reproduced; the annual phasing was not. Interestingly, the model did not capture the ammonia sediment flux accurately during the initial year of the model run, indicating a phase lag before the model could track the observations.

The model tended to underestimate the range of the observations, although the variability away from the extremes was well represented in a statistical sense. The discrepancies between the model and the observations were similar to those observed when the calibrated model is applied to a natural coastal setting. (Also, the calibration coefficients are similar between these natural settings and the mesocosm experiments, except when they relate to species-dependent phytoplankton parameters.) This suggests that the capabilities of existing coastal circulation models are generally sufficient as input to drive existing coastal eutrophication models.

In an attempt to model the system's recovery from excessive nutrient loading, the model was run for 15 years after nutrient loading ceased. Although loading was applied for less than 3 years, certain parameters of the system required the full 15 years to recover to pre-loading conditions. The implication for the needed long-term duration of monitoring programs is significant.

FIGURE 7-4. Results of comparison between WASP model results and experimental results from Marine Ecosystem Research Laboratory mesocosms. For chlorophyll *a* (Chl *a*), dissolved oxygen (DO), and nitrate (NO_3). Note the change in scale for the chlorophyll and nitrate plots as loading increases. The solid lines are the model results, while the circles and bars are monthly means and ranges of the observations. The dotted line in the dissolved oxygen plots are the saturation values (modified from DiToro et al. 2000). The "control" series shows nutrient additions to the mesocosm at the spatial average of nutrient additions to Narragansett Bay; 2× and 8× show results when adding twice and eight times this level, respectively, for each nutrient shown.

Figure 7-4 on next page

BOX 7-3 Continued

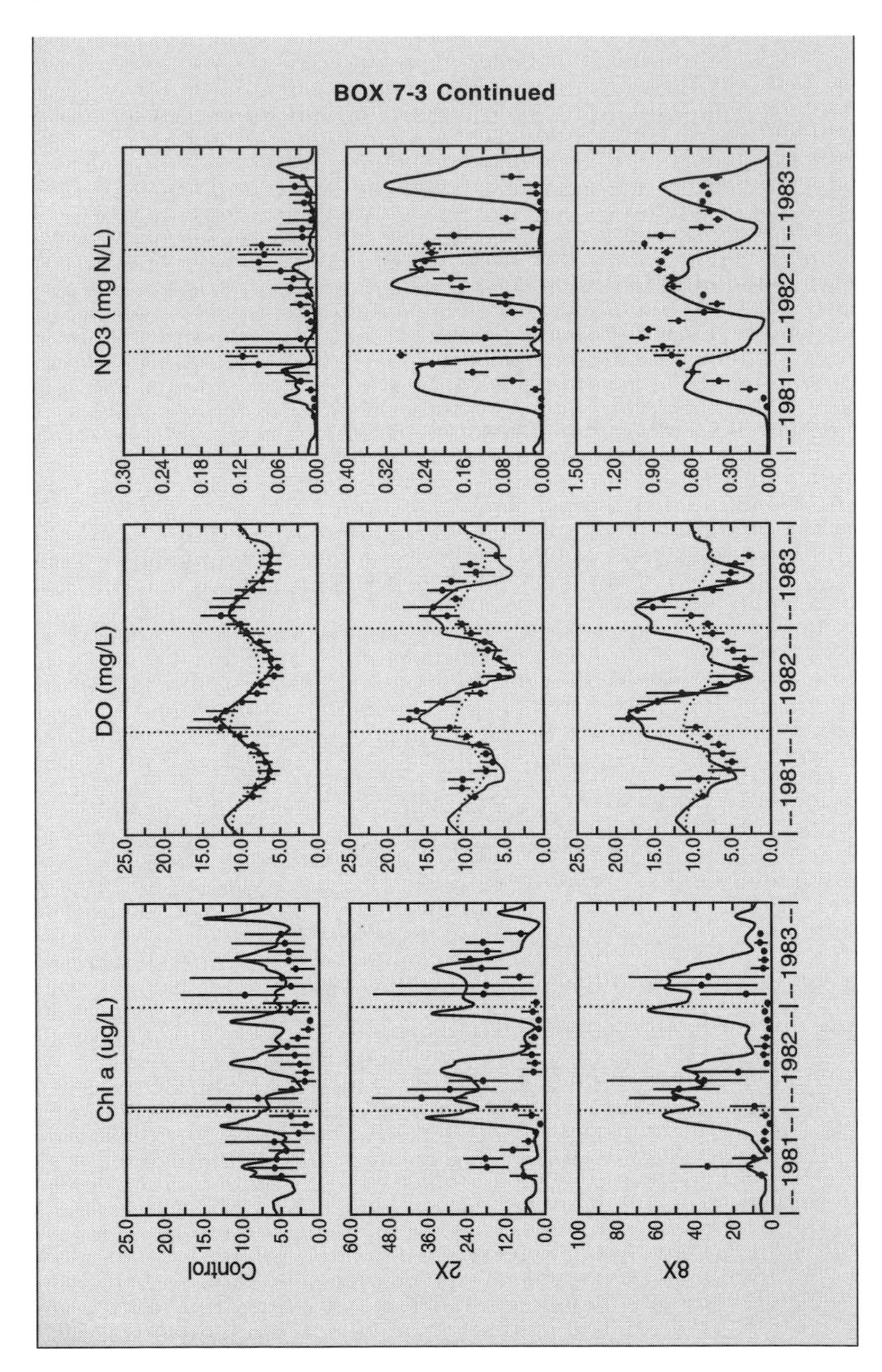

processes and not be valid over all forcing and loading conditions. For example, a change in higher trophic-level community structure will alter grazing and, consequently, affect biomass, organic carbon flux to the bottom, and related processes.

Most of the successful examples mentioned above have been able to reproduce the range of observed variability in parameters during a seasonal cycle, as well as the important processes inferred from the field data. They are not as good at capturing the phasing of all important seasonal variations or producing large-scale space-time means that match observations. For example, they often fail to reproduce important small-scale blooms. In spite of the effort and resources expended on the Chesapeake Bay model, for instance, it has been suggested that "three caveats need to be appreciated in interpretations of the watershed-water quality models: (1) the model predictions are very sensitive to several uncertain assumptions, (2) the models calculate 'average' conditions in a variable world, and (3) the models assume immediate benefits of source reductions in the Bay's tidal waters" (Boesch et al. 2000).

State-of-the-art models have difficulty reproducing the observed amplitudes and phases of observed nutrient cycles in large continental shelf domains (e.g., Radach and Lenhart 1995). Because of the reduced residence time of material, success is often greater in smaller estuarine systems. Particularly problematic are nonlinear processes, many of which are poorly understood even in estuarine situations. For instance, it has been suggested that the generation of hypoxic bottom waters over the shoal regions of the Chesapeake Bay may be reducing the rates of denitrification occurring near the seabed.

Simpler statistical and spreadsheet models are not without their problems, as well. Frequently, export coefficient models have been used in the management of estuaries to determine sources of inputs. These generally are very poor at dealing with atmospheric deposition of nitrogen onto the landscape, if they make any attempt at all. These export coefficient models seldom, if ever, have any independent verification.

The Long Island Sound nutrient input analysis illustrates the problems that can be encountered by applying export coefficient models in an area of high atmospheric deposition of nitrogen. The Long Island Sound study concludes that most of the nonpoint inputs of nitrogen to Long Island Sound come from natural sources not subject to human control (CDEP and NYSDEC 1998). The core approach for estimating nitrogen loads from nonpoint sources used in the Long Island Sound study is fairly simple: using a spreadsheet approach, land use in the watersheds is divided into three categories (urban/suburban, agriculture, and forest), and a nitrogen export from each land type is assumed. Similarly, an export value for nitrogen from an undisturbed "pre-colonial" landscape of forested land is assumed.

The problem with these estimates is that such export coefficients are poorly known, and can be greatly in error. For the Long Island Sound study, the pre-colonial export of nitrogen is assumed to be 920 kg N km^{-2} yr^{-1} (CDEP and NYSDEC 1998). This is a high value, not seen anywhere on Earth in undisturbed landscapes of the temperate zone (Howarth et al. 1996). The analysis of the International Scientific Committee on Problems of the Environment (SCOPE) Nitrogen Project concluded that for regions surrounding the North Atlantic Ocean (both in Europe and in North America, but excluding the tropics), it is likely that the flux of nitrogen from landscapes prior to human disturbance would be of the order of 133 kg N km^{-2} yr^{-1}, and is unlikely to be greater than 230 kg N km^{-2} yr^{-1} (Howarth et al. 1996; Howarth 1998) (Chapter 5). Indeed, data from a century ago for the Connecticut River show fluxes into Long Island Sound of only a little more than 100 kg N km^{-2} yr^{-1} (Jaworski et al. 1997). Most of the watershed for Long Island Sound is forested (CDEP and NYSDEC 1998), but the forested landscape is more affected by human activity than is assumed by the Long Island Sound study. The high export of nitrogen from forests there probably reflects the high level of atmospheric deposition of nitrogen. As stated in the 1994 the National Research Council report, "It is important that the watershed models under development be calibrated with accurate and detailed data from each region. It is equally essential that these calibrated regional watershed models be verified with other data. . . ." (NRC 1994).

Several topics in the previous discussion highlight uncertainty in model computations. Uncertainty arises in connection with an imperfect representation of the physics, chemistry, and biology of the real world, caused by numerical approximations, inaccurate parameter estimates and data input, and errors in measurements of the state variables being computed. Whenever possible, this uncertainty should be represented in the model output (e.g., as a mean plus a standard deviation) or as confidence limits on the output of a time series of concentrations or flows. The tendency described earlier for decisionmakers to "believe" models because of their presumed deterministic nature and "exact" form of output must be tempered by responsible use of the models by engineers and scientists so that model computations or predictions are not over-sold or given more weight than they deserve. Above all, model users should determine that the model computations are reasonable in the sense of providing output that is physically realistic and based on input parameters that are within accepted ranges. When model results are presented to managers, they should be accompanied by estimates of confidence levels.

Watershed models and estuarine and coastal models have developed independently because of the scales involved, the connectivity, and the dominant processes in each system. The following discussion briefly discusses characteristics of each type. More detailed characterizations of specific models are presented in Appendix D.

WATERSHED MANAGEMENT MODELS

Models that simulate the runoff and water quality from watersheds are categorized in several ways, but for purposes of this brief review they are segregated into three groups:

1. Models that explicitly simulate watershed processes, albeit usually conceptually. These models typically involve the numerical solution of a set of governing differential and algebraic equations that are a mathematical representation of such processes as rainfall-runoff, build-up and wash-off of surface pollutants, sorption, decay, advection, and dispersion.
2. Models that rely on land use categorization (sometimes through linkage to a geographic information system evaluation) coupled with export coefficients or event mean concentrations. These models are sometimes called spreadsheet approaches, but they actually can be highly sophisticated. These models rarely, if ever, involve solution of a differential equation and almost always rely on simple empirical formulations, such as the use of a runoff coefficient for generation of rainfall runoff.
3. Statistical models involve regression or other techniques that relate water quality measures to characteristics of the watershed. These models range from purely heuristic regression equations (e.g., Driver and Tasker 1990) to relatively sophisticated derived-distribution approaches for estimation of the frequency distribution of loadings and concentrations (e.g., DiToro and Small 1984; Driscoll et al. 1989; Smith et al. 1997).

In addition to these categories, the simplest modeling techniques involve the use of constant concentrations applied to measured or simulated flows, or alternatively, export coefficients, in the form of mass/area-time. Constant concentrations are usually obtained from measurements based on land use and other parameters and are in the form of flow-weighted averages or event mean concentrations. EPA's Nationwide Urban Runoff Program, for example, provides a good basis for these numbers for urban areas (EPA 1983), as well as storm event sampling from hundreds of cities around the United States that have submitted

National Pollutant Discharge Elimination System permit applications for their stormwater and combined sewage. Unfortunately, much sampling data languishes in state agency or consultant files; a coordinated effort on the part of EPA is sorely needed to publish and analyze the tens of thousands of samples collected as part of the National Pollutant Discharge Elimination System permitting process. Export coefficients may be derived from event mean concentration values, if runoff volumes are known, and this is a common method for obtaining these somewhat less common parameters. Both event mean concentrations and export coefficients fit easily into spreadsheet formats for watershed loading estimates. An advantage of event mean concentrations is that they may be coupled with any hydrologic simulation model to produce loads.

The committee recognizes that, especially in the urban environment, there is no coordinated effort to maintain a database of samples collected under the National Pollutant Discharge Elimination System and similar nationwide monitoring efforts. Such a database would be of inestimable value for developing loading estimates to receiving waters. Consequently, as the agency responsible for implementing National Pollutant Discharge Elimination System legislation, EPA should develop and maintain a current nationwide database of urban and other surface runoff samples for use in nonpoint source water quality analyses and modeling. Additional effort should be made to analyze such information in a manner similar to that of Driver and Tasker (1990), for purposes of developing simplified relationships between concentrations, loads, and causative factors.

There are many ways to characterize watershed runoff models, such as transient versus steady-state and lumped versus distributed. Most hydrologic simulation models are transient models in the sense that they produce a hydrograph (flow versus time) that is based on a time series input of precipitation. An especially useful additional categorization of such models is whether they can generate a continuous (long-term) hydrograph (e.g., for a period of many years) or whether they are event models (e.g., just for one storm event). Continuous models (sometime referred to as period-of-record models) may use a longer time step (often one hour to correspond to hourly precipitation data) and must rely on a statistical screening of very long time series of flows and quality parameters. The output may be the basis for a frequency analysis in the absence of long-term measured data from which design events may be selected for more detailed analysis (Bedient and Huber 1992). Continuous models are especially suited to planning and frequency-based analysis since the output time series is based on historic precipitation data and is representative of climatological extremes that influence the basin and its runoff and loadings. Event models typically use more detailed schemes (i.e., the level of detail in characterizing the watershed) and shorter time steps, and are executed for a single storm. The output hydrograph and pollutograph (concentration versus time) can be viewed graphically, and a model run in this mode is often used for detailed design (e.g., of a hydraulic structure or for a best management practice). Several watershed models can be run in either mode, and fast computers with extensive memory make the distinction between degree of schematization and time step less of an issue.

Growing interest in the application of statistical models is taking place. Such models can vary in complexity from simple regression models such as used in the International SCOPE Nitrogen Project, to more complex models such as the the Spatially Referenced Regressions on Watersheds (SPARROW) model developed by USGS. This class of models represents useful tools for understanding the relative roles various sources may contribute to the overall nutrient load delivered to a receiving body from a complex or extremely large watershed where insufficient observational data are available to initialize or verify a process based model.

ESTUARINE AND COASTAL MODELS

Development of process models for estuaries and open coastal systems is still in its infancy. While it is clear that transport by both advection and diffusion is important for controlling the final distribution of nutrients and carbon in a system, existing models generally uncouple the hydrodynamics from the biological and chemical kinetics. One justification for this is that the time step necessary to accurately describe the hydrodynamics of the system is much smaller than that assumed necessary to describe the biological and chemical processes. However, this assumption has not been carefully examined for the full extent of potentially important situations, and efforts are under way to examine its validity for shallow estuaries (e.g., Inoue et al. 1996). The solution techniques used in most existing hydrodynamic models encourage acceptance of this assumption.

The number of published model studies of physical-biological interactions in the coastal zone is increasing. Generally, they can be divided into two groups: (1) qualitative process models designed to increase understanding of the interactions observed in nature (e.g., Chen et al. 1997), and (2) prognostic models designed to enhance management decisions. The former often limit the size of the state space simulated (i.e., the number of independent variables). For example, many models of the lower trophic levels model generic categories termed phytoplankton, zooplankton, and nutrients. This ignores the significant differences in the interactions between subgroups of these three categories (e.g., diatoms and flagellates respond differently to various nutrient loadings and different classes of zooplankters prefer different phytoplankters as food sources). This, though, is a two-edged sword as there is evidence that increasing the number of dependent variables being modeled promotes the development of wildly varying (chaotic) solutions (Nihoul 1998).

The predictive models used as management tools often are limited in the level of sophistication used in developing the dynamics that force the final biological and chemical kinetics module. On the other hand, these biological and chemical kinetics are generally more sophisticated than those appearing in the process study models.

Perhaps the most challenging region to model is the continental shelf, where knowledge of the conditions along important open boundaries is generally absent. Our ability to successfully model such flow fields is limited and is only gradually improving (e.g., Herring et al. 1999). The biological variables need further differentiation, and additional chemical variables are required. The phytoplankton must be differentiated, at the very least, into diatoms and silico-flagellates. This will require the addition of a state variable for silica. The production of large versus small diatoms should influence grazing rates and efficiency. This will require development of more complex grazing models or differentiation of any zooplankton state variable.

If adequate observational data exist to initialize and verify them, calibrated process models of estuarine water quality can be useful forecasting (extrapolation) tools. Under such circumstances, they have many advantages over simpler formulations (such as statistical and spreadsheet models) because they tend to include a greater representation of the physics, chemistry, and biology of the system being simulated. Statistical and spreadsheet models should be limited to use strictly in the range of their calibration. However, in appropriate settings, the simple statistical or spreadsheet approaches may be perfectly adequate. Because these simple approaches do not require large, site specific, observational data sets or complex computer facilities or expertise, they may represent a cost-effective option. Furthermore, the utility of these simple water quality models could be greatly expanded. For example, because they employ similar conceptual formulations for modeled processes, databases of typical model parameters would greatly help in application of such models (e.g., functions to describe buildup and wash-off of surface pollutants, partition coefficients, and decay coefficients) as well as for hydrologic and hydraulic parameters. Parameters used to model best management practices (e.g., removal efficiencies) are even sparser. Agencies that sponsor watershed and water quality models should also sponsor development of databases of typical modeling parameters and case studies (bibliographic databases) of modeling efforts. Such databases would facilitate the modeling of new locations.

Estuarine and nearshore coastal models generally consist of hydrodynamic and water quality components. Although the hydrodynamic component is generally independent of the water quality component, water quality depends on transport processes. As a result, estuarine models are classified according to the temporal and spatial complexity of the hydrodynamic component. Most existing applications of such models have been to estuarine situations. The open coastal situation is complicated theoretically by the presence of open boundaries and practically by the large domain size and consequent requirements for extensive field measurements to initialize, force, and verify the model. Notable exceptions are the efforts to model the North Sea ecosystem (Baretta et al 1995) and the Louisiana inner shelf (Bierman et al. 1994; Chen et al 1997). Significant further effort in this important area is needed. As with the watershed models described above, estuarine and coastal models can be segregated into a small set of model types (EPA 1990).

1. The lowest level of complexity consists of desktop screening methodologies that calculate seasonal or annual mean total nutrient concentrations based on steady state conditions and simplified flushing time estimates. These models are designed for relatively simple screening-level analyses. They can also be used to highlight major water quality issues and important data gaps in the early stage of a more complex study.
2. The next level of complexity includes numerical steady state or tidally averaged quasi-dynamic simulation models, which generally use a box or compartment-type network. Tidally averaged models simulate the net flow over a tidal cycle, and can estimate slowly changing seasonal water quality with an effective time resolution of two weeks to one month.
3. Numerical one-dimensional and quasi-two-dimensional dynamic simulation models simulate variations in tidal heights and velocities throughout each tidal cycle. One-dimensional models treat the estuary as well mixed vertically and laterally. Quasi-two-dimensional models employ a link-node approach that describes water quality in two dimensions (longitudinal and lateral) through a network of one-dimensional nodes and channels. Tidal movement is simulated with a separate hydrodynamic package. These models are used when the data or modeling resources necessary for more complex models are unavailable.
4. At the highest level of complexity, two-dimensional and three-dimensional dynamic simulation models, dispersive mixing and seaward boundary exchanges are treated more realistically than in one-dimensional models. These models are almost never used for

routine nutrient assessment, because of the high level of resources required. Although recent advances in computer design have made such models more accessible, the field data required to initialize and drive them remain impressive.

OTHER RELEVANT MODELS

The watershed and estuarine models discussed here can be effective in describing the transport and fate of nutrients in water. However, other models may be needed to help assess management scenarios and best management practices. Most estuarine water quality models are formulated around phytoplankton-based primary production, but in some instances it may be important to focus on macrophytes. Special models have been developed to treat such situations (Box 7-4). Within the water-

BOX 7-4
Seagrass Models

Process-based simulation models of seagrasses and seagrass ecosystems have been developed for a variety of reasons. Some synthesize and guide research, or identify areas of inadequate information, while others generate research hypotheses or predict the effects of nutrient additions. Recent models attempt to guide restoration efforts. Models designed to understand the effects of nutrient enrichment on seagrasses range from physiological to landscape levels. As an example, Richard Wetzel and colleagues at the Virginia Institute of Marine Sciences have been using models to understand the processes responsible for seagrass disappearance, especially in the lower Chesapeake Bay (Wetzel and Neckles 1986; Wetzel and Buzzelli 1997; Buzzelli et al. 1995, 1999; Buzzelli and Wetzel 1996; Buzzelli 1998). Through this work, researchers learned that the principal factors governing eelgrass growth and survival were largely light related. Important interactions identified in the model were epiphytes growing on plant leaves, water column particulates and phytoplankton, grazers on epiphytes, and the indirect effects of inorganic nitrogen on phytoplankton and epiphyte growth. Overall, these models have proven to be a reliable predictor of the lower Chesapeake Bay seagrass survival and depth distribution, and have been useful for testing habitat criteria in relation to restoration goals. They are presently being integrated with the Chesapeake Bay three-dimensional water quality-hydrodynamic model.

The coupled seagrass-water quality model for Chesapeake Bay is a landscape-level model with certain cells focusing on the littoral zone (Wetzel 1996). The water quality model provides boundary conditions for water quality in the littoral zone and the ecosystem process models provide estimates of littoral zone transformations and exchange with the adjacent non-shoal or deeper areas represented

shed, biogeochemical models have been developed to describe the response of the plant community and soil to different management protocols. Such models are largely based on regression relationships, although development of models based on more fundamental relationships is progressing.

CENTURY (Parton et al. 1988) is an example of such a biogeochemical model. It was developed to describe soil organic matter changes on the North American Great Plains, but has been applied to a variety of other agricultural and grassland settings with varying degrees of success (e.g., Paustian et al. 1992; Carter et al. 1993; Gijsman et al. 1996; Vallis et al. 1996; Gilmanov et al. 1997). The model considers carbon, nitrogen, phosphorus, and sulfur. Carbon is distributed among pools with different turnover rates, which depend on the plant lignin content. The time step of the original model was one month, so short-scale meteorological varia-

by the water quality model. Four characteristic littoral zone habitats are represented in the model: (1) nonvegetated subtidal, (2) seagrass, (3) nonvegetated intertidal with microalgae, and (4) marshes. The relative distribution of each varies with water level and light penetration. The model has been used to address such questions as:

- What is the effect of halving or doubling the inorganic nutrient concentrations?
- What is the effect of varying open bay concentrations of phytoplankton and suspended particulate matter (an indication of eutrophication)?
- What is the effect of a doubling or halving the distribution of seagrass habitat on total ecosystem production, sediment microalgal production, and phytoplankton production?

Other modeling efforts focus on seagrass restoration, the role of seagrasses in sediment stabilization and water column turbidity, and the linkage between seagrass habitat quality and higher trophic levels. Kremer and Deegan (J. Kremer, University of Connecticut, personal communication) are developing a model that statistically links water quality to seagrass distribution and vitality, with the goal of linking eutrophication to declines in fish community structure and productivity.

The next generation of seagrass models may be particularly useful for planning restoration efforts because they address the feedback between seagrass biomass and sediment stabilization. Once seagrasses start to decline in the littoral zone, suspended solids often increase, thereby decreasing light penetration, which has a further negative impact on the distribution of remaining seagrasses. Reversing this trend through management actions directed at nutrient reduction will also have to address the issue of sediment stabilization and water clarity.

tions were missed. Furthermore, the original model did not include runoff, although this process could be added with additional effort (e.g., Probert et al. 1995). More recent developments (Parton et al. 1998) have shortened the model time step to one day, and added runoff processes and multiple layers to the soil model. The original model reproduces multi-year trends in soil nitrogen change and annual statistics, although events are frequently missed. The recent improvements reproduce short-term variability on some occasions and fail to reproduce reality on others (Parton et al. 1998). They clearly need further development.

Sixteen such models, applied to a long-term study of a forest-soil-atmosphere system, are described by Tiktak and van Grinsven (1995). They conclude that most such models are poorly documented for community use and include significant simplification, such as ignoring dynamic feedbacks, ignoring short time scale events, aggregating state variables, simplifying and ignoring boundary conditions, and simplifying and ignoring processes. Furthermore, many of the models were unbalanced in their treatment of the processes included. Of significance to the present discussion is the lack of an agreement among the models as to description of the nutrient cycle. Tiktak and van Grinsven (1995) recommended the development of lumped parameter models suitable for scenario assessment, but, because the necessary data are not available to verify such models, they need to be compared to detailed mechanistic models. The development of the latter was strongly recommended. Nitrate modeling by 11 of these models (Kros and Warfvinge 1995) was not very successful despite the complexity of approaches used. None of the models could reproduce the seasonal soil nitrate variations and only two could reproduce observed extreme values. These failures were attributed to soil heterogeneity, unresolved litter layer hydrologic processes, and complex microbial transformations.

Some of these models estimate the loss of nitrogen to the atmosphere as trace gases. While the different models adequately reproduce the flux of N_2O, they fail to accurately predict the flux of NO or NH_3 (Frolking et al. 1998), the gases that are most important in driving nitrogen deposition on coastal watersheds.

The committee recommends that biogeochemical models be developed because they are so important for understanding a watershed's response to best management practices and other proposed management scenarios. This is proposed even though this type of model has not yet performed at a uniformly reliable level in all environments (e.g., forests, plains, and agricultural fields). Detailed mechanistic models must be developed for comparison with field results and used to develop simple calibrated lumped-parameter models for regional analyses.

Because of the importance of atmospheric deposition of nitrogen to coastal eutrophication, it would be helpful to have models of the atmospheric transport and deposition of this nutrient available to managers. While a number of such models exist, it seems that they must be calibrated to particular regions of the globe to accurately predict deposition (Holland et al. 1999), and that their spatial resolution is so large that deposition at the scales of ecosystems is poorly resolved (Holland et al. 1997). For example, as discussed in Chapter 5, the ability of forests to store nitrogen is limited. Once a forest is saturated with respect to nitrogen, losses both to the atmosphere and to downstream ecosystems can increase rapidly. Some evidence indicates that the process whereby forests switch from retaining nitrogen to exporting nitrogen as they become nitrogen saturated can be self-accelerating due to related changes in biogeochemical cycling and ecosystem decline (Schulze et al. 1989; Howarth et al. 1996). Successfully capturing such complex ecosystem behavior through time should be an important component of efforts to develop the next generation of process-based models.

A related issue is linking models that were developed for different purposes. For instance, it is not necessarily a simple task to input a time series of atmospheric deposition values into a watershed loading model or a receiving water quality model. Nor is it straightforward to input the time series of flows and loadings from a watershed model into a receiving water model. To facilitate interfacing of different models, EPA or other relevant agencies should develop standards for storage and manipulation of hydrologic, hydraulic, and water quality time series. This will make it much easier to link models that may not have been developed for similar purposes, but may usefully provide input from one to another.

Since people are so important a component of the environment contributing to eutrophication, it will be important to incorporate socioeconomic variables into models that purport to predict landscape variability and its results. Initial efforts in this direction are being made using transition probability matrixes for future land use based on socioeconomic indicators (e.g., Berry et al. 1996). The effectiveness of such models has yet to be demonstrated. At present, the best approach to account for long-term changes in climate, land use, and related factors is to run the same model under different scenarios or forcing. This is similar to running coupled global ocean-atmosphere models under the assumption of doubled atmospheric carbon dioxide content to infer the potential system response to continued fossil fuel burning.

RECOMMENDATIONS

How should a prospective modeler select models? As can be seen from the models highlighted in Appendix D, there are few federal agency sources of useful off-the-shelf models. For watershed models, these are found primarily at USDA, EPA, USGS, and the U.S. Army Corps of Engineers' Hydrologic Engineering Center. For receiving water quality models, EPA, and the U.S. Army Corps of Engineers' Waterways Experiment Station are the primary federal sources. Most software provided by federal agencies is essentially free, but may include minimum costs for support and training. All these models are supplemented by proprietary and other software developed by American and European companies and universities. Proprietary models may be relatively costly ($5,000 to $30,000), but typically they include sophisticated interfaces and extensive user support. In some cases, models may only be available through the purchase of the services of the model developer. Thus, a client may decide on a model based as much on the qualifications of the sponsoring firm or agency as on the properties of the software.

Because modeling continues to be partially an art, the expertise and experience of the modeler is of consequence. It is a good idea to rely on competent technical personnel and allow them to decide on the choice of a model. If the model will be provided to a particular manager for his long-term use, then the model itself must be well documented and understandable by its future users. Data requirements also enter into the choice of a model, since some process models may require more information than can be affordably provided. This tends to drive the model selection toward less sophisticated techniques that require fewer data. Verification and checks for reasonableness of estimates provided by simple models are especially important in order to lend credibility to such estimates.

Since model choices remain somewhat limited, a manager may be tempted to develop an in-house model. While this option may be feasible, in-house models tend to be specific to individual problems and locations and are seldom subject to peer review and the experience of a variety of users. Apart from simple statistical or spreadsheet models, this approach is usually not cost effective relative to acquisition of federal models or use of the services of model providers.

It is clear that a number of numerical modeling codes of varying degrees of sophistication are available to the scientist and manager interested in eutrophication. While many of these models are accepted and used successfully in general practice, the level of quantitative verification, post-audit, and skill assessment that has been applied to them is highly variable. Most have only been subjected to qualitative comparison to field data sets. In many cases, these models probably overestimate nutrient inputs prior to European settlement. As a result, they underestimate the extent to which human activity has accelerated inputs of nutrients. Development of fully three-dimensional, verified, and skill-assessed water quality models should be encouraged for synthesis and management.

This implies that well designed monitoring programs be effectively linked to the iterative development of models so that both the data and model syntheses can be used in management decisions and policy.

It is expected, however, that input parameter estimates for process models should also fall within accepted ranges for such values. Verification and checks for reasonableness of estimates provided by simple models are especially important for lending credibility to such estimates. Whenever possible, uncertainty in the model output should be represented (e.g., as a mean plus a standard deviation) or as confidence limits on the output of a time series of concentrations or flows.

To facilitate interfacing of different models, EPA, USGS, or other relevant agencies should develop standards for storage and manipulation of hydrologic, hydraulic, water quality, and atmospheric deposition time series. This will make it much easier to link models that may not have been developed for similar purposes, but may usefully provide input from one to another.

Models are excellent tools for synthesis of our understanding of systems. As management tools, they can only begin to describe the variables specified per se in the model. Such management concepts as sustainable, healthy ecosystems are not quantifiable and cannot be a variable predicted from a numerical model. Great care must be given to identifying the appropriate parameters to estimate and the measures to be applied to these parameters (Huggins 1963). The assumptions that enter into this definitional process are often as important and interesting as the science involved.

All models benefit from continuous improvement. Federal agency support for widely used models that the agency has developed or sponsored should include maintenance, improvements, help with parameter estimates, and a feedback mechanism. The latter could conveniently be accomplished through discussion groups on the internet.

The numerical values of the often very large number of required input parameters are as important as the model formulations themselves. Data sets of input parameters for calibrated models should be provided by the agencies charged with oversight of the models. Agencies that sponsor watershed and water quality models should also sponsor development of databases of typical modeling parameters and a bibliographic database of case studies of model applications. Such databases would enormously ease the effort in modeling new locations.

Agencies responsible for monitoring of parameters useful for calibration and verification of models should also develop and maintain databases containing this information. For example, as the agency responsible for implementing the National Pollution Discharge Elimination System legislation, EPA should develop and maintain a current nationwide database of urban and other surface runoff samples collected during this program for use in nonpoint source water quality analyses and modeling. Additional effort should be devoted to characterization and statistical analysis of such data.

Managerial concern with the impacts of nutrient over-enrichment often is concerned with the perceived effects that nutrient loading will have on higher trophic levels in the system (e.g., the loss of commercial and recreational fisheries). These linkages are not always clearly demonstrable, and modeling of such cause and effect relationships is in its infancy. To further complicate the situation, the phytoplankton appear to be readily modeled as a continuum, while higher trophic levels often are characterized using individual models. The European regional seas ecosystem model (Baretta et al. 1995) is a suite of interconnected models that attempt to model the entire North Sea ecosystem up through the fish communities and including the benthos and the microbial loop. The present lack of knowledge concerning the connections among nutrient loadings, phytoplankton community response, and higher trophic levels implies a disconnection between the estimates used to evaluate management scenarios and the goals for which management is taking place. Therefore, the development of heuristic models using comparative ecosystem approaches is needed to identify and better understand key processes and their controls in estuaries.

8

Water Quality Goals

KEY POINTS IN CHAPTER 8

This chapter discusses how the setting of water quality goals can be used for combating nutrient over-enrichment problems in coastal areas and finds:

- To design effective approaches to mitigating nutrient over-enrichment, decision-makers must understand the physical and ecological processes at work, outline clear management goals, set specific targets, and develop a range of possible policy approaches or management tools that are suitable to the site and its problems.
- To make efficient use of available resources, managers should adopt policies that ensure that targets will be met at the lowest possible cost. In many cases, control costs will vary across sources and, if equally effective, the total cost of meeting the target will be lowest if the lowest cost sources are controlled first.
- In designing policies to achieve nutrient reductions, decision-makers will need to choose between voluntary approaches and mandatory controls or financial penalties. Each approach has advantages and disadvantages, and managers must assess how successful a given approach is likely to be in their specific context.
- Voluntary changes in behavior can be difficult to motivate. Providing information and education is not always effective, but it is relatively inexpensive and non-controversial. Providing subsidies designed to reduce nutrient inputs can be effective, but requires funds that are generally raised via taxes, which may impose a cost on society. Also, subsidies can inadvertently encourage pollution, because polluters are not required to pay the full costs of their activities, which in turn can lead to lower product costs, then higher product demand, and ultimately increased pollution to meet that demand.

- Some of the shortcomings of both regulatory and tax-based approaches can be overcome with the use of marketable permits. A careful examination of the effectiveness of this approach where it has already been implemented should be undertaken.
- In many instances, managers may find that a well-formulated mix of incentives (voluntary approaches) and disincentives (mandatory or punitive approaches) works better than either approach alone. Managers might increase the likely success of a voluntary approach by making it clear that, if the voluntary approach is not successful, an approach based on disincentives will be adopted.
- An information clearinghouse should be established that provides local managers with information about the cost and effectiveness of alternative source control methods, the effectiveness of alternative policy options, and the policy experiences that other managers have had in attempting to control nutrient over-enrichment. This information should emphasize the role of site characteristics in determining effectiveness and costs.

In developing an effective strategy for mitigating the effects of nutrient over-enrichment one must understand the physical and ecological relationships that determine the extent and causes of nutrient over-enrichment, along with societal objectives and behavioral responses. Societal objectives determine goals that a management strategy will strive to achieve and the benchmark against which it will be evaluated. Behavioral responses ascertain how the various parties contributing to nutrient over-enrichment are likely to respond to different policies designed to affect that behavior. Managers must anticipate and understand these responses as they choose among policy alternatives, since the responses will determine the effectiveness of any given policy.

This chapter discusses issues that arise both in setting goals for nutrient over-enrichment management strategies and in choosing among policy alternatives. The appropriate set of policies for any given estuary will depend on the nutrient sources for that estuary. For example, if the main nutrient source is agriculture, a set of policies designed to promote the adoption of best management practices is required. These can be implemented at the local, regional, or national level. Alternatively, if atmospheric deposition is the main source of nutrients, policies that reduce atmospheric emissions of nitrogen are needed. Since the source of atmospheric nitrogen is often outside the local jurisdiction governing the estuary, policies to combat this nutrient source must be implemented at the regional or national level. Because both the susceptibility and the specific policy needs of any given waterbody are site-specific, this chapter does not attempt to prescribe specific water quality goals or policy choices for adoption by local managers. Rather, it is intended to provide managers with an improved understanding of the factors that should be considered in setting goals and making policy choices. With this understanding, managers will then be able to begin the process of crafting a management strategy that addresses the needs and challenges of waterbodies in their jurisdiction.

SETTING GOALS

Choosing Targets

The most common approach to setting water quality goals to combat nutrient-related problems is to select target levels of ambient concentrations in the receiving body, nutrient loadings to the receiving body, or resource stocks (e.g., acres of submerged aquatic vegetation [SAV] or marine populations). Concentration levels are not always useful indicators of the eutrophic state of an estuary, since they reflect both nutrient inputs to the estuary and the response of the estuary. For instance, nutrient concentrations can be low in a highly eutrophic estuary if the nutrients are rapidly taken up by phytoplankton. Conversely, nutrient concentrations can be high in a non-eutrophic estuary if factors such as short residence times or low light availability (from deep mixing and high turbidity) make the estuary non-susceptible to eutrophication. For this reason, targets based on primary productivity (Nixon 1995), chlorophyll, or phytoplankton biomass (NRC 1993a) are likely to be better indicators of an estuary's eutrophic state and hence better measures of whether overall water quality goals (i.e., desired reductions in eutrophication) are being achieved.

Setting policy goals involves not only choosing a target indicator but also setting a target level for that indicator. There are a number of different bases that can be used to set target levels. One possibility is to seek to increase/decrease the indicator value by some arbitrary amount (e.g., 25 percent) or to restore it to its level during some previous period. For example, for Chesapeake Bay, an arbitrary goal of reducing controllable sources by 40 percent was set. Similarly, managers might seek to restore SAV acreage to some historical level. Alternatively, the target level might be determined by a specific use that is desired (e.g., a "fishable/swimmable" criterion). Setting target levels in either of these two ways typically focuses on the benefits of improvements, or, similarly, on what level of improvement is deemed "feasible." In such cases, the target level is chosen without an explicit regard to the cost of achieving the improvement, although that cost is implicit in the definition of "feasibility." Similarly, the benefits gained by meeting the target based on "feasibility" are not always clearly articulated.

Another way is to consider explicitly both the costs and benefits of alternative targets. For example, a target level of SAV acreage could be based not on an historical level but on a comparison of the cost and benefit of achieving the restoration. Such a comparison requires a measure of the benefit of restoration or a measure of the cost of degradation. Unfortunately, these costs can be difficult to estimate. Although it may be relatively easy to calculate the commercial value of a resource that is harvested and sold in the market (e.g., a commercial fish stock), the non-commercial (or non-market) value of natural resources is inherently difficult to estimate. Nonetheless, techniques exist for valuing non-market goods such as wetlands, water quality, and wildlife populations, and these techniques have been applied to the valuation of estuaries (Chapter 4). With estimates of both the benefits and costs of achieving improvements in water quality or resource stocks, an economically efficient target level for improvement can be identified. Improvements would be sought up to the point where the cost of any additional improvement would exceed the benefit of the additional improvement. However, since the ability to estimate benefits of improving winter quality remains imperfect and imprecise, the use of this approach alone can be difficult.

Establishing Criteria and Standards

Targets usually fall into two categories: water quality *criteria* and water quality *standards*. McCutcheon et al. (1993) differentiate the two as follows:

> A water quality *criterion* is that concentration, quality, or intensive measure (e.g., temperature) that, if achieved or maintained, will allow or make possible a specific water use. [For the toxic substance,] a criterion may be a concentration that, if not exceeded, will protect an organism, aquatic community, or designated use with an adequate degree of safety.[1] A criterion may also be a narrative statement concerning some desirable condition. While water quality criteria are often the starting point in deriving standards, criteria do not have a direct regulatory impact because they relate to the effect of pollution rather than its causes. A water quality *standard* is the translation of a water quality criterion into a legally enforceable ambient concentration, mass discharge or effluent limitation expressed as a definite rule, measure, or limit for a particular water quality parameter. A standard may or may not be based on a criterion.

Standards may differ from criteria for a variety of reasons, including natural impairment of water quality even in the absence of anthropogenic

[1] As noted above, nutrient concentrations do not alone provide adequate criteria for control of eutrophication in estuaries. However, the oxygen concentration of an estuary could be used, setting a concentration below which oxygen should not fall. Other appropriate criteria might include chlorophyll concentrations or nutrient inputs.

pollution, the perceived importance of a particular ecosystem, or the degree of safety required for a particular waterbody (McCutcheon et al. 1993). While criteria are typically defined relative to ambient water quality, standards may take different forms. Ambient standards often are based on the establishment of threshold values for a particular contaminant, and they consider the intended use of the waterbody, as well as its ability to assimilate wastes. Effluent standards limit the amount of material that may be discharged regardless of the size of the receiving waterbody or the intended use of its waters. Effluent standards are often technology-based and may be imposed even if the level of contamination is less than that required to achieve ambient water quality standards (McCutcheon et al. 1993). When a receiving waterbody is affected by a discharge, the standard, be it ambient or effluent, must govern the discharge or loading of wastes into the water. Even for large rivers, loadings over the entire watershed may impact the estuarine or coastal water quality.

Traditionally, water quality standards have been absolute numbers in the sense of a concentration or discharge of a toxic substance that may not be exceeded, or an oxygen concentration that must be maintained. However, water quality in a given waterbody can fluctuate as a result of random factors, such as weather and uncertainties in hydrologic processes. When natural variation is important, it is still possible to implement probabilistic standards—standards that, for example, specify that a given concentration of a toxic substance is not to be exceeded more than once, on the average, during a certain number of years (EPA 1991b).[2] Probabilistic standards account for extreme hydrologic events that rarely create excessive nonpoint source loads, and are more reasonable when receiving-water impacts are driven in large part by these factors.

Other impacts of nutrient enrichment could also lead to a standard, if criteria can be established through scientific studies and modeling. Eutrophication impacts were discussed in Chapter 4. They include increased primary productivity; increased phytoplankton biomass; reduction in water clarity; increased incidence of low oxygen events (hypoxia and anoxia); changes in the trophic structure, trophic interactions, and trophodynamics of phytoplankton, zooplankton, and benthic communities; damage to coral reefs; fish kills; reduced fisheries production; and decreased biotic diversity. In general, the public must accept the need for improved receiving-water quality and prevention measures. If the public does not perceive a problem, it is unlikely that elected officials will pur-

---

[2] EPA's total maximum daily load (TMDL) process allows the use of probabilistic standards (EPA 1991b). See further discussion below.

sue this issue or that agency staff will have the resources or authority to implement solutions.

Finally, as with any other complex water quality problem, many other factors must also be considered in setting nutrient-based water quality standards, including:

- Are damages subject to threshold effects or are they continuous? In some cases, impacts are gradual up to some critical loading, at which point the receiving water may react in a way that greatly increases algal growth. In other cases, impacts remain gradual throughout. (Susceptibility is discussed in Chapter 5.)
- During which seasons are impacts the greatest? Warmer water during the summer often leads to worsening of dissolved oxygen levels because of lower saturation concentrations and higher rate constants, combined with greater stratification and more light. If there is a low-flow season, dilution of wastes entering an estuary may be lower. Year-to-year variability in climate is also a driving force. Critical conditions can usually be determined by simulation over periods of at least a year.
- Are sudden discharges (e.g., due to a thunderstorm) controlled? In some settings, a heavy rainfall during dry weather may lead to intense, temporary nonpoint source loading.
- What are the flushing and mixing conditions in the estuary or coastal water? Exchange with the open ocean or sometimes with wetlands may mitigate the impact of heavier loadings. Effects will differ for every coastal water.
- What time scales are involved? Control efforts are likely to take a significant amount of time to become effective in reducing an existing negative impact. It is likely that improvements due to reduced nutrient loadings will be felt only over a period of several years because of nutrient storage in the system, especially in sediments, although in some cases some systems may show a rapid response.

Current Criteria and Standards

Most water quality criteria in the United States are based on the "Gold Book" (EPA 1987), in which criteria are given for over 100 constituents, most of which are heavy metals and organic chemicals. Criteria may differ for fresh and marine waters and may be based on toxicity to various aquatic organisms or hazard to human health (e.g., drinking water standards).

Another current approach to regulating water quality, defined in section 303(d) of the Clean Water Act, is an approach known as total maximum daily load (TMDL). This method, which was refined in later regulations, is a water quality-based standard. It strives to assure water quality through a series of steps that, in effect, require a watershed scale approach (EPA 1999a). For problems of eutrophication in estuaries, this emphasis on water-quality-based standards was recommended by a previous National Research Council (NRC) report (NRC 1993a) and is endorsed by this committee. Under proposed 1999 modifications to the TMDL process, a TMDL must contain the following minimum elements (EPA 1999a):

- the name and geographic location of the impaired or threatened waterbody for which the TMDL is being established;
- identification of the pollutant for which the TMDL is being established and quantification of the pollutant load that may be present in the waterbody and still ensure attainment and maintenance of water quality standards;
- identification of the amount or degree by which the current pollutant load in the waterbody deviates from the pollutant load needed to attain or maintain water quality standards;
- identification of the source categories, source subcategories, or individual sources of the pollutant for which the wasteload allocations and load allocations are being established;
- wasteload allocations for the pollutant to each industrial and municipal point source; for discharges subject to a general permit, such as storm water, combined sewer overflows, abandoned mines, or combined animal feeding operations; pollutant loads that do not need to be reduced to attain or maintain water quality standards; and supporting technical analyses demonstrating that wasteload allocations, when implemented, will attain and maintain water quality standards;
- load allocations, ranging from reasonably accurate estimates to gross allotments, to nonpoint sources of a pollutant, including atmospheric deposition or natural background sources; and supporting technical analyses demonstrating that load allocations, when implemented, will attain and maintain water quality standards;
- a margin of safety expressed as unallocated assimilative capacity or conservative analytical assumptions used in establishing the TMDL (e.g., derivation of numeric loads, modeling assumptions, or effectiveness of proposed management actions that ensure attainment and maintenance of water quality standards for the allocated pollutant);
- consideration of seasonal variations and environmental factors that

affect the relationship between pollutant loadings and water quality impacts;
- an allowance for future growth, if any, which accounts for reasonably foreseeable increases in pollutant loads; and
- an implementation plan, which may be developed for one or a group of TMDLs.

In most cases, item 4 (identification of source categories, etc.) of this process virtually mandates a watershed approach since waters are impaired by multiple dischargers and pollutants, and these derive, to a considerable extent, from nonpoint sources distributed over broad regions. The Environmental Protection Agency (EPA) recommends that TMDLs be developed on a regional basis, that is, by watershed. A virtue of the TMDL approach is that it is flexible and considers water quality to be a function of an extensive range of sources distributed across the landscape. It is so flexible that physical and biological stressors, like water temperature and habitat alteration, can be considered within the same management framework (NRC 1999a).

The TMDL approach is promising for control of pollution, including nutrients, but it can be hampered by data gaps. For instance, development of a TMDL presupposes that a waterbody has been classified as water quality impaired, that its condition has been ranked and prioritized with respect to other impaired waters within a state, and that standards for specific contaminants have been established. Development of TMDLs requires understanding of point and nonpoint sources; the processes that influence the magnitude, timing, transport, and attenuation en route of pollutants; and how those pollutants affect aquatic biota.

The TMDL procedure is fairly site-specific. Thus, the management approach tends to be tailored to the needs of a specific watershed or receiving body. Because of data gaps and limitations in knowledge of the structure and function of watershed ecosystems, and because the approach is still relatively new, development of TMDLs has proceeded slowly. Table 8-1 summarizes current nutrient criteria for coastal states by EPA region (EPA 1998c). As discussed in Chapter 7, information on the ambient concentration of nutrients in watersheds and rivers is important for calculating the load that downstream coastal water bodies will receive from these sources. However, resistance to concentration-based standards by some coastal states is understandable, given their limited utility as a measure of waterbody impairment. Ambient concentration of nutrients in receiving waters rarely reflects the degree to which the body has been impacted by nutrient over-enrichment. For example, if an estuary is nitrogen limited, primary productivity will be stimulated by nitrogen loading. This increased primary productivity will then remove nitrogen

TABLE 8-1

Region/State	Nitrate	Ammonia	Total Nitrogen	Total Phosphorus
Region 1				
Connecticut				3
Maine				7,3,8
Massachusetts	2	2		2
New Hampshire	2			
Rhode Island	3	3		3
Region 2				
New Jersey	2	2		9,3
New York				2
Puerto Rico	9	9		8,2
Virgin Islands				8,9
Region 3				
Delaware	2	2		2
District of Columbia				
Maryland				
Pennsylvania				
Virginia				4
Region 4				
Alabama				
Florida			2	7,2
Georgia	3			
Mississippi				
North Carolina	3			
South Carolina	2	2		3
Region 6				
Louisiana	2	2		2
Texas	2	2		2
Region 9				
American Samoa	2	2	1,9	1,9
California	1,5	5	1,2	1,6,7
Guam	2,7	5	2	2,7
Hawaii	1,9	1,9	1,9	1,9
Nevada	5	2,5	1,7,9	1,7,9
Northern Mariana Islands	7	5	7	7
Trust Territories of the Pacific Islands	2	2	7	7,9
Region 10				
Alaska				
Oregon				
Washington				

TABLE 8-1 Summary of existing water quality criteria and standards for nutrient over-enrichment for coastal states and territories (EPA 1998c). As of 1999, 21 states and territories had proposed water quality criteria for nutrients, with significant differences in the nutrients addressed and whether the criteria are narra-

continued

TABLE 8-1 Continued

tive or quantitative. Blank entries indicate that no criterion for the nutrient has been specified by the state. Key: (1) Site-specific numeric values for ambient nutrient levels; (2) narrative criteria related to natural conditions, eutrophication and nutrient over-enrichment for nitrate, ammonia, inorganic nitrogen, total nitrogen, or total phosphorus; (3) narrative criterion that is not related to natural conditions, eutrophication, or nutrient over-enrichment issues; (4) numeric values for effluent nutrient levels; (5) numeric values related to public health (nitrate) or aquatic toxicity (ammonia); (6) habitat-based numeric values for ambient nutrient levels; (7) water use classification- or water use designation-based numeric values for ambient nutrient levels; (8) state-wide numeric values for ambient nutrient levels; and (9) waterbody-based (streams, rivers, lakes, estuaries, coastal/oceanic waters) numeric values for ambient nutrient levels.

from the water column at a high rate and tie it up in organic matter. Thus, the ambient nitrogen concentration in the water column may never rise significantly, or remain elevated long enough to be observed, even as eutrophication of the body takes place (Box 8-1). The TMDL approach avoids this limitation by directly addressing nutrient loading.

As the TMDL approach evolves in support of controlling nutrient problems in coastal waters, it will need to recognize the variation among estuaries and follow a consistent classification scheme (Chapter 6). Criteria endpoints are likely to vary by type of estuary (e.g., some might address seagrass extent, others chlorophyll, and others dissolved oxygen).

EPA recently developed nutrient standards on a regional or watershed basis (EPA 1998e). The major elements of this strategy include:

- use of a regional and waterbody-type approach for the development of nutrient water quality criteria;
- development of waterbody-type technical guidance documents (i.e., documents for streams and rivers; lakes and reservoirs; estuaries and coastal waters; and wetlands) that will serve as user manuals for assessing trophic state and developing region-specific nutrient criteria to control over-enrichment;
- establishment of an EPA national nutrient team with regional nutrient coordinators to develop regional databases and to promote state and tribal involvement;
- EPA development of nutrient water quality criteria guidance in the form of numerical regional target ranges that EPA expects states and tribes to use in implementing state management programs to reduce over-enrichment in surface waters (i.e., through

the development of water quality criteria, standards, EPA's National Pollutant Discharge Elimination System permit limits, and TMDLs); and
- monitoring and evaluation of the effectiveness of nutrient management programs as they are implemented.

The major focus of this strategy is the development of waterbody-type technical guidance and region-specific nutrient criteria by the year 2000. Once the guidance and criteria are established, EPA will assist states and tribes in applying numerical nutrient criteria to water quality standards by the end of 2003.

CHOOSING A POLICY APPROACH

Once water quality or other resource-based targets have been set, managers must decide which policy approaches to use for achieving those targets and the details regarding implementation. Some approaches are based on voluntary action, while others involve mandatory controls or the use of financial or other penalties (e.g., taxes) to induce desired behavioral changes. In this section, the strengths and weaknesses of alternative approaches are discussed.

Evaluation Criteria

In choosing among alternative approaches to reducing eutrophication or other effects of nutrient over-enrichment, it is important to specify the criteria used to rank the alternatives. The type of criteria typically used (e.g., Bohm and Russell 1985; NRC 1993a) are:

1. cost-effectiveness;
2. dynamic adjustment (flexibility, adaptability, and innovation incentives); and
3. distributional impacts and fairness.

Cost Effectiveness

Given a predetermined target or goal for a given response (e.g., reduced eutrophication), cost effectiveness is achieved when that target is met at the lowest possible cost (Bohm and Russell 1985; Baumol and Oates 1988). Applying this principle requires both a definition of effectiveness and a mechanism for ensuring that, other things being equal, low cost sources are controlled first. In defining effectiveness, a number of issues arise. First, the criterion must be based on measures of water

BOX 8-1
Why Concentration Alone is an Inadequate Measure of Nutrient Over-Enrichment

The concentrations of inorganic nutrients (such as nitrate, ammonium, and phosphate) are often measured in estuarine monitoring programs, and nutrient data can provide important information on the functioning of the system. However, data on nutrient concentrations alone give little indication of whether or not an estuary is eutrophic. For instance, inorganic nutrient concentrations may be low in an estuary that is receiving high nutrient inputs and is eutrophic because the nutrients are largely assimilated by phytoplankton. On the other hand, inorganic nutrient concentrations may be high in another estuary that is not eutrophic, perhaps because high turbidity and light limitation prevent phytoplankton from growing and effectively using the nutrients. An example of this can be seen by comparing dissolved inorganic nitrogen concentrations in Chesapeake Bay and in Delaware Bay (Figure 8-1). These nitrogen concentrations tend to be higher in much of Delaware Bay than in the Chesapeake, even though eutrophication is a larger problem in the Chesapeake, because high turbidity in Delaware Bay limits phytoplankton production and allows the nutrient concentrations to remain high (EPA 1998d).

Rather than relying on inorganic nutrient concentrations as an index of water quality, the extent of eutrophication in an estuary is best determined from measurements of primary productivity (Nixon 1995) or from measurements of chlorophyll or other measures of phytoplankton biomass (NRC 1993). Similarly, nutrient problems in estuaries are best managed by reducing the rate of input of nutrients into the system rather than by setting a standard for an acceptable nutrient concentration within the estuary.

FIGURE 8-1 Comparison of the concentrations of dissolved inorganic nitrogen measured during the summer months in surface waters of Chesapeake Bay and Delaware Bay. Nutrient levels are higher than optimum in most of the rivers and upper bays. "Poor" refers to nitrogen levels greater than 0.45 mg l^{-1}, "fair" is for nitrogen levels between 0.15 and 0.45 mg l^{-1}, and "good" nitrogen levels are less than 0.15 mg l^{-1} (EPA 1998d).

Figure 8-1 on next page

BOX 8-1 Continued

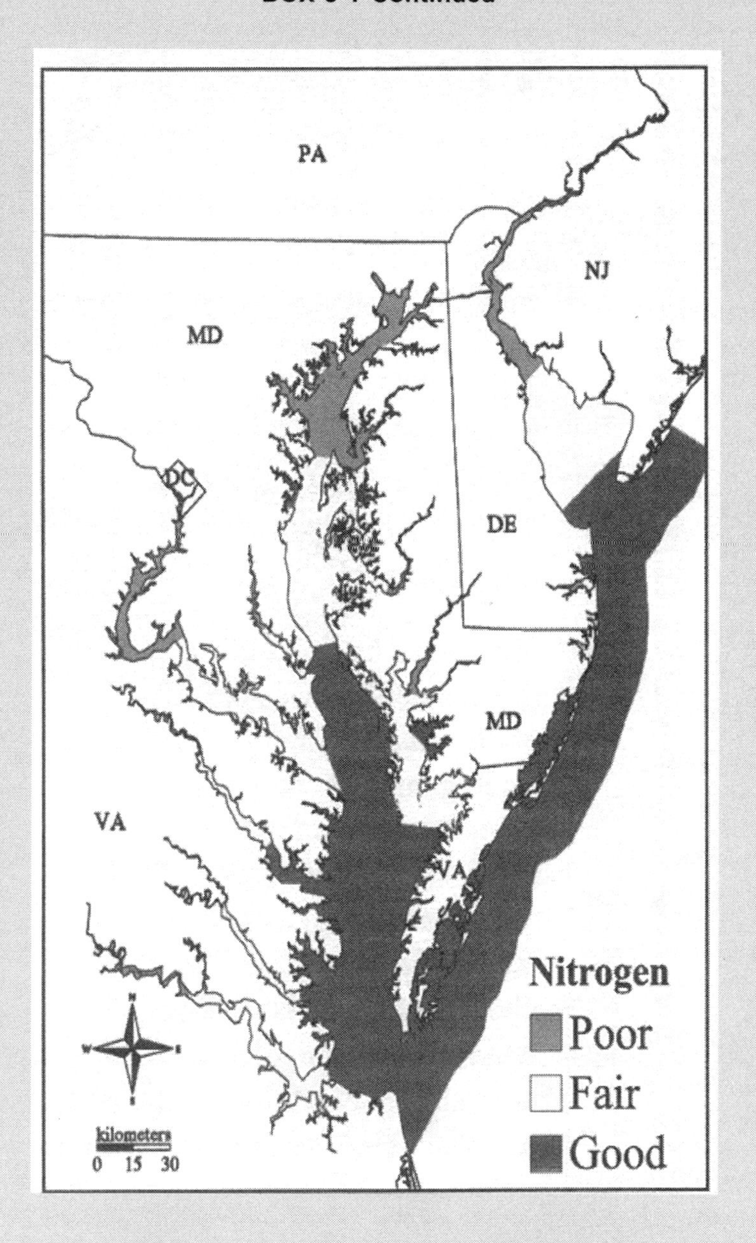

quality or living resources rather than action indicators (e.g., number of projects or number of acres enrolled). It is important to recognize the spatial, temporal, and seasonal variability in water quality measures and the resulting variation in "effectiveness." Measuring effectiveness can be difficult when the effect is primarily through groundwater recharge, because of the potentially long lag times that result when groundwater moves slowly or when the recovery of the estuary is slow because of nutrient retention in sediments or other "memory" effects. Nonetheless, reliable measures of effectiveness are necessary to ensure that policies are having the desired impact.

Given equal effectiveness, an evaluation of eutrophication reduction policies can be based on a comparison of the cost of alternatives, both to private parties and to government agencies. Private costs include direct expenditures on pollution abatement services and equipment, as well as losses in profits due to alterations in production processes or changes in the products that are produced and sold. For example, changes in crop mix to reduce nitrogen runoff can reduce farm revenues, while investments in manure storage facilities directly increase farm costs. Both of these costs reduce profits for the farmer, and impose direct or indirect costs on the private parties. Government costs include direct public expenditures or investments, such as investment in improved sewage treatment, and expenditures on hydrologic or biologic treatment programs.

The costs of meeting a given water quality target include direct compliance or investment costs (both private and public), associated administrative costs, and other monitoring and enforcement costs. These costs can be borne by both the public agency responsible for administering the policy and the affected private parties. The magnitude of these costs is generally related to both the amount of information necessary to determine compliance (including the reporting requirements) and the complexity of the policy. All else being equal, policies that involve low information and monitoring costs are preferred.

Policies can generate uncertainty for private parties, and this uncertainty can be costly. For example, if taxes are used as a mechanism to push industry to make investments in pollution control, and industry makes investments in response to the tax, but the changes do not bring about the needed level of improvement and the tax is adjusted, the businesses may find that they have under- or over-invested. If polluters are held liable for damages that result from their discharges, uncertainty regarding those future liabilities can create significant risks. If polluters have no mechanism for spreading these risks (through, for example, the purchase of liability insurance), these risks impose costs.

Given this broad definition of costs, cost effectiveness requires that the environmental goal be met at the lowest possible cost. In many cases, control costs will vary across sources and, if equally effective, the total cost of meeting the target will be lowest if the lowest cost sources are controlled first. Some policies (such as marketable permits) are designed to ensure that the target is met at the least cost, while others (such as uniform emission limits) are not.

Dynamic Adjustment

Since environmental and economic conditions can change rapidly, it is desirable to have a policy that is flexible enough to respond to such changes or to new information as it becomes available. For example, policies that embody marketable permits can automatically accommodate economic growth without an associated increase in pollution (since new firms would simply be required to buy existing permits), while regulatory policies based on allowable discharges cannot (Tietenberg 1985a).

Because pollution control technologies can change over time, policies should be designed to provide incentives for innovation and the design and development of improved control or response strategies. Policies under which private parties realize the gains from innovation provide incentives for investment in pollution control research and development (Prince and Milliman 1989; Carraro and Siniscalco 1994; Laffont and Tirole 1994, 1996; Jaffe and Palmer 1997).

Distributional Impacts

While the above criteria relate to reducing nutrient over-enrichment as cheaply and effectively as possible, policies can also be evaluated on the basis of how they distribute the costs (and benefits) of eutrophication control. The costs will be distributed both regionally and sectorally (e.g., agriculture, electric utilities, households, public sector). There are at least two alternative principles for determining an appropriate cost allocation. The first is "the polluter pays principle," under which polluting parties bear the costs of pollution control (Tobey and Smets 1996). Taxes (such as carbon taxes, energy-use taxes, land-use taxes, and pesticide or fertilizer-use taxes) and, to a lesser extent, regulatory approaches are based on this principle. The second principle is "the beneficiary pays principle," under which those who benefit from pollution control bear the associated costs. Subsidies and public investment in pollution control are based on this principle, as are water treatment strategies (for example, Safe Drinking Water Act) that raise water prices to consumers. When polluters and beneficiaries are separated sectorally and regionally, the two principles imply very different distributions of costs. The alternative distributions imply different assumptions about property rights. The polluter pays principle implies that the public has a right to a clean environment, and polluters must bear the cost of maintaining a clean environment or pay for degradation that interferes with the enjoyment of that right. In contrast, the beneficiary pays principle implies that the polluters have a right to use (or abuse) the environment as they choose, and the public must pay to ensure or restore a clean environment. The choice between the two principles is essentially a choice about the appropriate allocation of property rights.

Alternative Policy Approaches

NRC (1997b) identified a number of general policy approaches or management tools that can be used to improve marine management: (1) command and control or direct regulation, (2) moral persuasion, (3) liability and compensation, (4) direct production of environmental quality, (5) education, (6) economic incentives, and (7) tools that affect the underlying dynamics of the marine system. Of these, both the direct production of environmental quality (through, for example, improved sewage treatment plants or water treatment systems) and modifying the dynamics of the natural system are purely public management options because they are undertaken and financed directly by public authorities. The remaining tools are aimed at private source reduction and involve government efforts to force or induce private parties to reduce pollution.

Historically, the tools for private source reduction have often been divided into regulations and other mandatory controls (command and control) and economic incentives (Bohm and Russell 1985; NRC 1993a). Command and control policies set limits on allowable emissions of pollutants or dictate allowable or required production practices (e.g., installation of specific pollution control equipment). Failure to comply triggers enforcement actions. Incentive policies, on the other hand, try to induce rather than dictate changes in private behavior. Incentive policies can be based on either positive incentives (a carrot approach) or negative incentives (a stick approach). Positive incentives include subsidies for land use changes (for example, under the Conservation Reserve Program), subsidies for reductions in soil erosion (under cross-compliance provisions), and cost sharing for investment in best management practices. Negative incentives include fertilizer or pesticide taxes (designed to reduce use), land use taxes, and legal liability for groundwater contamination.

Another way to think about management tools is whether they are voluntary or mandatory (Stranlund 1995; Segerson and Miceli 1998; Segerson and Li 1999). Voluntary approaches include both information- and moral persuasion-based policies and subsidies that induce voluntary reductions in pollution. Mandatory policies include both regulatory or command-and-control policies and negative incentives (e.g., taxes, fines, penalties, liability). The crucial distinction is that under voluntary programs, polluters have a choice regarding whether to participate in the program, while under a mandatory policy, compliance with the regulation or payment of a penalty or tax is not voluntary. Under a voluntary approach, managers cannot impose involuntary net costs on polluters, while under the mandatory approach, they can. This voluntary-mandatory dichotomy mirrors the choice faced by resource managers, who often must decide whether to require pollution abatement or encourage voluntary abatement. Each approach has its advantages and disadvantages.

Voluntary Approaches

Voluntary approaches to pollution control can be divided into three types (Carraro and Léveque 1999; Segerson and Li 1999): (1) unilateral initiatives or commitments, (2) public voluntary schemes, and (3) negotiated agreements. Unilateral initiatives are actions initiated by individual polluters or groups seeking to establish standards for themselves or to self-regulate; regulatory agencies or resource managers are not actively involved in these initiatives. Public voluntary schemes, on the other hand, are designed by managers or policy-makers; the manager designs the program, including eligibility criteria and the rewards and obligations of participation, and then seeks participation. With negotiated agreements, the obligations of the involved parties are determined through negotiation among the parties (Box 8-2).

Regardless of the approach used, the effectiveness of voluntary measures depends on participation. There are a number of reasons why a polluter might participate in a voluntary program or voluntarily undertake abatement (Segerson and Miceli 1998; Segerson and Li 1999), including:

- a strong commitment to environmental improvement or stewardship;
- a personal benefit, as, for example, when the polluter is also a user of the polluted resource (e.g., groundwater);
- a perceived payoff in the marketplace, as, for example, when a firm feels it will benefit directly or indirectly from having a "green" image or marketing a "green product";
- a sufficiently large financial inducement or subsidy; and
- fear that failure to participate will lead to more stringent mandatory controls in the future (a regulatory threat).

Managers can increase the likelihood of participation by affecting one or more of these motivating factors. For example, through information or

BOX 8-2
Tampa Bay Nitrogen Management Consortium: A Public-Private Partnership

An example of a voluntary, public-private partnership can be seen in Tampa Bay. As part of its involvement in the Tampa Bay National Estuary Program, coastal managers decided to set nitrogen-loading targets for Tampa Bay based on the water quality and related light requirements of turtle grass, *Thalassia testudinum*, and other native seagrass species. Monitoring data indicate that light levels can be achieved at necessary levels by maintaining existing nitrogen loadings. However, that goal may be difficult to achieve because human population in the watershed is expected to increase 20 percent over the next 20 years, which is projected to increase nitrogen loadings by 7 percent. The sources of the nitrogen are varied, which means there is no single way to combat the problem. External nitrogen sources and their relative contribution to existing loads include industrial wastewater (4 percent), municipal wastewater (10 percent), fertilizer spillage (7 percent), residential runoff (10 percent), commercial and industrial runoff (5 percent), intensive agriculture (6 percent), pasture and rangeland (13 percent), undeveloped land (7 percent), mining (4 percent), groundwater and springs (5 percent), and atmospheric deposition from a combination of point, mobile, and natural sources (29 percent).

Long-term management of this disparate mixture of nitrogen sources called for a partnership of many interests, and a group called the Nitrogen Management Consortium was formed. This group is composed of representatives of the local electric utility, local industries, agricultural interests, local governments, and regulatory agencies. The consortium is developing an action plan to set the target load reduction needed to maintain nitrogen at 1992-1994 levels, estimated to be a total reduction of 17 tons of nitrogen per year through 2010.

Projects planned and implemented by the consortium are expected to reduce existing nitrogen loads by 140 tons yr^{-1} by the year 2000, which meets and exceeds the agreed reduction goal. Even now, before the action plan has been fully implemented, seagrass extent, the environmental indicator of success for this program, is expanding at the rate of about 500 acres per year. If that rate is maintained, Tampa Bay will meet its long-term restoration target in about 25 years.

moral persuasion, managers can try to increase the public's environmental stewardship and recognition of the benefits of pollution abatement. Managers can also design financial incentives to encourage participation.[3]

[3] Although managers can also increase participation by applying a regulatory threat, the response in this case is not really voluntary, since polluters are simply choosing the lesser of two evils.

Use of Information or Moral Persuasion

In some cases, voluntary source reduction can be achieved simply through education or technical assistance (NRC 1993a). To the extent that "win-win" situations exist (i.e., opportunities for changing behavior in a way that benefits not only the environment but also the individual), making people aware of these opportunities can cause voluntary improvements. For example, under some conditions, adoption of conservation tillage may simultaneously reduce erosion and runoff of nutrients and increase on-farm profitability. Scaling back applications of fertilizers, pesticides, and feeding rations or increasing irrigation efficiency may generate this type of double benefit as well. Similarly, improvements in energy efficiency can decrease costs by decreasing fossil fuel use and the associated emission of pollutants such as NO_x. This is the goal of Energy Star, a set of voluntary programs jointly managed by EPA and the Department of Energy. These programs are designed to improve the energy performance of commercial products and buildings through education and technical assistance. Even when adoption of an environmentally friendly practice entails some costs–if those costs are modest–a combination of information and persuasive appeal to environmental stewardship ethics may induce adoption. An example of such a program is the Florida Yards and Neighborhood Program, a program of the University of Florida Cooperative Extension Service, which provides information regarding the impact of yard care and design on the environment and guidance for reducing that impact.

Evidence on the effectiveness of information and moral persuasion approaches to environmental protection is mixed. For example, Lohr and Park (1995) found that information variables (such as contact with agricultural agencies) were significant determinants of participation in a voluntary program, although the variables that were significant varied across the sites. Bosch et al. (1995) found information and education to be important in the adoption of nitrogen testing of the soil. However, information, education, and cost sharing information are not necessarily successful in motivating landowners to change production practices (Napier and Johnson 1998). A similar picture emerges from evidence in the industrial and commercial sectors. For example, two recent EPA studies (EPA 1997, 1998e) reported significant energy savings and economic gains from Energy Star. However, other voluntary EPA programs appear to have been less effective (for example, Davies and Mazurek 1996).

While the effectiveness of information-based policies is uncertain, the advantage of this approach is that it relies on only voluntary changes in behavior. In addition, it can be relatively inexpensive (compared to capital investments) and is generally less controversial than other approaches to water quality improvement (NRC 1993a).

Subsidies and Cost Sharing

Subsidies designed to induce reductions in loadings can take a number of forms. For example, farmers can be paid to take land out of production, such as occurs under U.S. Department of Agriculture's (USDA) Conservation Reserve Program (CRP).[4] Alternatively, cost-sharing funds can be provided for certain production practices, such as best management practices (BMPs). (See Chapter 9 for a discussion of these programs.) The intention is to increase pollution abatement by reducing or reimbursing polluters for the associated cost.

The use of subsidies to induce pollution abatement suffers from at least two important weaknesses. The first is the need to raise the funds necessary to finance the subsidy. In most cases, the taxes that are used to raise the necessary revenue distort other sectors of the economy (Atkinson and Stiglitz 1980), and may thus impose a net cost on society. Second, polluters who receive subsidies do not pay the full costs of their activities. As a result, they tend to engage excessively in those activities. For example, even though a polluter receiving a subsidy might be induced to invest in pollution control, the existence of the subsidy makes his production activity more profitable than it would otherwise be. This can either induce more firms to enter the industry or reduce the number who would otherwise move into alternative production activities. In addition, the firm's costs are lower with the subsidy than they would otherwise be, and, as a result, the price of its product is lower as well. With a low price, demand for the polluting product tends to be high.

The net result is that, even though pollution per polluter has been reduced as a result of the subsidy, it is possible that by enhancing the demand for the polluting product and the profitability of the polluting industry, the overall level of pollution may actually increase (Baumol and Oates 1988). For these reasons, subsidies may be economically inefficient.

Despite their inefficiencies, agricultural pollution control policy historically has been based on technical assistance and subsidies (Reichelderfer 1990; Ribaudo 1998). Agricultural legislators, administrators, and interest groups have historically been very effective at fighting any effort to impose regulatory restrictions aimed at reducing the environmental impact of agriculture (Reichelderfer 1990). In addition, the agricultural sector

[4] Although CRP initially was designed as an erosion control policy, its successor, the Environmental Quality Incentives Program uses a more general definition of environmental improvement.

has historically benefited from federal commodity price support programs, which encourage production and in many cases exacerbate environmental degradation (Just and Bockstael 1991). Subsidies in the form of tax credits have also been used to promote energy conservation and conversion to renewable energy sources.

Although inefficient, empirical research has shown that cost sharing and other subsidy-based policies can be effective in inducing voluntary pollution abatement. For example, Cooper and Keim (1996) surveyed farmers to determine whether they would adopt specific farm management practices to improve water quality if they were paid a fixed amount per acre. Lohr and Park (1995) used a similar methodology to predict participation in a program to encourage the use of filter strips, as part of CRP. Both studies found that participation rates were sensitive to financial incentives (i.e., increases in payments increased the probability of participation). However, previous studies of the CRP have shown that participation is also affected by non-economic factors, such as education and land quality (Esseks and Kraft 1988; Konyar and Osborn 1990; McLean-Mayinsse et al. 1994; Parks and Schorr 1997).

Both the theoretical and the empirical literature on the use of voluntary approaches to pollution control reveal that these approaches can be effective, but that success is not guaranteed (Segerson and Li 1999). As expected, the programs are most likely to be successful if there is a strong inducement for individuals to participate. Based on experience with voluntary agricultural programs, Ribaudo (1998) concludes that voluntary approaches are most likely to succeed when:

- farmers recognize that agriculture contributes to severe local or on-farm pollution problems, such as groundwater contamination;
- practices that lead to pollution reduction also generate higher returns;
- education, technical, and financial assistance are offered in a coordinated fashion;
- local research can provide information on the economic and physical performance of recommended practices;
- there is interaction with non-USDA agencies, organizations, and local businesses in the watershed; and
- voluntary programs are supported by regulations that clarify goals and provide an impetus for participation in voluntary programs.

Similarly, voluntary approaches to promote reductions in other nitrogen sources, such as fossil fuel combustion, are most likely to be effective when there are strong incentives for firms to participate. These incentives are likely to be greatest when firms (1) are aware of the technical feasibility of reducing loadings, for example, through improved energy efficiency, and such improvements are cost-effective; (2) produce products where demand is sensitive to their environmental characteristics or energy efficiency; (3) perceive public relations benefits from participation in voluntary programs; or (4) fear the imposition of mandatory controls if voluntary approaches are not successful in achieving desired reductions in loadings or fossil fuel use.

Mandatory Policies

While information-based policies and subsidies rely on voluntary changes in polluting behavior with no long-term net cost to polluters, mandatory policies dictate behavioral changes or payments based on polluter choices. Irrespective of whether the mandatory policy takes the form of command-and-control regulation, taxes, or fees, it puts the burden and the associated cost of pollution control on the polluters (Box 8-3).

Command-and-Control Regulations

Command-and-control regulations can take a number of forms, including mandatory limits on emissions of a pollutant (e.g., NO_x emission caps or nitrogen effluent limits), required investment in pollution control equipment (e.g., use of best available control technologies), or required use of specified production practices (e.g., reduced tillage).

To be cost effective, regulations must be designed to ensure that pollution reductions are achieved in the least costly way. Historically, regulations have not always been designed with this goal in mind, and they have been criticized for their high costs (Hahn 1994). Environmental regulations have relied heavily on the use of technology standards, which require installation of a particular type of pollution control equipment and are generally not cost effective. This standardized, "one-size-fits-all" approach deters firms from developing and taking advantage of alternative, less costly technologies and methods of reducing emissions.

More recently, the nation has moved toward greater reliance on performance standards, which grant polluters the flexibility to meet standards in a variety of ways, and this is expected to lead to greater cost effectiveness (Besanko 1987; Burtraw 1996) and encourage innovation. With technology standards, firms have no incentive to develop less costly approaches to pollution reduction, since the regulation does not allow them to benefit from such improvements. With a performance standard, any reduction in the cost of meeting the standard (through, for example, an innovation in pollution control techniques) generates direct benefits

BOX 8-3
Maryland Tries Mandatory Nutrient Management

In 1998 Maryland passed its Water Quality Improvement Act (WQIA), perhaps the most comprehensive farm nutrient control legislation in the country. The law marks a transition from voluntary to mandatory nutrient management, and it brings new attention to phosphorus as a nutrient of concern.

Under WQIA, by 2005 all agricultural operations with annual incomes greater than $2,500 or more than eight animal units (one animal unit equals 1,000 pounds of live weight) must implement nutrient management plans that consider both nitrogen and phosphorus application rates. In the past, when animal manure or sludge was applied, the amount of recommended materials was based on crop nitrogen needs. However, because the amount of phosphorus in manure is generally high relative to nitrogen and the nutrient needs of growing crops, this practice resulted in substantial excess application of phosphorus. Although it was long thought that controlling erosion controlled phosphorus loss, research has shown that, even without erosion, runoff from soils with excessive phosphorus levels can contain high levels of dissolved phosphorus.

The law allows at least three approaches to phosphorus control. Farmers can test their soil and follow recommendations to match agronomic and environmental needs, although this approach might greatly restrict phosphorus application on soils with optimum to slightly excessive levels without considering other site-specific factors that affect phosphorus loss. Farmers also can establish "critical" soil test values that limit phosphorus application, meaning that a level could be established at which only as much phosphorus as the crop removes could be applied, while for soils at some higher level no additional phosphorus could be applied. Scientists have objected to both approaches, since their research indicates that many site-specific factors influence the potential for phosphorus loss. Instead, they have proposed the use of a phosphorus site index.

This phosphorus site index is a generalized national index that has been developed and is now being adapted by the University of Maryland for possible use in Maryland. The index evaluates slope, runoff potential, proximity to surface water, soil phosphorus levels, and fertilizer and manure application rates and methods; it thus allows site-specific assessments and comprehensive evaluation of potential environmental impacts without restricting phosphorus application to low-risk sites.

To help farmers meet WQIA requirements, Maryland has committed $800,000 per year for at least three years for agricultural research and education programs, which could include research and extension programs on composting, analysis of the pilot transport program, animal nutrition management, development of a phosphorus index, and phosphorus dynamics in soils (EPA 1999b).

for the firm in the form of reduced compliance costs. Thus, firms have an incentive to innovate and adopt new, more efficient techniques.

Although greater reliance on performance standards rather than technology standards should lead to lower costs for individual polluters, achieving an aggregate target level of water quality in a watershed involving multiple polluters at least cost is more complex. Each source must meet its required reduction at least cost and have the required reductions allocated efficiently across sources. The total cost of meeting an aggregate abatement target is minimized when the required reductions are allocated so that each source faces the same incremental cost from additional abatement (NRC 1993a). Unfortunately, if abatement costs differ across sources, this implies different required reductions for different sources. Such differential regulation can be both administratively complex and politically difficult to implement.

Who pays to compensate for environmental damages is another difficult issue. Under regulations, polluters pay only for the cost of complying with regulations and not for the damages that any remaining pollution causes. As a result, the price of their products does not reflect all the associated costs of production, including both market and non-market costs. With the product price artificially low, consumption of those products tends to be high. For example, if agricultural producers comply with regulations, but nutrient runoff still occurs, that runoff could still generate costs for society (e.g., increased eutrophication) that the farmer did not consider when making production and pricing decisions. If these costs were reflected in the product prices, prices would rise and the demand for those products would adjust to reflect the full cost of production. Thus, those who bear the environmental costs of the production would no longer be implicitly subsidizing consumers of agricultural products. However, higher agricultural prices could cause hardship both to marginal farmers who might be forced out of business and to low-income consumers. The use of regulation is consistent with the polluter pays principle, to the extent that polluters pay for compliance with the regulations. However, as noted earlier, they do not pay for any damages that result despite that compliance.

There have been numerous studies of the impact of nutrient-based regulation, particularly in the context of agriculture. The type of regulation (e.g., mandated reduction in excess application or limiting animal densities) strongly affects the burden and effectiveness of regulation (McSweeny and Shortle 1989). Many studies have found that regulation is more efficient when aimed at areas or farms with the greatest pollution contributions, but this increases the administrative cost of regulations substantially, and these administrative costs may outweigh the efficiency gains from varying regulations across sources according to their pollution contributions (Mapp et al. 1994; Moxey and White 1994; Carpentier et al. 1998; Yiridoe and Weersink 1998).

Taxes and Fees

In contrast to regulatory approaches, which mandate certain changes in behavior, taxes and fees (negative economic incentives) are designed to induce (rather than force) those changes using financial incentives. They take a variety of forms, including effluent charges, user charges, product charges, administrative charges, tax differentiation, non-compliance fees, performance bonds, and legal liability for damages (NRC 1993a). For example, the state of Florida has a coastal protection tax of two cents per barrel that is charged for pollutants (petroleum products, pesticides, chlorine, and ammonia) produced in or imported into the state. The revenue from this tax goes to the Coastal Protection Trust Fund, which is used by the Florida Department of Environmental Resources for cleaning up spills (NRC 1997b).

In principle at least, taxes and fees are the negative counterpart to the positive incentives created by subsidies or cost sharing. With positive inducements, polluters receive payment for voluntarily undertaking desired behavior or investment. With negative inducements, they are forced to pay for undesirable behavior.

A common feature of economic incentives is that they put a price on environmental degradation. Whether in the form of forgone subsidy or explicit tax payment, polluters pay for "consuming" (or reducing) environmental quality just as they pay for the use of other inputs, such as labor and capital. Economic incentives thus put environmental inputs on a par with other inputs used in production. As with other inputs, polluters have an incentive to use environmental inputs only up to the point where the polluter's benefit from increased use equals the price the polluter must pay for that use. As that price rises, they face an increased incentive to reduce use of environmental inputs.

One of the main advantages of pollution taxes over regulatory policies is that they are generally thought to be more cost effective. Since polluters directly benefit from any cost savings, each polluter is encouraged to reduce its emissions in the least costly way. Polluters with low abatement costs have an incentive to reduce emissions more than high-cost polluters. As a result, the allocation of emission reductions will not be uniform across sources but will be more heavily borne by low-cost sources, as required for overall cost effectiveness. However, high-cost polluters will discharge relatively more and therefore bear larger total tax burdens. In addition, polluters will have an incentive to innovate, since they will benefit directly from any resulting cost savings.

Although, in principle, pollution taxes induce efficient abatement, their actual effectiveness is likely to be uncertain, at least initially. When setting a regulatory standard, authorities can be reasonably certain of the resulting level of emissions (assuming polluters comply). However, when setting a tax level, regulators often cannot predict with certainty how polluters will react and what the resulting level of environmental quality will be. While the level of the instrument can be adjusted over time to ensure that targets are met, such adjustments can be costly and can generate strategic behavior by polluters (Livernois and Karp 1994). In addition, it may be costly to adjust the level of the instrument in response to changes over time in economic conditions and in the demand for improvements in environmental quality.

Under tax-based instruments, polluters pay not only for the costs of any abatement undertaken but also for the remaining discharges. The resulting cost allocation is hence consistent with the polluter pays principle. Because polluters have to pay both the costs of abatement and the tax, the total cost to polluters is higher under a tax policy than it would be under a regulation leading to the same level of total discharges. While this ensures that product prices reflect the full social cost of production, the total cost may create considerable hardship both for marginal firms and low-income consumers who would be hard hit by the associated price increases. This is particularly true when, in many cases, relatively high taxes would be required to induce the desired change in behavior.

The magnitude of the tax increase that would be needed to induce a reduction in discharge depends on how responsive polluters are to the tax. Numerous studies have shown that, in the case of agricultural fertilizers, farmers are not very responsive to price increases (i.e., the demand for fertilizers is generally inelastic) and hence that relatively large tax rates would be needed to ensure that environmental objectives are met (McSweeny and Shortle 1989; Heatwole et al. 1990; Johnson et al. 1991; Helfand and House 1994; Pan and Hodge 1994; Weersink et al. 1998). For example, a simulation done by Giraldez and Fox (1995) found that an ad valorem tax rate of 55 percent would have to be applied to nitrogen to induce farmers to reduce nitrogen use to satisfy drinking water standards. Even though, from an economic efficiency perspective, taxes are desirable instruments for reducing nitrogen application, the large impact on farmers may limit their appeal (Moxey and White 1994; Helfand and House 1995). Similarly, while gasoline taxes may promote reductions in fuel use and hence emissions of NO_x, gasoline demand tends to be relatively unresponsive to price increases and hence large tax increases would be needed to have a significant impact on fuel consumption, particularly in the short run (Espey 1998). Such tax increases are generally viewed as politically unappealing.

Marketable Permits

Some of the shortcomings of both regulatory and tax-based approaches to environmental protection can be overcome with the use of marketable permits (Tietenberg 1985b). A marketable permit system starts with an allocation of allowable emissions across sources, as under a regulatory approach. However, by allowing sources to trade their allocations, the final allocation (after all trades have occurred) will be such that low-cost avoiders will undertake more abatement than high-cost avoiders, and the aggregate emission reduction will have been achieved at least cost. Low-cost avoiders will have an incentive to reduce discharges below their allocation and will sell their excess permits on the market. Similarly, high-cost avoiders can buy additional permits rather than incur their high costs of pollution control. If there are a sufficient number of buyers and sellers, the resulting market for permits establishes a price for emissions that reflects the total allowable emissions (i.e., the supply of permits) and the costs of pollution abatement for all polluters (i.e., the demand for permits). Unlike other economic incentives that also establish a price, the total impact of marketable permits on environmental quality is known since the total number of permits is fixed. Thus, the use of marketable permits combines the certainty of the regulatory approach with the cost effectiveness of economic incentives.

The use of marketable permits also allows a regulator to achieve any desired distribution of total costs by altering the way permits are allocated initially. Economic growth is possible without changing the total level of emissions, because new firms can simply be required to purchase permits from the market. The result is an increase in permit prices, but no increase in aggregate emissions.

Numerous economic studies have shown the potential for cost savings when polluters are allowed to trade pollution permits (Tietenberg 1985b; Klaassen 1996). The success of the sulfur dioxide emissions-trading program established under the 1990 Clean Air Act amendments (for example, Joskow et al. 1998) has heightened interest in this efficient pollution control tool. While this trading program was targeted toward sulfur dioxide emissions, to the extent that it is coupled with overall reductions in fossil fuel consumption, it would also help promote reductions in NO_x and the associated atmospheric deposition of nitrogen. Consideration has also been given to the use of trading programs for surface sources of water pollution, including trading between point source and nonpoint source. Such trading allows point sources to sponsor implementation of nonpoint source controls rather than further cutting back on their own emissions. Assuming nonpoint source loadings are significant and the marginal costs of nonpoint source reductions are lower than the costs of additional point source pollution controls, ambient water quality goals could be met at a lower cost by substituting nonpoint source for point source reductions (Crutchfield et al. 1994). Note, however, that the trading of point and nonpoint loadings requires the establishment of an appropriate trading ratio, as well as a means of quantifying the diffuse nonpoint loadings.

There have been several studies of the issue of pollution abatement trading between point and nonpoint sources. Letson (1992) has provided an economic analysis of the issue, illustrating the appeal as well as the difficulties in application of such a policy. Among the difficulties cited by Letson are monitoring, use of market power to manipulate permit price, and the distribution of the financial burden of loadings reductions. In addition, the rate at which nonpoint source abatement can be substituted for point source abatement must be established. The appropriate value of this trading ratio is uncertain because of qualitative differences between the two classes of sources. The optimal trading ratio will depend on the relative costs of enforcing point and nonpoint reductions and on the uncertainty associated with nonpoint loadings (Malik et al. 1993).

Crutchfield et al. (1994) isolated several practical circumstances that facilitate source abatement trading and developed an empirical protocol to determine the extent to which they exist in coastal watersheds. Their nationwide screening analysis was not designed to locate "good" candidates for trading programs. Their goal was to rule out many coastal watersheds, thus allowing researchers and planners to better focus their water quality efforts.

Several efforts are under way to implement point source and nonpoint source trading programs to improve water quality. Connecticut has applied one such program to Long Island Sound. A nitrogen-trading plan has been established to achieve reductions in nitrogen discharges cost effectively and expeditiously. The Connecticut Department of Environmental Protection anticipates that this plan will reduce the statewide bill for nitrogen removal by more that $200 million (CDEP 1998).

A similar program to limit phosphorous loads exists in the Cherry Creek basin in Denver, Colorado. The Cherry Creek Trading Program involves two types of trades: authority pool and in-kind trades. In authority pool trading, phosphorous reduction credits from Cherry Creek Basin Water Quality Authority projects are allocated to a trading pool. A qualified discharger may apply to the Authority for the purchase of trade credits from the trading pool for its wastewater treatment plant. For in-kind trades, non-Authority owners of independent nonpoint source pollutant reduction facilities receive credits to be used for their own wasteload allocation or to be transferred to a wastewater treatment facility (Sandquist and Paulson 1998).

The Tampa Bay National Estuary Program uses a cooperative approach that resembles a watershed trading program. No actual trades take place, but some sources make pollutant load reductions that they otherwise would not have been required to make in order to offset increases occurring at other sources. This approach to watershed management may be applicable to areas where trading is technically or politically inappropriate or unnecessary (Bacon and Greening 1998).

STEPS IN DEVELOPING EFFECTIVE WATER QUALITY GOALS

The first step in designing an effective approach to combat the effects of nutrient over-enrichment is to understand the physical and ecological processes at work. Next, decision-makers at local or regional levels must outline clear management goals, set specific targets to achieve the overall goals, and develop a range of possible policy approaches or management tools that are suitable to the site and its problems. Targets can be based on various measures or indicators of nutrient over-enrichment or estuarine health and can take the form of general water quality criteria or more specific water quality standards. Federal development of many of the resources and research efforts called for in Chapter 2 would greatly facilitate these efforts.

Once water quality goals or targets are set, managers must choose among a variety of policy approaches or management tools. To make efficient use of available resources, managers should strive to adopt policies that ensure that targets will be met at the lowest possible cost. In many cases, control costs will vary across sources and, if equally effective, the total cost of meeting the target will be lowest if the lowest cost sources are controlled first. Thus, when control costs vary, managers should not seek to achieve uniform reductions across all sources. Rather, they should target first the sources where reductions can be made at relatively low cost.

In designing policies to achieve these reductions, a fundamental choice must be made between the use of a voluntary approach and the use of mandatory controls or financial penalties. There are advantages and disadvantages of each approach, and managers must assess how successful a given approach is likely to be in their specific context. In many instances, managers may find that a well-formulated mix of incentives (voluntary approaches) and disincentives (mandatory or punitive approaches) works better than either approach would work alone.

Voluntary approaches that rely on moral persuasion, information, technical assistance, and possibly financial subsidies can be effective if there are sufficiently strong incentives for participation. While participation can be increased through financial incentives, local managers are not likely to have the local resources to finance subsidies for participation. In addition, even if financial incentives were available, subsidies of this type are generally inefficient because of both the need to raise the funds to finance the subsidy and the inefficient product prices that result. Thus, voluntary approaches at the local level will generally have to rely on other participation incentives (e.g., appealing to local commitment to water quality improvement). Managers considering reliance on a voluntary approach should evaluate how likely it is that people would participate in the program, since this will be a key determinant of the effectiveness of a purely voluntary approach. Without these incentives, a purely voluntary approach may not provide sufficient protection. Managers may be able to increase the likely success of a voluntary approach by making it clear that, if the voluntary approach does not appear to be working, an approach based on regulation and/or financial penalties will be adopted.

A mandatory approach based on regulations or taxes places a greater burden on the pollution sources, but if compliance can be ensured, it can be more effective in achieving water quality goals than a purely voluntary approach. However, when the costs of control vary across sources, uniform regulations will not meet those targets at the lowest possible cost. Cost-effective reduction can be achieved by allowing loading allocations to be traded. For example, allowing trades between point and nonpoint sources can generate significant cost reductions. In addition, managers can distribute initial permits in a variety of ways (e.g., uniformly across sources) without affecting the cost-effectiveness of the program. Marketable permit systems can, however, involve substantial administrative and information costs, and they may not work well if the number of sources that could participate in the permit market is small. The likely gains (in the form of cost reductions) must be weighed against the likely costs of using such a system. A careful examination of the effectiveness of trading in settings already employing it should be undertaken so managers have a better understanding of when this approach should be used.

The choice among alternative policy instruments will depend on the nature of the available control options and the characteristics of the watershed. For example, for estuaries where a primary nutrient source is agricultural production within the watershed, managers can choose among policy instruments designed to reduce nutrient runoff, such as the provision of technical or financial assistance for the adoption of best management practices (see Chapter 9), regulations requiring adoption of those practices, mandatory soil testing, taxes on fertilizer use, and land use taxes. Alternatively, if a primary nutrient source is atmospheric deposition, local managers will need to work with regional or federal officials to develop strategies that reduce nutrient inputs to the estuary. While the same types of policy options exist (e.g., technical or financial assistance for energy conservation or other reductions in NO_x emissions, regulations limiting allowable emissions, energy use taxes, and emission taxes), these policies would generally have to be implemented regionally or nationally to combat atmospheric sources of nutrients.

Because the appropriate choice of both the water quality target and the choice of policies to achieve that target are site-specific, a national recommendation regarding policy design is inappropriate. However, as part of a national strategy aimed at helping local managers reduce nutrient over-enrichment, the Committee recommends that a web-based clearinghouse for information relating to nutrient over-enrichment be developed. One component of that clearinghouse should be the compilation of three types of information that would aid local managers in developing nutrient management strategies that are appropriate for their estuaries:

- The first type of information would be a summary of and guide to research on the economic impacts of alternative source reduction methods, with particular emphasis on the role of site-specific characteristics in determining those impacts. This information would allow a local manager to determine which source reduction methods are likely to be more effective and cost-efficient, given the characteristics of the watershed and estuary of concern. For example, a manager of a local estuary with excessive nutrient inputs from agriculture could find information on the cost and effectiveness of various agricultural best management practices (see Chapter 9).
- The second type of information would be a summary of and guide to research on the effectiveness of alternative policies in achieving the most effective forms of source reduction, again given local circumstances. For example, if particular best management practices are identified as effective for a given watershed, this second type of information would provide a manager with information on the likely effectiveness of alternative policies in promoting increased adoption of those practices.
- Thirdly, the clearinghouse should contain documented case studies of both successful and unsuccessful attempts by local managers to combat nutrient over-enrichment in different types of estuaries. This would include not only attempts based on local policy implementation, but also documentation of attempts by local managers to work with regional or federal officials to combat nutrient loadings that originate outside the watershed, such as those from atmospheric sources.

By providing meaningful and easily understood information about both the results of scientific research on source control and policy design, and on what has or has not worked in practice in different settings, local managers can increase their understanding of the likely effectiveness of alternative policies and hence make informed decisions about which policy approaches are most appropriate for them.

9

Source Reduction and Control

KEY POINTS IN CHAPTER 9

This chapter reviews what is known about management options for reducing nutrient supply to coastal environments. It finds:

- Nutrient loads to coastal areas can be reduced by a variety of means, including improvements in agricultural practices, reductions in atmospheric sources of nitrogen, improvements in treatment of municipal wastewater (including tertiary treatment in some cases), and better control of diffuse urban nutrient sources such as runoff from streets and storm sewers (including both structural and passive controls). Regional stormwater control facilities, use of wetlands as nutrient sinks, better forest management to limit nitrogen export, enhancement of circulation in coastal waterways, and biological treatment also offer promise in some settings.
- Options to minimize nutrient export from agricultural areas include manure management strategies, careful estimation of native nutrient availability and crop requirements, and supplemental fertilizer application timed to meet crop demand. Watershed-scale implementation of best management practices needs to be targeted to ensure maximum reduction in nitrogen and phosphorus export. Post-implementation monitoring should be done to assess effectiveness.
- Lasting reductions in nutrient export from agriculture can be encouraged by focusing on consumer-driven programs and education, as well as on-farm production. Farmers' decisions are often influenced by regional or even global economics. At these scales, farmers have little or no control over these economic pressures and the resulting changes in nutrient flows and distribution. New ways of using incentives to help farmers implement innovative source reduction and control are needed.

- A positive side effect from regulatory initiatives to reduce NO_x emissions, targeted to minimize ozone and acid rain, is a reduction in the atmospheric contribution to nutrient loading in estuaries. The need to minimize coastal eutrophication should be a component of air pollution control strategies. Unfortunately, current NO_x emission efforts are aimed principally at control during the summer because of emphasis on ozone and smog formation; for eutrophication, year-round emission controls are necessary.
- A wide variety of methods, with variable effectiveness, are available to reduce urban point and nonpoint sourced nutrients. Natural options (enhancement of coastal wetlands) are one of a range of management tools.

Many factors contribute to nutrient over-enrichment, and thus there are many avenues by which the associated loads might be reduced. The effectiveness of any method depends, in part, on how large a contribution the source in question makes: minor improvements to major sources can sometimes offer more overall improvement than eliminating some minor nutrient source. Accurate information about relative contributions is essential if policymakers are to prioritize control efforts. Again, the federal actions called for in Chapter 2 would greatly strengthen efforts by local, state, and regional decisionmakers to successfully prioritize control efforts.

Nutrient over-enrichment in coastal waters is inextricably linked to human activities within estuarine areas as well as upstream, which in turn are tied to management and policy decisions. Conversely, physical, chemical, and biological impacts can be reduced by more effective control of anthropogenic inputs to the watershed, for instance by reducing loadings from agricultural, urban, or atmospheric sources. Figure 9-1 illustrates the significant effects that changes in tillage practices can have on nitrogen and phosphorus in a watershed. This chapter explores management strategies designed to reduce nutrient inputs. Because agricultural runoff is one of the greatest challenges in nutrient control, considerable attention is focused on control of agricultural sources, followed by control of atmospheric sources, urban sources, and control by other mechanisms.

AGRICULTURAL SOURCES

The goal of efforts to reduce nitrogen and phosphorus loss from agriculture to water is to increase nutrient use-efficiency. To do this, farmers attempt to balance the input of nutrients into a watershed from feed and fertilizer with outputs in crop and livestock produce, and also to manage the level of nitrogen and phosphorus in the soil. Reducing nutrient loss in agricultural runoff can be achieved by both source and transport control measures (Table 9-1). In general, there are reliable ways to reduce the transport of sediment-bound phosphorus from agricultural land by controlling erosion, and, to a lesser extent, there are methods to control nitro-

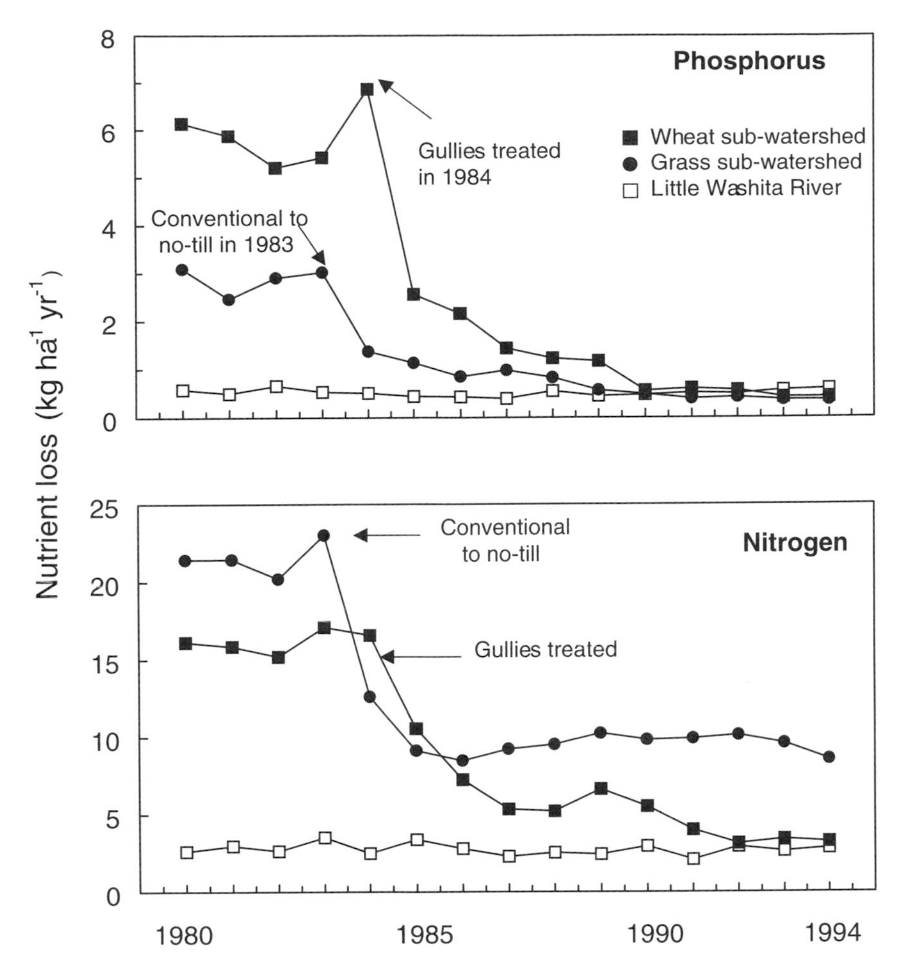

FIGURE 9-1 Annual nitrogen and phosphorus loss into the Little Washita River from a wheat-dominant and grass-dominant sub-watershed. Note the reduction in both nutrients after the eroding gullies in the grass watershed were treated in 1984 and conventional tillage was replaced with no-till in 1983 in the watershed growing wheat (modified from Sharpley and Smith 1994; Sharpley et al. 1996).

gen leaching to groundwater and the transport of dissolved phosphorus in runoff. However, less attention has been directed toward source management of nutrients because controlling nutrients at the source typically requires significant extra labor and thus is an economic burden to the farmer.

TABLE 9-1

Practice	Description
Source Measures	
Feed additives	Enzymes increase nutrient utilization by animals
Crop genetics	Low phytic-acid corn reduces phosphorus in manure
Manure management	Compost, lagoons, pond storage; barnyard runoff control; transport excess out of watershed
Rate added	Match crop needs
Timing of application	Avoid autumn and winter application
Method of application	Incorporated, banded, or injected in soil
Crop rotation	Sequence different rooting depths
Manure amendment	Aluminum reduces NH_3 loss and phosphorus solubility
Soil amendment	Flyash, iron oxides, gypsum reduce phosphorus solubility
Cover crop and residues	If harvested can reduce residual soil nutrients
Plowing stratified soils	Redistribution of surface phosphorus through profile
Transport Measures	
Cultivation timing	Not having soil bare during winter
Conservation tillage	Reduced and no-till increases infiltration and reduces soil erosion
Grazing management	Stream exclusion, avoid overstocking
Buffer, riparian, wetland areas, grassed waterways	Removes sediment-bound nutrients, enhances denitrification
Soil drainage	Tiles and ditches enhance water removal and reduce erosion
Strip cropping, contour plowing, terraces	Reduces transport of sediment-bound nutrients
Sediment delivery structures	Stream bank protection and stabilization, sedimentation pond
Critical source area treatment	Target sources of nutrients in a watershed for remediation

TABLE 9-1 Best management practices designed to control nonpoint sources of agricultural nutrients (unpublished table from A. Sharpley).

Source Management

Animal Feeding of Nitrogen and Phosphorus

Both nitrogen and phosphorus are important dietary nutrients for animals and have a key role in various metabolic functions (NRC 1989). Most feedstuffs do not contain adequate nitrogen and phosphorus to meet the needs of growing animals; thus additional nutrient supplements are brought onto the farm. The nutritional goal is to feed adequate nitrogen and phosphorus to meet the animal's requirements while minimizing excretion. Recommended dietary requirements of nitrogen and phosphorus have been established by the National Research Council (NRC) and are routinely updated (e.g., NRC 1989). Although studies show these requirements to be accurate, many farms routinely over feed nitrogen and phosphorus (Shaver and Howard 1995; Wu and Satter 1998). Because about 70 percent of the nitrogen and phosphorus in feeds is excreted, routine overfeeding of nitrogen and phosphorus animals contributes to nutrient surpluses on farms (Isserman 1990; Morse et al. 1992; Wadman et al. 1987).

It is common to supplement poultry and pig feed with mineral forms of phosphorus because of the low digestibility of phytin, the major phosphorus compound in grain. This supplementation contributes to phosphorus enrichment of poultry manures and litters. Enzyme additives for livestock feed that will increase the efficiency of uptake from grain during digestion are now being tested. Development of such enzymes would be cost-effective in terms of livestock weight gain and it is hoped that lowering mineral phosphorus supplementation of feed would reduce the phosphorus content of manure. One example is the use of phytase, an enzyme that allows the digestive systems of chickens and hogs to absorb phosphorus from grains. Ertl et al. (1998) showed a 23 percent reduction in excretion of phosphorus by poultry fed "low-phytic acid" grain compared to those fed normal corn grain.

Another approach to balance farm phosphorus inputs and outputs is to increase the quantity of phosphorus in corn that is available to poultry and pigs. Corn can be genetically engineered to decrease unavailable phytate-phosphorus, which contributes as much as 85 percent of phosphorus in corn grain. Ertl et al. (1998) manipulated the genes controlling phytate formation in corn and showed that phytate-phosphorus concentrations in "low-phytic acid" corn grain were as much as 51 percent less than in normal grain. There was a 23 percent reduction in excretion of phosphorus by poultry fed the "low-phytic acid" grain compared to those fed the "wild type" corn grain. Thus, the use of low-phytate corn in poultry and pig feed can increase the assimilation of phosphorus and other phytate-bound minerals and proteins.

Reducing Off-Farm Inputs of Nitrogen and Phosphorus

The accumulation of nutrients on many animal feeding operations where on-farm crop production is supplemented by feed inputs is generally not as great as in other operations where the animals depend primarily on off-farm feed. The distinguishing feature among these animal operations is the breakdown between the amount of crops produced on a farm (and the potential nutrient utilization by those crops) and the animal numbers on the farm, because the manure applied to crop fields will include both nutrients coming from those fields and from imports of off-farm feeds. The application of imported nutrients to crop fields can compensate for those lost in marketed products and manure handling operations and help to sustain the farm, but the additional nutrients can also be the source of excess nutrient loading. Nevertheless, restricting feed purchases to maintain the balance of nutrients can limit herd size and economic return if all manure from the herd must be applied to the farm cropland (Westphal et al. 1989).

Improving Nutrient-Use Efficiency

Management practices that improve nutrient-use efficiency are vital for minimizing losses to ground and surface waters. Specific best management practices (BMPs) for nitrogen and phosphorus vary from region to region due to large-scale differences such as climate, geology, depth to water, and irrigation or drainage practices, and also due to small-scale differences such as soils, cropping systems, and past field history. Therefore, BMPs for nitrogen and phosphorus will necessarily be site-specific and should be prescribed by a technical advisor who has a good knowledge of local nutrient cycles. This task often is assigned to state and federal extension agencies or soil conservation agencies.

Nitrogen

Nitrogen sources and reduction control strategies for Mississippi River Basin inputs to the Gulf of Mexico have been outlined by the National Oceanic and Atmospheric Administration (NOAA 1999b). Even though the major inputs of nitrogen and phosphorus to agricultural systems in this region are via fertilizer rather than manure, the principles of increasing nutrient-use efficiency are appropriate to other regions of the United States. Loadings to surface waters in the Upper Mississippi River and Ohio River sub-basins occur primarily by infiltration of water beyond the crop rooting zone into deeper soil layers, where it is collected by subsurface tile drains. In other basins, the primary pathway for nitrate loading to surface waters is groundwater seepage and irrigation return flow. Reduction of nitrate loading to surface waters in the Mississippi River basin can be achieved by reducing nitrate sources and controlling drainage (NOAA 1999b).

Although the selection of BMPs for nitrogen must depend on the specific hydrologic setting, field, and source of nitrogen, there are some basic nitrogen management principles that apply if the goal is to minimize nitrogen losses to ground or surface water. The most fundamental principle is to supply only the nitrogen needed to meet the needs of the next crop, and to apply it in synchrony with crop use. Steps in applying this principle include:

- estimating the nitrogen requirement of the next crop (expected yield);
- evaluating nitrogen available from native sources (soil nitrogen mineralization, residual soil nitrate, irrigation water, etc.);
- subtracting the available native nitrogen from the crop nitrogen requirement to estimate supplemental nitrogen needs;
- determining the most appropriate source of supplemental nitrogen (manure, fertilizer, crop residues); and
- determining the most efficient and practical management practice for the specific source of supplemental nitrogen (rate, time, and placement of the nitrogen).

Crop nitrogen requirement

Selection of the expected yield goal is one of the most critical BMPs for nitrogen, because most fertilizer and/or manure application rates are based directly on anticipated yield. Several studies have shown that farmers, or those advising them, often have unrealistic yield expectations and that resultant over-fertilization with nitrogen can be directly related to long-term increases in groundwater nitrate.

The most direct way to integrate overall site-specific factors is to calculate the average yield of the specific soil-crop system over the past 3 to 5 years. One can then adjust the average yield for unusual conditions (eliminating unusually wet or dry years), for current conditions (stored soil moisture, planting date, tillage practices, etc.), or for new technologies (new varieties, new irrigation, etc.), and then calculate a final estimate of expected yield. In any case, it is important to base the estimated yield on "real world conditions" (i.e., actual field yields) to avoid excess nitrogen applications.

Native nitrogen availability

The second step is to evaluate nitrogen available from native sources (i.e., sources that are not directly manageable by the farmer). These sources include nitrogen present in the root-zone as inorganic nitrate, nitrogen released through organic matter decomposition (mineralized nitrogen), nitrogen contributed through water sources (irrigation), and nitrogen from atmospheric inputs. The most recent tools for including native sources of nitrogen are the pre-sidedress nitrate test and the leaf chlorophyll meter (Magdoff et al. 1984; Meisinger et al. 1992; Schepers et al. 1992). The chlorophyll meter essentially measures the "greenness" of a specific leaf, from which the need (or lack of need) for more fertilizer nitrogen or potential for over fertilization can be estimated.

Management of nitrogen source (rate, placement, and timing)

The above steps produce an estimate of the appropriate nitrogen rate for a realistic yield of the next crop, which is the basic principle behind efficient nitrogen use. The final step is to manage the selected nitrogen source in a manner to supply nitrogen in phase or balance with crop demand. For fertilizer nitrogen, this is a relatively easy task because it can be applied just before the period of rapid crop nitrogen uptake.

Applying nitrogen when needed does not itself ensure adequate control of losses. For instance, one study showed that losses from unfertilized cereal crops were not much less than losses where fertilizer had been applied at the economic optimum input, with both resulting in nitrate concentrations in excess of the European Community limit of 10 mg nitrate-nitrogen l^{-1} (Sharpley and Lord 1998). This is because nitrate leached during winter is largely derived from that mineralized by the soil during late summer and autumn, when crop uptake is small especially in arable systems. This mineralization is affected little by fertilizer inputs. It is only when inputs exceed crop uptake capacity (usually close to the economic optimum) that excess fertilizer nitrogen contributes directly to losses. The economic optimum application of nitrogen for most crops exceeds offtake, resulting in a small positive balance. Within grazing systems, nitrogen surpluses are often a large proportion of the total fertilizer input, because most of the nitrogen consumed in grazing is redeposited as urine and dung. This nitrogen is not efficiently recycled because some of it is not immediately available and it is deposited unevenly over the field. Thus for nitrogen, enforcing a balance of inputs against removals could seriously reduce productivity, cause significant economic penalties, and would not in itself solve the nitrate problem.

Specific improvements in management may include: (1) reducing rates of nitrogen fertilizer by following fertilizer guidelines developed by land-grant universities, (2) switching from fall to spring or split applications, (3) changing the form of fertilizer nitrogen from anhydrous ammonia to slow-release urea fertilizers, (4) switching from broadcast to banded or incorporated application methods, (5) calibrating fertilizer application equipment, and (6) applying nitrification inhibitors (CENR 1997).

Phosphorus

The long-term use of commercial fertilizers and manures has increased the phosphorus status of many agricultural soils to optimum or excessive levels. This, of course, was the intended goal of phosphorus fertilization, to remove soil phosphorus supply as a limitation to agricultural productivity. However, for many years actions taken to achieve this goal did not consider the environmental consequences of phosphorus loss from soil to water. The constraint on phosphorus buildup in soils from commercial fertilizer use was usually economic, with most farmers recognizing that soil tests for phosphorus were an accurate indicator of when to stop applying fertilizer phosphorus. Some "insurance" fertilization has always occurred, particularly in high value crops, such as vegetables, tobacco, and sugar cane. However, the use of commercial fertilizers alone would not be expected to grossly over fertilize soils because farmers would cease applying fertilizer phosphorus when it became unprofitable. Today's concerns with phosphorus are caused by the realization that soils that are considered "optimum" in soil test phosphorus (or perhaps only slightly over fertilized) from a crop production perspective may still provide environmentally significant quantities of phosphorus in surface runoff and erosion.

Basing manure application on estimates of soil phosphorus and crop removal of phosphorus can reduce the buildup of soil phosphorus but can present several technical and economic problems to many farmers. A soil test phosphorus-based strategy could eliminate much of the land area with a history of continual manure application from further manure additions, as several years are required for significant depletion of high soil phosphorus levels. This would force farmers to identify larger areas of land to use the generated manure, further exacerbating the problem of local land area limitations. In addition, farmers relying on manure to supply most of their crop nitrogen requirements may be forced to buy fertilizer nitrogen to supplement foregone manure nitrogen.

As phosphorus is relatively immobile in soil compared to nitrogen, timing of application is less critical in BMP development for phosphorus than nitrogen. However, methods of phosphorus application are important. Rotational applications of phosphorus designed to streamline fertilizer operations may leave large amounts of available phosphorus in the surface, and should be avoided in areas of the landscape at risk of erosion or surface runoff. Efficient management of phosphorus amendments to soils susceptible to phosphorus loss involves the subsurface placement of fertilizer and manure away from the zone of removal in surface runoff, and the periodic plowing of no-till soils to redistribute surface phosphorus accumulations throughout the root zone. Both practices may indirectly reduce the loss of nitrogen and phosphorus by increasing crop uptake and yield, which affords a greater vegetative protection of surface soil from erosion. However, these measures are often unrealistic for a farmer to implement. For example, subsurface injection or incorporation in rocky soils may be difficult, and without manure storage, farmers who contract out the cleaning of poultry houses will have little flexibility for when manure or litter is applied.

Manure Management

As discussed in Chapter 5, animal wastes are a major part of the nutrient over-enrichment problem, and management efforts are complicated by the long distances that feedstocks are transported. Managing nutrients from manure is often more difficult than from fertilizer, due to uncertainties in initial composition (e.g., ration, animal age, etc.), losses during storage or handling (e.g., ammonia volatilization), uncertainties of application rates (e.g., uncalibrated spreaders, uneven applications), difficulty in spreading manure to a growing crop without causing crop damage, greater gaseous nitrogen losses with manure after application, and time pressures producers face because of weather uncertainties.

It is also important for manure management to know the approximate decomposition rate of the organic nitrogen, so as to minimize nitrogen loss in groundwater. This is generally estimated as a decay series for the particular type of manure. An example of a decay series for solid beef manure would be 40 percent mineralized the first year, 25 percent of the remaining nitrogen the next year, 6 percent the next year, and so on (Gilbertson et al. 1979).

The last step is to calibrate the manure spreader. Obviously it does little good to know the crop nitrogen need, the manure composition, the likely ammonia loss, and the decomposition rate, if one cannot apply the calculated rate of manure accurately. Manure spreader calibration programs in Maryland and Pennsylvania frequently find that farmers are applying two to five times more manure than they originally estimate. Educational materials for spreader calibration can significantly improve manure nitrogen utilization, and further improvements could be obtained with monitoring or incentives.

Farm advisors and resource planners now recommend testing manure for nitrogen and phosphorus, and soils for phosphorus, prior to land application of manure. However, nitrogen-based manure management plans are still based on crop needs. Without these determinations, farmers and their advisors can underestimate the fertilizer value of manure. Soil test results can also demonstrate the positive and negative long-term effects of manure use and the time required to build-up or deplete soil nutrients. For instance, they can help a farmer identify the soils in need of fertilization, those where moderate manure applications may be made, and those fields already containing excess nitrogen and phosphorus where manure should not be applied.

Commercially available manure amendments, such as slaked lime or alum, can help in manure management. Such amendments can decrease ammonia volatilization, which can significantly affect export to estuaries (Chapter 5), and at the same time lead to improved animal health and weight gains. Amendments can also decrease the solubility of phosphorus in poultry litter by several orders of magnitude and decrease dissolved phosphorus, metal, and hormone concentrations in surface runoff at least 10 fold (Moore and Miller 1994; Moore et al. 1995; Shreve et al. 1995; Nichols et al. 1997). Perhaps the most important benefit of manure amendments for both air and water quality would be an increase in the nitrogen:phosphorus ratio of manure, by reducing nitrogen loss because of ammonia volatilization. An increased nitrogen:phosphorus ratio of manure would more closely match crop nitrogen and phosphorus requirements.

One approach to better manure management would be to establish a mechanism to facilitate movement of manures from surplus to deficit areas. At present, manures are rarely transported more than 10 miles from where they are produced. But mandatory transport of manure from farms with surplus nutrients to neighboring farms where nutrients are needed would face several significant obstacles. First, it must be shown that manure-rich farms are unsuitable for manure application, based on soil properties, crop nutrient requirements, hydrology, actual nutrient movement, and proximity of sensitive water resources. Second, it must be shown that the recipient farms are more suitable for manure application. The greatest success with re-distribution of manure nutrients is likely to occur when the general goals of nutrient management set by a national (or state) government are supported by consumers, local governments, the farm community, and the livestock industry involved. This may initially require incentives to facilitate subsequent transport of manures from one area to another.

This may be a short-term alternative if nitrogen-based management is used to apply the transported manures. If this happens, soil phosphorus in areas receiving manures eventually may become "excessive." To date, however, large-scale transportation of manure from producing to non-manure producing areas is not occurring. The main reasons for this are the high transportation costs and concern that avian diseases will be transferred from one farm (or region) to the next. Consequently, there is a need to develop a means to ensure the biosecurity of any manure transportation network that is developed, and in general to seek ways to overcome existing obstacles to better manure management (e.g., incentives and disincentives).

Composting is another potential management tool to improve manure distribution. Although composting tends to increase the phosphorus concentration of manures, the volume is reduced and thus transportation costs are reduced. Additional markets are also available for composted materials. Finally, there is interest in using some manures as sources of bioenergy. For example, dried poultry litter can be burned directly or converted by pyrolytic methods into oils suitable for use to generate electric power. Liquid wastes can be digested anaerobically to produce methane that can be used for heat and energy. As the value of clean water and cost of sustainable manure management is realized, it is expected that alternative entrepreneurial uses for manure will be developed, become more cost-effective, and, thus, create expanding markets. Research is needed to speed the development of these types of technologies and approaches.

Transport Management

Once water and sediment begin to move in the landscape, taking with them the phosphorus originally applied as fertilizer and/or manure, the quantities that reach the stream can be reduced by any feature that slows flow and/or encourages infiltration or sediment trapping. Such transport management measures include terracing, contour tillage, cover crops, buffer strips, riparian zones, and impoundments. These transport measures are generally more efficient at reducing particulate phosphorus rather than dissolved phosphorus. However, such approaches only work where subsurface pathways of phosphorus loss are unimportant. Furthermore, by encouraging infiltration of surface runoff, which may be enriched with phosphorus, the problem is simply translated from surface to subsurface delivery. While uptake by plant roots and adsorption onto soil particles may delay the delivery of phosphorus to surface waters, such mechanisms may be ineffective in soils with a high hydraulic conductivity (e.g., sands) or where macropore or drainflow is important.

For nitrogen, losses can also be reduced by improved water management, including adoption of controlled drainage or sub-irrigation methods, switching from furrow irrigation to surge irrigation or sprinkler irrigation with fertigation, and the use of irrigation scheduling techniques (Skaggs and Gilliam 1981). Nitrate losses can also be reduced by control of water table depth by managing tile drain spacing and depth and by control structures on the tile drain outlets, to limit tile flow when the potential for nitrate may be greatest (Gilliam et al. 1979; Kladivko et al. 1991; Zucker and Brown 1998). Moving the runoff through anoxic sediments (e.g., in wetlands) can also help remove nitrogen through denitrification.

Careful selection of the type and sequencing of crops in a rotation, taking into account the timing and position in the soil profile of residual nitrate, rooting depth, and soil-water movement, can maximize nitrogen-use efficiency and minimize nitrate leaching potential (Sharpley et al. 1992). Crop cover during the period of agricultural runoff will therefore reduce losses of both nitrogen and phosphorus. For nitrogen, the critical period is autumn, to allow plants to take up nitrogen mineralized in soils after harvest. In temperate regions, autumn-sown cereals usually do not take up much nitrogen before winter. However cover crops established immediately after harvest (and killed in winter or early spring) can be highly effective (Shepherd and Lord 1996). Data from farms within the United Kingdom Nitrate Sensitive Areas Scheme show that such crops are compatible with commercial farming systems, relatively inexpensive to manage, and can reduce nitrate losses by about 50 percent (Sharpley and Lord 1998).

For phosphorus, crop cover at any time during the period when agricultural runoff can occur will help protect against total phosphorus loss but will be less effective against losses of dissolved phosphorus (Sharpley and Smith 1991). Crop residues can be as effective as crop cover in reducing erosion or surface runoff, and hence protecting against phosphorus loss. Equally, anything that keeps the surface rough, such as plowing, can be effective. However creation of fine seedbeds, as are required for winter cereals, has been shown to increase erosion. This is especially the case where cultivations are up-and-down slopes.

Cultivation may promote mineralization of nitrate, especially where plant residues with high nitrogen content are present. Thus, delaying autumn plowing can reduce leaching losses. Cereal straw, conversely, because of its high carbon:nitrogen ratio, may actually reduce leaching losses slightly by using soil nitrate in the early stages of decomposition.

Riparian zones play an important role in reducing non-point sources of nitrogen and phosphorus, can increase wildlife diversity and numbers, and improve aquatic habitat and diversity via shading. In addition to acting as physical buffers to sediment-bound nutrients, plant uptake captures nitrogen and phosphorus, resulting in a short-term accumulation of nutrients in non-woody biomass as well as a long-term accumulation in woody biomass (Peterjohn and Correll 1984; Fail et al. 1986; Correll and Weller 1989; Groffman et al. 1992). Even more importantly, denitrification of nitrogen in riparian zones is a significant mechanism for decreasing nitrogen (Jacobs and Gilliam 1985; Pinay et al. 1993). Denitrification rates of 30 to 40 kg N ha^{-1} y^{-1} have been measured for natural riparian forests in the United States. Most denitrification occurs in the top 12 to 15 cm of the soil. Within a riparian zone, the fastest rates occur at the riparian-stream boundary where nitrate enriched water enters organic surface soil (Cooper 1990).

The effectiveness of riparian zones as nutrient buffers can vary significantly. For instance, the route and depth of subsurface water flow paths though riparian areas can influence nutrient retention. Riparian zones are most efficient when sheet flow occurs, rather than channelized flow. The key to successful denitrification in riparian wetlands is for the water to flow (not too quickly) through the surface layer of waterlogged, wetland soils, where denitrification is fed by the high input of organic carbon from the wetland plants and where oxygen is low to zero.

In several locations of the coastal plain of the Chesapeake Bay watershed, average annual terrestrial boundary nitrate concentrations of 7 to 14 mg NO_3-N l^{-1} decrease to 1 mg NO_3-N l^{-1} or less in shallow groundwater near streams (Lowrance et al. 1995). However, in the same area, a single site with a nitrate concentration of 25 mg NO_3-N l^{-1} at depth had a concentration of 18 mg NO_3-N l^{-1} in shallow groundwater at the stream. Lowrance et al. (1984b), who estimated annual denitrification rates to average 31 kg N ha^{-1} yr^{-1} in the top 50 cm of soil, measured denitrification rates between 1.4 kg N ha^{-1} yr^{-1} in a riparian zone adjacent to an old field (which received no fertilizer) to 295 kg N ha^{-1} yr^{-1} under conditions of high nitrogen and carbon subsidies. Such results illustrate the potential for denitrification in riparian zones, the high spatial variability that can be expected, as well as the importance of carefully managing riparian areas.

Usually, farm nitrogen inputs can be more easily balanced with plant uptake than can phosphorus, particularly where confined animal operations exist. In the past, separate BMP strategies for nitrogen and phosphorus have been developed and implemented at farm or watershed scales. Because of differing chemistry and flow pathways of nitrogen and phosphorus in soil and through the watershed, these narrowly targeted strategies often are in conflict and lead to compromised water quality remediation. For example, basing manure application on crop nitrogen requirements to minimize nitrate leaching to groundwater increases soil phosphorus and enhances potential phosphorus surface runoff losses. In contrast, reducing surface runoff losses of phosphorus via conservation tillage can enhance nitrate leaching (Sharpley and Smith 1994).

Nitrogen and phosphorus transport management strategies may differ because nitrogen losses can occur from any location in a watershed, while areas prone to surface runoff contribute most to phosphorus loss. Nitrogen also volatilizes to the atmosphere, whereas, phosphorus does not. Hence, remedial strategies for nitrogen may be applied to the whole watershed, whereas the most effective phosphorus strategy would be a combination of simple measures over the whole watershed to avoid excessive nutrient buildup, thereby limiting losses in subsurface flow, and more stringent measures to the most vulnerable sites to minimize loss of phosphorus in surface runoff. Thus, BMPs must consider both nitrogen and phosphorus sources and export pathways at farm and watershed scales.

Implementing Remedial Measures

Since the early 1980s, several studies have investigated the long-term (7 to 10 yr) effectiveness of BMPs to reduce nitrogen and phosphorus export from agricultural watersheds (National Water Quality Evaluation Project 1988; USDA and ASCS 1992; Goldstein and Ritter 1993; Richards and Baker 1993; Bottcher and Tremwell 1995). These studies quantified nutrient loss prior to and after BMP implementation or attempted to use untreated watersheds as control. Overall, these studies showed BMPs reduced nutrient export. However, it is evident that several factors are critical to effective BMP implementation. These factors include targeting watersheds that will respond most effectively to BMPs, identifying critical source areas of nutrient export, as well as accounting for both watershed and estuary response time and equilibration (capacity to buffer added nitrogen and phosphorus).

The time of watershed or estuary response to BMP implementation is particularly important for phosphorus, due to its long residence time in ecosystems, compared to nitrogen. Watersheds may become saturated with phosphorus where animal feeding operations are concentrated (Lander et al. 1997; USDA and EPA 1999). Studies have shown that even where phosphorus applications are stopped, elevated soil phosphorus can take up to 20 years to decline to levels at which crops will respond to applications (McCollum 1991). Also, internal recycling of phosphorus in estuarine sediments can supply sufficient phosphorus to maintain eutrophic conditions in phosphorus-sensitive waters.

Watershed Identification and Cost-Effectiveness

Because resources are limited, local decisionmakers often focus management efforts on those watersheds that will provide the greatest reduction in nitrogen and phosphorus loss following BMP implementation. Otherwise, overall nutrient inputs to an affected waterbody may not be decreased sufficiently. Similarly, farmers make decisions about on-farm management approaches based on cost-effectiveness, which is affected by many things.

Model simulation and field studies provide data illustrating that the cost-effectiveness of BMPs varies (Table 9-2). Although protection of

TABLE 9-2

Best Management Practice	Phosphorus Loss (kg ha^{-1} yr^{-1})	Cost-Effectiveness ($ kg P $saved^{-1}$)
None	10.0	
Contour cropping	6.3	1.7
Terraces	3.2	4.7
Conservation tillage	3.9	0.8
Vegetative buffer areas	2.5	1.1
Manure management	2.8	3.3
All BMPs	1.8	4.9

TABLE 9-2 Cost-effectiveness of BMPs for reducing phosphorus losses from continuous corn with a 5 percent slope and 140 kg P ha^{-1} yr^{-1} manure broadcast. Conservation tillage achieves significant phosphorus control with good cost-effectiveness, while manure management is less cost-effective (modified from Meals 1990).

riparian areas with buffers and manure management can reduce runoff phosphorus more than tillage management, conservation tillage is often a more cost-effective measure. These generalized examples emphasize the need to determine the load reduction required for a given watershed and waterbody to select appropriate BMPs. Clearly, construction of terraces, which are initially expensive, may in some cases be a viable option. However, careful selection and integration of different practices can improve overall cost-effectiveness. Cost-effectiveness includes the cost of land taken out of production.

In another example, Meals (1990) evaluated the effect of several manure BMPs on phosphorus export from two watersheds in the LaPlatte River basin draining into Lake Champlain, Vermont. BMPs included barnyard runoff control, milkhouse waste treatment, and construction and use of manure storage facilities. Phosphorus losses were lower than before BMPs. For both watersheds, barnyard runoff control resulted in the greatest reduction in phosphorus export and was the most cost-effective BMP (Table 9-3). The results of this simple cost-effectiveness analysis have important implications for formulating remediation strategies. If a watershed project was being developed with limited funding, the cost-effectiveness analysis could help target a watershed that would provide the greatest impact for the money invested (Meals 1990). For instance, if a choice had to be made between the two Vermont watersheds shown in Table 9-3, watershed 1 would have been selected based on better cost-effectiveness ratios.

TABLE 9-3

	Watershed 1		Watershed 2	
Management	Phosphorus Reduction (kg)	Effectiveness ($ kg P^{-1})	Phosphorus Reduction (kg)	Effectiveness ($ kg P^{-1})
Barnyard runoff control	311	4	78	14
Milkhouse waste treatment	34	12	11	32
Waste storage facility	154	269	14	1,963
Total	567	77	103	282

TABLE 9-3 Comparison of cost-effectiveness of animal waste management BMPs for two watersheds in the LaPlatte River Basin project, Vermont, 1980 to 1989 (modified from Meals 1990). Simple cost-effectiveness analysis can be key to helping farmers implement strategies that contribute the greatest benefits to watershed protection, and for targeting actions in watersheds where they can have the most impact.

If regional assessments are to identify critical source areas within large regions, results obtained from experimental sites as well as model estimates have to be scaled up. The accuracy of regional estimates depends on how good our experimental results or models are and how reliable available regional data are describing the factors governing nutrient transport. In addition to regional assessments, models can be used to make comparative studies on the effectiveness of different remedial measures.

Targeting Within a Watershed

Once an area has been selected for remediation, the next step is selection of appropriate BMPs. Using cost-effectiveness ratios like those outlined in Table 9-3, BMP implementation can be prioritized. For the example of the two Vermont watersheds in Table 9-3, the most effective BMP installation priority would be barnyard runoff control, followed by milkhouse waste treatment, and then animal waste storage facilities. Without careful targeting of critical nutrient source controls within a watershed, BMPs may not produce the expected reductions in nitrogen and phosphorus export.

The importance of targeting BMPs within a watershed or basin is shown by several studies in the Little Washita River watershed (54,000 ha) in central Oklahoma (Sharpley and Smith 1994). Nutrient export from two subwatersheds (2 and 5 ha) were measured from 1980 to 1994, while

BMPs were installed on about 50 percent of the main watershed. Practices included construction of flood control impoundments, eroding gully treatment, and conservation tillage. Following conversion of conventional-till (moldboard and chisel plough) to no-till wheat in 1983, nitrogen export was reduced 14.5 kg ha^{-1} yr^{-1} (3 fold) and P loss 2.9 kg ha^{-1} yr^{-1} (10 fold) (Figure 9-1; Sharpley and Smith 1994). A year later, eroding gullies were shaped and an impoundment constructed in the other subwatershed. Both nitrogen and phosphorus loss decreased dramatically (5 and 13 fold, respectively) (Sharpley et al. 1996). There was no effect of BMP implementation, however, on phosphorus of nitrogen concentration in flow from the main Little Washita River watershed (Figure 9-1). A lack of effective targeting of BMPs and control of major sources of phosphorus and nitrogen export in the Little Washita River watershed contributed to no consistent reduction in watershed export of phosphorus or nitrogen.

Apparently, there is a minimum threshold level of implementation that must be achieved before a significant response to BMPs occurs. For instance, in the LaPlatte River Basin, Vermont example (Meals 1990), animal waste control measures were implemented during the early 1980s. These BMPs included control of barnyard runoff, milkhouse waste treatment, and construction of waste storage facilities. However, there was no apparent reduction in either dissolved or total phosphorus concentration in runoff with increasing percent of animals in a watershed under a BMP (dashed lines; Figure 9-2). If the runoff phosphorus values for watersheds where less than 50 percent of the animals were under BMPs are not considered, then both dissolved and total phosphorus in runoff were decreased significantly. The low values of implementation (less than 42 percent) represent the initial years of land treatment when BMP implementation was incomplete.

Selecting a Best Management Practice

The cost-effectiveness of BMPs for reducing nitrogen and phosphorus loss varies with both types of practice and among watersheds. Remediation strategies are ongoing processes, in which BMP selection and operation should be reevaluated regularly to optimize nutrient export reductions.

Research into the effectiveness of BMPs has shown that they can be successful in reducing overall loads, but will not necessarily be adequate during extreme snowmelt or rainfall events. For instance, Meals (1990) studied BMPs in an agricultural watershed leading to Lake Champlain, which is sensitive to phosphorus loadings. Surprisingly, this analysis showed an increase in phosphorus export from the Mud Hollow Brook watershed following BMP implementation (Figure 9-3). Further analysis

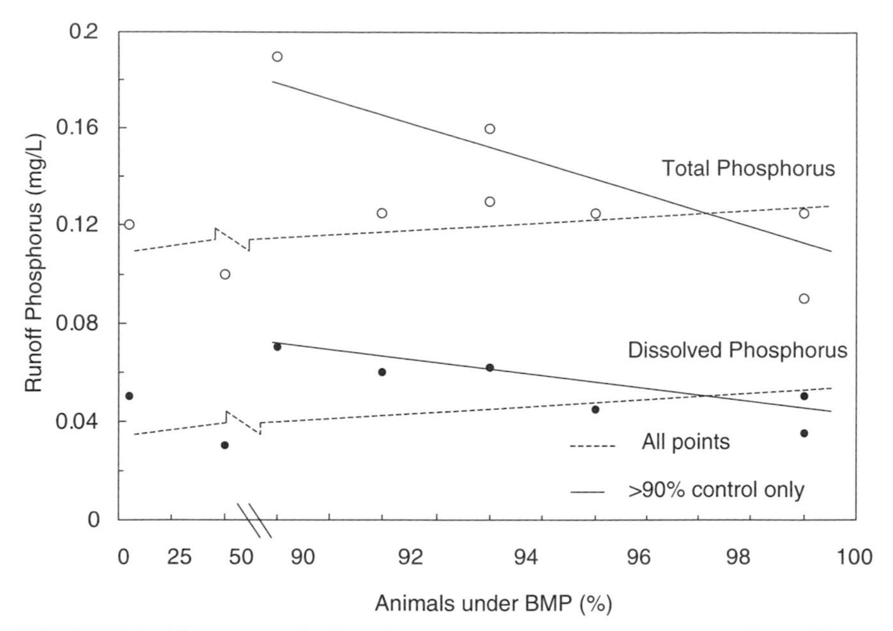

FIGURE 9-2 Mean annual phosphorus concentration in watershed runoff as a function of the percentage of watershed animals under BMPs in the LaPlatte River basin, Vermont (modified from Meals 1990).

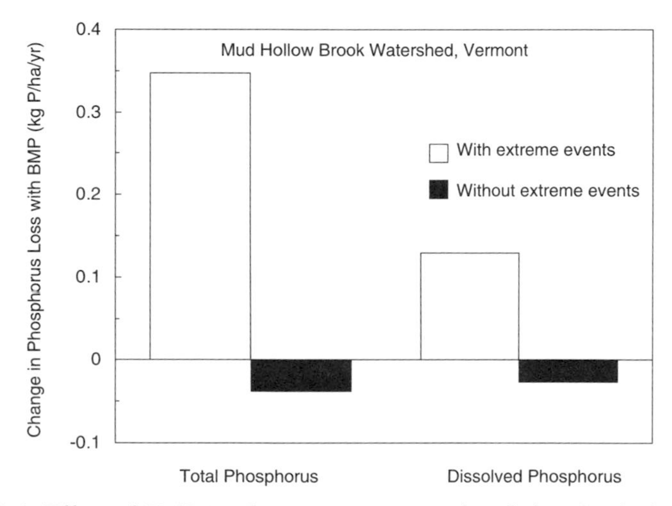

FIGURE 9-3 Effect of BMP implementation on total and dissolved phosphorus loss from Mud Hollow Brook watershed, Vermont, 1985 to 1989, with and without inclusion of extreme flow events. Positive values indicate an increase in phosphorus loss and negative values and decrease in phosphorus loss with BMP implementation (modified from Meals 1990).

revealed that annual phosphorus export was dominated by one or two extremely high flows (greater than the 95th percentile of all recorded stream flows) that were generally associated with snowmelt or intense rainfall events (Meals 1990). These few extreme events had a dramatic effect on the overall assessment of BMP impact on phosphorus export. When these events were not included in the analysis, BMPs reduced dissolved and total phosphorus export as anticipated. Effective remediation strategies should consider such extreme events in situations where they can dominate phosphorus export.

Incentives for BMP Adoption

As noted in Chapter 8, policies to promote source reduction can be based on either a voluntary approach or a regulatory approach. A voluntary approach has the advantage of promoting a more cooperative environment for encouraging adoption of, for example, BMPs. However, if the voluntary approach involves significant subsidies or cost-sharing to induce adoption, it will require the regulator or manager to raise funds to finance the subsidies and will generally result in product prices that do not fully reflect the total cost of production (including environmental costs). Nonetheless, voluntary approaches have been used successfully to promote source reduction and recent experience suggests these approaches could play an important role in reducing nutrient loadings.

There are several sources of technical assistance and financial programs to help defray the costs of constructing or implementing BMPs (EPA 1998e). Some of these sources are Conservation Technical Assistance, Conservation Reserve Program, Environmental Quality Incentives Program, Special Water Quality Incentives, Wetland Reserve Program, and Wildlife Habitat Incentive Program. Also, stakeholder alliances have been developed among state, federal, and local groups, producers, and the public to identify problems, focus resources, and implement BMPs in Chesapeake Bay and the New York City watershed, for example.

ATMOSPHERIC SOURCES

As discussed in Chapter 5, atmospheric deposition is a significant source of nitrogen loadings to some estuaries and other coastal waters, particularly in the northeastern United States. The deposition can be directly to the surface of the waterbody or onto the watershed with subsequent export to the estuary. This nitrogen deposition onto the watershed with export to downstream waters is more important than deposition directly onto the waterbody for many estuaries where the ratio of watershed area to estuary area is high. Both deposition of ammonia/

ammonium and of oxidized nitrogen compounds can be important. (NOTE: Deposited oxidized nitrogen compounds are denoted NO_y as opposed to NO_x, which denotes emitted compounds.) In the United States, most of the ammonium deposition comes from agricultural sources, particularly from animal wastes but also from volatilization of fertilizers, while the oxidized nitrogen comes principally from the combustion of fossil fuels.

For nitrogen deposition directly onto the water surface of an estuary, reductions in the sources of nitrogen to the atmosphere are the only possible approach to control. For nitrogen deposition onto the watershed, reductions in sources to the atmosphere may still be the easiest and most effective approach for control, but other options are possible, such as encouraging denitrification in riparian wetlands (discussed earlier in the context of controlling nitrogen runoff from agricultural fields), managing the composition of tree species in forests so as to reduce nitrogen export, and treating runoff from urban streets (where atmospheric nitrogen deposition is particularly high).

The reduction of atmospheric ammonium deposition requires better management of animal wastes to decrease the volatization of ammonia to the atmosphere, as is discussed earlier in this chapter. For the deposition of oxidized nitrogen compounds, the emission of nitrogen oxides (NO_x) must be controlled. In tropical areas, agricultural practices can also contribute greatly to the emissions of these compounds, but in the United States, most NO_x emissions originate from the burning of fossil fuels. These emissions originate from both point (stationary) and nonpoint (mobile) sources. The primary stationary sources are electric utilities and industrial facilities, while the primary mobile sources are motor vehicle emissions. Nationally, NO_x emissions rose rapidly with post-war economic activity. However, due primarily to regulation of NO_x emissions under the Clean Air Act, national emissions have been roughly constant since 1980 (EPA 1999c).

Reduction in atmospheric nitrogen deposition is directly related to reductions in NO_x emissions. NO_x emissions can be reduced either by burning less fossil fuel or by removing NO_x from the combustion exhaust. For mobile sources, reductions in fossil fuel consumption can result from a number of different mechanisms (e.g., Calvert et al. 1993; Krupnick 1993). One obvious mechanism is a reduction in the number of vehicle miles driven. Incentives to reduce mileage can be created through promotion of mass transit or car pooling. Individuals can be expected to make transportation mode choices by comparing a number of factors, including convenience and price. The more expensive car travel is, the less individuals are likely to use it. Thus, incentives for either reductions in overall travel or for switching to alternative transportation modes can be created through increases in the price of gasoline (Haughton and Sarkar 1996; Goldberg 1998). Gasoline prices can be increased through increases in gasoline taxes. However, the effectiveness of a tax depends on the responsiveness of demand to price increases. Empirical evidence suggests that demand is relatively unresponsive to price increases in the short run, although the responsiveness is greater in the long run (Espey 1998).[1] Thus, while gasoline taxes may generate significant revenue that can be used to finance pollution abatement, it is likely that the tax increase would have to be large in order to induce a significant change in driving behavior. Such a large price increase could have undesirable distributional consequences, since gasoline taxes are generally thought to be regressive (Chernick and Reschovsky 1997).

A second potential mechanism for reducing fuel consumption is increasing miles per gallon of fuel (i.e., increasing the fuel efficiency of cars). This is the aim behind the Corporate Average Fuel Efficiency (CAFE) standards (Goldberg 1998). Overall reductions in the cost of meeting these standards can be achieved by allowing firms to average emissions across their fleets or to trade emissions with other firms (Kling 1994).[2] A requirement that all vehicles meet a common emission standard in terms of grams per mile suggests that improvements in fuel economy per se may not generate reductions in vehicle emissions (NRC 1992). Although this is true for new cars (Khazzoom 1995), a recent study by Harrington (1997) finds that, as cars age and the emission control equipment breaks down, better fuel economy is strongly associated with low emissions, at least for CO and HC. If vehicle emissions in fact vary considerably by year, make and size (Kahn 1996), reductions in total emissions can be achieved by changing the composition of the existing automobile fleet. Incentives can be provided for early retirement of high-emission vehicles (Alberini et al. 1995, 1996). Reductions of this type may, however, be offset by emission increases due to the increased popularity of high-emission sport utility vehicles, which are currently subject to less stringent emission restrictions than passenger cars (0.7 grams per mile versus 0.4 grams per mile for passenger cars).

Reductions in fossil fuel consumption can also result from the development and use of vehicles that rely on alternative power sources, such as methanol (Krupnick and Walls 1992; Michaelis 1995; Hahn and Borick 1996; Kazimi 1997) or ethanol (Michaelis 1995; Rask 1998). Given current markets conditions, these alternatives do not appear to be very cost-

---

[1]The greater long run responsiveness is due to the impact that higher prices have on automobile purchases. As consumers replace their automobiles, higher gasoline prices can encourage them to purchase more fuel efficient cars.

[2]See Chapter 8 for a discussion of the economic implications of allowing firms to trade pollution allowances.

effective and hence are unlikely to be adopted voluntarily. If these sources are preferred from a social point of view, based on a full cost accounting that includes the environmental costs of alternative fuels, some mechanism for inducing private adoption would be required.

An alternative to reducing NO_x emissions through reductions in fuel consumption is to remove NO_x from vehicle exhaust through the use of catalytic converters. However, the continued effectiveness of these devices requires inspection and maintenance, and existing vehicle inspection programs have recently been subject to criticism for failure to provide consumers with incentives to maintain vehicles to minimize emissions (Hubbard, T. 1997, 1998).

For stationary sources, emissions from fossil fuel consumption can be reduced by reducing output (e.g., reducing the amount of electricity generated), or by switching to alternative fuels (e.g., nuclear or renewable sources such as hydroelectric power). To avoid shortages, reductions in electricity output would have to be accompanied by reductions in demand, either through price increases or other energy conservation measures. As with gasoline demand, the effectiveness of these programs depends on how responsive consumers are to electricity price increases and incentives for conservation (Silk and Joutz 1997; Wirl 1997). Alternatively, emissions can be reduced through the use of pollution control equipment such as flue-gas desulfurization equipment (scrubbers).

Since there are a number of different ways in which stationary or point sources of NO_x can be reduced, cost-effective policies must have the flexibility to allow firms to achieve reductions with the least cost. This type of flexibility is not provided when regulations mandate the installation of certain pollution control devices (e.g., scrubbers). As a result, technology standards of this type do not achieve emission reductions at least cost. Fullerton et al. (1997) estimated that the cost of compliance under a "forced scrubbing" policy is almost five times the minimum that is possible under the Clean Air Act Amendments. Increased flexibility can be provided through the use of emission taxes, performance standards, or tradable permits (Baumol and Oates 1988; Burtrow 1996; see related discussion in Chapter 8). However, in designing flexible policies, the impact of other market distortions, such as imperfect competition or public utility regulation, must be considered. Failure to account for these distortions can result in flexible policies that are actually less economically efficient than more rigid regulations (Besanko 1987; Fullerton et al. 1997).

To date, policies designed to reduce NO_x emissions have stemmed not from concerns about excess nutrient loadings to waterbodies, but rather from the other environmental impacts of NO_x emissions. For example, NO_x is a precursor to the formation of ground level ozone or photochemical smog (NRC 1991a; U.S. Congress 1991). Ozone is formed when nitrogen oxides and volatile hydrocarbons react in the presence of sunlight. In addition, NO_x is thought to be a contributor to acid rain (NRC 1986). Because of concerns about the health effects of ozone, considerable attention has been focused on efforts to reduce NO_x emissions. While not designed specifically to reduce nutrient loadings to waterbodies, these efforts can be expected to have eutrophication-related benefits as well.

Concerns about the transboundary nature of ground level ozone led in 1995 to the formation of the Ozone Transport Assessment Group, with representatives of 37 eastern states, the District of Columbia, Environmental Protection Agency (EPA), and stakeholder groups. The Ozone Transport Assessment Group recommended the need for further NO_x reductions to reduce ozone, and, in response, EPA finalized the NO_x State Implementation Plan Call in 1998. This regulation limits summer NO_x emissions for 22 states and the District of Columbia, and requires states to amend their State Implementation Plans to ensure that NO_x budgets are met. As part of this, EPA advocated the adoption of a nitrogen trading program. EPA also noted the potential need for Federal Implementation Plans if states fail to meet their obligations under the State Implementation Plan provisions. The 1990 Clean Air Act Amendments required additional NO_x control as well. Under the Amendments, both existing and new or modified sources are required to meet certain technology standards, which differ depending on whether the facility is located in an area that meets the current ozone standards. As noted above (Chapter 8), in the absence of other market distortions, technology standards are generally not cost-effective policies, since they do not allow firms flexibility to meet emission reduction goals in a least cost way. In addition, since these regulations target summer emissions, their impact on atmospheric deposition of nitrogen is limited to that period. Efforts to control emissions on a year-round basis are necessary for eutrophication-reduction benefits, particularly because of time delays between when nitrogen is deposited onto a watershed and when it is exported downstream to estuaries.

EPA has also sought further reductions in NO_x emissions from new or reconstructed sources. In September of 1998, EPA was forced to finalize a standard for fossil-fuel fired utilities and industrial boilers for which construction or modification was begun after July 9, 1997. EPA projects that this standard will reduce NO_x emissions from new sources by approximately 42 percent (EPA 1999c).

Federal initiatives have also had a significant impact on motor vehicle emissions of NO_x. The effect comes through regulations relating both to fuel economy and to tailpipe emissions. As noted above, the Corporate Average Fuel Economy program was created to establish vehicle manufacturers' compliance with fuel economy standards set by Congress in

1975 (EPA 1991c). Manufacturers' cars and light trucks must meet minimum miles per gallon standards or face monetary damages from the Department of Transportation (see Section 508 of the Motor Vehicle Information and Cost Savings Act). The Corporate Average Fuel Economy values are sales-weighted averages of fuel economy test results. In addition, motor vehicle emissions standards (i.e., limits on allowable grams of emissions per mile) are set by the Clean Air Act. The current standard for NO_x for cars is 0.4 grams per mile (EPA 1994), with a higher standard for light duty trucks. EPA's Tier 2 Report to Congress found the need for more stringent standards in order to meet and maintain the National Ambient Air Quality Standards for both ozone and particulate matter (EPA 1998a).

In summary, atmospheric sources of nitrogen constitute a significant (in some cases, the major) source of nutrient loadings to many estuaries. There are a number of federal and regional initiatives currently under way that are designed to reduce emissions of NO_x from both stationary and mobile sources. The impetus for these policies has come primarily from concerns about the contribution of NO_x to other pollution problems, such as ground level ozone and acid rain. Nonetheless, to the extent that these policies are effective in reducing NO_x emissions, they will reduce eutrophication as well. This additional benefit of NO_x emission reductions strengthens the case for stringent NO_x control. However, controls that target only summer emissions will be less effective in reducing atmospheric deposition and eutrophication than year-round controls. A recognition of the importance of atmospheric deposition as a nitrogen source to coastal and other water bodies would provide a rationale for year-round controls.

URBAN SOURCES

Urban sources of nutrients can be significant in some watersheds and coastal water bodies. These urban sources, and particularly wastewater discharges, were thoroughly discussed in an earlier NRC report (NRC 1993a). Therefore, in this report we only briefly discuss the control of these discharges.

Urban Point Sources

Treated Municipal Waste

Standards for treated municipal wastewater set goals for biochemical oxygen demand and total suspended solids. In most cases, these goals are achieved through biologically-based secondary treatment processes.

However, the 1977 Clean Water Act recognized the argument made by coastal cities that these effluent standards might be too high for settings where ocean currents disperse pollutants and result in minimal oxygen depletion. For such coastal systems, nutrient releases are of much more significance than biological oxygen demand. Thus, Section 301(h) of the Clean Water Act established a waiver process by which municipalities could avoid constructing full secondary treatment facilities if, on a case-by-case basis, they could demonstrate compliance with a strict set of pollution control and environmental protection requirements (NRC 1993a). Some cities, such as San Diego, have been able to use the waiver process. Others, such as Boston, have not.

The goal of secondary treatment is primarily to reduce solids and organic oxygen demand (Table 9-4). Communities that have bans on phosphate detergents typically enjoy a reduction of approximately 32 and 39 percent on total phosphorus and orthophosphate concentrations, respectively, in domestic wastewater (WEF 1998). Biologically-based secondary treatment typically reduces total nitrogen by approximately 31 percent and total phosphorus by approximately 38 percent (NRC 1993a), although national survey data indicate a wide range may be observed in practice (zero to 63 percent and 10 to 66 percent, respectively, for total nitrogen and total phosphorus) (NRC 1993a). Hence, additional (tertiary) treatment for removal of nutrients may be required for nutrient control. Although less common, and rare in coastal areas, tertiary treatment is

TABLE 9-4

Constituent	Before Sedimentation	After Sedimentation	Biologically Treated
Total solids	800	680	530
Total volatile solids	440	340	220
Suspended solids	240	120	30
Volatile suspended solids	180	100	20
Biological oxygen demand	200	130	30
Ammonia nitrogen as nitrogen	15	15	24
Total nitrogen as nitrogen	35	30	26
Total nitrogen generation (kg/cap-yr)	3.4-5.0		
Soluble phosphorus*	7	7	7
Total phosphorus*	10	9	8

*Before ban on phosphate detergents.

TABLE 9-4 Approximate composition of average domestic wastewater, mg l^{-1} before and after sedimentation after biological treatment, showing relative reductions in components (Viessman and Hammer 1998).

implemented in many locations in the United States. This usually takes the form of additional phosphorus removal in freshwater systems and nitrogen removal for nitrogen-limited coastal systems.

As described in NRC (1993a), nitrogen removal may be accomplished by an extension of the conventional biological system to incorporate the biochemical processes of nitrification and denitrification. Nitrification is the oxidation of ammonia and organic nitrogen to nitrate nitrogen. The process is mediated by the activity of a specialized class of bacteria that can be grown in conventional activated sludge biological systems by extending the biological solids residence time resulting in more complete biodegradation of organic matter. Nitrogen removal is subsequently obtained by denitrification whereby the nitrate nitrogen is reduced to nitrogen gas and then released into the atmosphere.

Biological phosphorus removal can be accomplished through the selection of high phosphorus content microorganisms, resulting in a greater mass of phosphorus in the excess biological solids removed. Biological phosphorus removal systems are more capital cost-intensive and less operations and maintenance cost-intensive than alternative chemical phosphorus removal systems, and their efficiency can vary depending on a number of factors. Consequently, biological phosphorus removal systems typically incorporate some degree of chemical addition (usually for polishing). Additional physico-chemical mechanisms for nutrient removal are also summarized in NRC (1993a).

Depending on the process employed, up to 97 percent of total nitrogen and 99 percent of total phosphorus may be removed from the waste stream (Table 9-5). Biological treatment is more costly both from a capital and from an operational point of view, and is implemented only for water-quality limited water bodies for which municipal loadings constitute an important nutrient source (e.g., Tampa Bay for nitrogen, the Potomac Estuary for phosphorus, and the Chesapeake Bay for both). Effects on eutrophication can be dramatic, although as for most loading reduction practices, several years may be required for the effects to become evident.

Treated Industrial Waste

Industries that discharge directly to receiving waters are required to meet the same EPA National Pollutant Discharge Elimination System requirements as municipalities. Nitrogen concentrations vary widely depending on the industry (Table 9-6), with most of the nitrogen in the form of organic nitrogen (WEF 1998). The seafood industry, typically located in the coastal environment, can have high concentrations of total nitrogen in its waste stream. Phosphorus concentrations are too dependent upon the particular source to permit generalization in a table. Rela-

TABLE 9-5

	Conventional Primary (%)	Chemically Enhanced Primary (CEPT) (%)		Conventional Biological Secondary Preceded by Conventional Primary (%)	Biological Secondary Preceded by CEPT (%)	Nutrient Removal Preceded by Conventional Biological Secondary and Conventional Primary (%)	Reverse Osmosis (%)
		Low Dose	High Dose				
Suspended Solids							
as mg l^{-1} TSS	41-69	60-82	86-98	89-97	88-98	94	99
BOD							
as mg I^{-1} BOD_5	19-41	45-65	67-89	86-98	91-99	94	99
Nutrients							
as mg l^{-1} TN	2-28	26-48	0-63	NA	NA	80-88	97
as mg l^{-1} TP	19-57	44-82	10-66	83-91	83-91	95-99	99

TABLE 9-5 Typical removal capability percentages for a range of wastewater treatment processes (NRC 1993a). NA = not available; BOD = biological oxygen demand; CEPT = chemically enhanced primary treatment.

TABLE 9-6

Industry	Total Nitrogen Mean (mg/L)	Total Nitrogen Range (mg/L)
Meat Packing		52-115
Fish and Seafood		
Catfish	34	21-51
Crab	100	62-147
Shrimp	222	168-276
Tuna	38	
Combined Poultry		15-300
Fruit and Vegetable Processing		
Apples	3.9	
Citrus fruits	21	
Potatoes	5.5	
Winery Wastewater	20	5-40
Chemical Industries		
NH_{3}-N (mg/L)	1270	
NO_{3}-N (mg/L)	550	

TABLE 9-6 Nitrogen concentrations in wastewater derived from a range of industries (WEF 1998).

tively few industries employ advanced waste treatment for nutrient removal. The focus at urban industries is often on pre-treatment for targeted pollutants, such as heavy metals or other toxic constituents. The wastewater is then sent to the publicly owned treatment works for further treatment, where tertiary nutrient removal may or may not occur.

Septic Tanks

Septic tanks for individual household treatment serve approximately 29 percent of the U.S. population (Novotny and Olem 1994) and often result in wastewater discharged directly to the groundwater system. Septic tanks in the riparian zone of a coastal waterbody (e.g., from homes along a beach) will discharge directly to the affected waterbody. Discharge from septic tank systems is commonly estimated as 280 liters per capita per day (l cap^{-1} day^{-1}) with typical effluent concentrations of 40 to 80 mg l^{-1} for total nitrogen and 11-31 mg l^{-1} for total phosphorus (Novotny et al. 1989). Although a well-functioning septic system typically removes most organic pollutants and phosphorus, the ammonia is rapidly oxidized in the soil to nitrate-nitrogen that is not adsorbed by soils and readily moves into the groundwater (Novotny et al. 1989). Hence, septic systems can contribute significant quantities of nitrate to riparian zones of coastal waters in addition to their contributions upstream in the watershed.

Improvements to septic tank design to provide for nitrification-denitrification can reduce nitrogen loadings. For instance, whereas conventional septic systems remove 10 to 45 percent of total nitrogen, anaerobic up-flow filters remove 40 to 75 percent, and recirculating sand filters remove 60 to 85 percent (EPA 1993a). These options typically involve a recirculation loop or tanks in series, and it is possible to retrofit conventional systems to improve denitrification performance. Other factors that affect the degree of nitrogen removal include temperature and the density of the soil in the septic tank fields.

Urban Diffuse Source Discharges

Urban diffuse nutrient sources include both runoff from streets and storm sewers and combined sewer overflows. Because of the enormous volume of runoff associated with rainfall, control of low concentrations of nutrients in stormwater runoff may lead to improvements in nutrient levels. In urban areas, overflows in older cities with combined sewer systems have the potential for large, concentrated loadings at overflow points. A mixture of stormwater and dry-weather sewage, combined sewage generally contains higher nutrient concentrations than does stormwater (Table 9-7). Hence, it is usually logical to control combined sewer overflows before urban stormwater, although both types of discharges are regulated under National Pollutant Discharge Elimination System regulations (stormwater currently only for cities of greater than 50,000 population).

Sources of nutrients in urban runoff include automobiles, atmospheric deposition (Chapter 5), erosion, deterioration of pavement and structures, fertilizer application, and miscellaneous wastes. Excessive fertilizer application can create especially high loads. For example, residential stormwater runoff is thought to contribute as much as 30 percent of total nitrogen loads to Sarasota Bay, Florida, largely related to excessive use of lawn fertilizers (Camp, Dresser, and McKee, Inc. 1992; Sarasota Bay National Estuary Program 1995).

A very large number of control options exist to help manage urban nonpoint source runoff (Schueler 1987; Camp et al. 1993; EPA 1993a; NRC 1993a; Urbonas and Stahre 1993; Horner et al. 1994; Novotny and Olem 1994; ASCE and WEF 1998), and their efficiency for nutrient removal varies widely. Passive treatment controls (typically structural options) include:

TABLE 9-7

			Concentrations (mg/L)		
Constituent	Reference	Source	Median (50th Percentile)	90th Percentile	Coefficient of Variation
TKN	*a*	Stormwater	1.50	3.30	0.50-1.00
NO_2-N+NO_3-N	*a*	Stormwater	0.68	1.75	0.50-1.00
Total N	*a*	Stormwater	2.20	5.00	0.50-1.00
TKN	*b*	Combined sewage	6.50	10.30	0.60
NH_3-N (ammonia)	*b*	Combined sewage	1.90	3.90	0.80
NO_2-N (nitrite)	*b*	Combined sewage	0.10	0.10	0.60
NO_3-N (nitrate)	*b*	Combined sewage	1.00	4.50	0.50
Total N	*b*	Combined sewage	7.60	14.90	0.60
Soluble P	*a*	Stormwater	0.12	0.21	0.50-1.00
Total P	*a*	Stormwater	0.33	0.70	0.50-1.00
Ortho-P	*b*	Combined sewage	0.80	1.10	0.40
Total P	*b*	Combined sewage	2.40	7.90	0.70

TABLE 9-7 Nutrient characteristics of urban runoff and combined sewage. Definitions: TKN = Total Kjeldahl Nitrogen = organic-N + NH_3-N. Total nitrogen = TKN + NO_2-N + NO_3-N. Coefficient of variation = standard deviation/mean. *(a)* is EPA 1983 and *(b)* is Driscoll and James 1987.

- vegetated areas (filter strips, swales, riparian vegetation);
- extended detention basins (dry between storm events);
- wet ponds (continuously filled);
- infiltration devices (ponds, trenches, swales, porous pavement);
- filters (biofilters, compost filters, sand filters); and
- constructed wetlands.

Pollutants in particulate or solid form are most amenable to treatment (e.g., nitrogen or phosphorus organic forms). For example, retention ponds ("wet ponds" with continual water storage) might remove 30 to 40 percent of total nitrogen and 50 to 60 percent of total phosphorus, with removal increasing as the dissolved forms of nitrogen and phosphorus decrease (EPA 1983) (Table 9-8). Extended detention facilities dry out between storms and are not effective for removal of dissolved nutrients by biological mechanisms. However, such facilities can sometimes serve multi-purposes as recreational areas when dry. Design guidelines for storage and other passive treatment control devices are provided in references such as Schueler (1987), Camp et al. (1993), Urbonas and Stahre (1993), and American Society of Civil Engineers and Water Environment Federation (1998). Additional information may be found in Novotny and Olem (1994), EPA (1993b), Debo and Reese (1995), and Novotny (1995).

Caution should be exercised when comparing urban BMPs on the basis of removal efficiencies. It is often found that while influent quality varies considerably, effluent quality exhibits a much smaller range (Strecker et al. 1999). Thus, high removal efficiencies may result purely from the fact that influent concentrations are high. Hence, BMPs might be better characterized simply by effluent quality. Alternatives for determining urban BMP performance effectiveness, based on a review of the most complete data set currently available, are provided by Strecker et al. (1999).

Natural wetlands are protected waters by law in the United States and generally cannot be used for waste treatment, except sometimes for a "polishing" purpose. Nonetheless, they provide many functions that enhance water quality (NRC 1991b), including acting as a sink for phosphorus, and facilitating denitrification by converting nitrate to nitrogen gas. Their capacity for nutrient removal can be considerable (Mitsch and Gosselink 1986). Hence, protection of coastal wetlands and tidal exchange is an important water quality consideration for coastal waters. Many factors can act to impair the natural functions of wetlands, including the drainage of wetlands for additional cropland, overgrazing, construction of highways, channelization of an adjoining waterway, deposition of dredged material, and excavation for ports and marinas (EPA 1993a).

TABLE 9-8

Type of Pond	Total Suspended Sediments	Nitrogen	Phosphorus
Extended detention basins	70-80	0 (dissolved) 20-30 (total)	0 (dissolved) 20-50 (total)
Retention ponds	70-80	50-70 (dissolved) 30-40 (total)	50-70 (dissolved) 50-60 (total)

TABLE 9-8 Comparison of nutrient removal percentages from well-designed extended detention basins and retention ponds (EPA 1983).

Measures for protection include: acquisition, protective zoning, application of water quality standards to wetlands, education and training, inclusion in comprehensive watershed planning, and restoration. Restoration measures include: maintenance of a natural hydrologic regime, restoration of native plant species, reduction of nonpoint source or other pollutant inflows, and maintenance of historic wetlands sites (EPA 1993a).

Constructed wetlands function similarly to storage devices, and their nutrient removal effectiveness depends upon the characteristics of the inflow as well as hydraulic properties (e.g., avoidance of short-circuiting) and vegetation types (Schueler 1992; Strecker et al. 1992; Urbonas and Stahre 1993; Strecker 1996). Schueler (1992) indicates projected removal rates for total phosphorus and total nitrogen on the order of 40 to 60 percent and 20 to 30 percent, respectively. Actual monitoring of constructed wetland removal efficiencies yields a very large variability (e.g., from –4 percent to 62 percent for NH_3 and –4 percent to 90 percent for total phosphorus) (Strecker et al. 1992; Urbonas and Stahre 1993; Strecker 1996). Nutrient removal efficiency depends strongly on the loading rate, percentage solid material, site conditions (such as soils), and hydraulic conditions that might lead to short-circuiting or scour—which might account for occasional negative removal efficiencies. Furthermore, Harper et al. (1988) point out that systems designed for removal of nutrients should avoid long detention times and stagnant conditions, both of which can decrease oxidation reduction potential and pH and reduce the efficiency of phosphorus removal. There is no clear advantage of constructed wetlands over storage ponds for nutrient control apart from the public appeal of wetland systems (Urbonas and Stahre 1993). Schueler (1992) presented extensive design guidelines for constructed wetlands.

Maintenance is a critical concern for all stormwater management facilities. When maintenance is poor, both quantity and quality control effectiveness can be greatly diminished. It is important that operation and maintenance costs be included during the planning and design of BMPs. Robustness of a design is also a factor (ASCE and WEF 1998). High robustness implies that when all the design parameters are correctly defined and quantified, the design has a high probability of performing as intended. For instance, wet retention ponds have a high robustness for removal of particulates and solids, but only a low to moderate robustness for removal of dissolved constituents (ASCE and WEF 1998). The robustness of extended detention basins and wetlands is moderate to high for particulates. Wetlands have a low-moderate robustness for removal of dissolved constituents while extended detention has none to low.

An additional number of structural controls exists for combined sewers, most of which are designed to store sewage during a storm for eventual treatment at the treatment plant or to divert only the cleanest water to receiving waters during a storm event (WPCF 1989; Moffa 1990; EPA 1992, 1993b). The use of structural controls for combined sewer overflow management often involves significant costs when applied in dense, highly developed, older urban areas.

Most urban BMPs are designed to control parameters other than just nutrients, such as heavy metals, solids, and oxygen demanding constituents. Hence, their design will also be predicated upon removal of these other constituents, as well as for management of stormwater quantity.

A large number of nonstructural BMPs are available (ASCE and WEF 1998). In general, nonstructural BMPs emphasize source controls and "good housekeeping". Many such options are routinely implemented as part of stormwater master plans in cities, although their effectiveness has not generally been quantified, especially for nutrient control. Examples include:

- public education;
- use of alternative products;
- vehicle use reduction;
- storm drain system signs (e.g., "dump no waste, drains to stream");
- spillage control;
- control of illicit connections to storm sewers;
- street cleaning and catchbasin cleaning;
- general maintenance;
- control of leaking sanitary sewers; and
- land use controls.

Overall, post-construction monitoring data is lacking so there is little concrete evidence of the effectiveness of urban BMPs. Management practices are often implemented under the assumption that they will be effective in reducing the load of targeted pollutants, without any follow up on how well they actually perform. Hence, good design information is lacking. In response to the need to assemble and evaluate available effectiveness data, the American Society of Civil Engineers is conducting such a study for EPA (Strecker et al. 1999), due to be completed sometime in 2000 (Box 9-1). When finished, this study will provide a definitive statement about the effectiveness of urban BMPs.

OTHER MITIGATION OPTIONS

Regional Stormwater Control Facilities

Location as well as type of BMPs play an important role in control of nonpoint source runoff. For example, stormwater and combined sewer

BOX 9-1
National Stormwater Best Management Practices Database

In response to the need for a centralized, easy-to-use, scientifically-sound tool for assessing the appropriateness of stormwater runoff BMPs under various conditions, the Urban Water Resources Research Council of the American Society of Civil Engineers has entered into a cooperative agreement with the EPA to develop a National Stormwater BMP Database (http://www.bmpdatabase.org). The long-term goal of the project is to promote technical design improvements for BMPs and to better match their selection and design to the local stormwater problems being addressed. The database, which was released in late 1999, contains data from BMP evaluations conducted over the past 15 years (ASCE and EPA 1999). Database updates will be made available as additional BMP evaluation data are gathered.

overflow controls can be distributed at critical points throughout the watershed, such as at discharge locations. Another option is to collect nutrient-laden stormwater and control/treat it at a more centralized or downstream location. Such a regional facility may have advantages such as lower capital costs, reduced maintenance, and greater reliability (Stutler et al. 1995). Alternatively, a regional collection system may be better from the standpoint of the location of the receiving water discharge. For instance, the City of San Francisco collects combined sewage in large storage tanks placed along the Bay-side of the city. The combined sewage is then pumped across the dividing hills for treatment and discharge into the Pacific Ocean in the southwest corner of the city. In this way, combined sewer overflow loadings to San Francisco Bay are avoided, except for very high storm events.

Hydrologic/Hydraulic Alterations in the Watershed

The history of human development is one of encroachment upon wetlands and waterways, and loss of wetlands in coastal areas is substantial. An outstanding example is the Kissimmee River system of central Florida, which drains to Lake Okeechobee and ultimately, through the Everglades system to Florida Bay. Although not a coastal system, Lake Okeechobee has reacted to loss of upstream wetlands and attendant nutrient filtering with massive eutrophication problems since flood control facilities that straighten, narrow, and reduce the length of the river were built in the 1960s (Koebel 1995; SFWMD 1998; Koebel et al. 1999). The South Florida Water Management District is now working with the Corps of Engineers to restore wetlands for purposes of phosphorus removal from the predominant inflow, the Kissimmee River. Restoration of the wetlands themselves, as well as enhancement of hydraulic connections, can provide a nutrient removal function that helps mitigate the overall trend of increased loadings from tributary watersheds and growing urban areas. Restoration of non-tidal wetlands as a management strategy has received much attention for the Mississippi Basin (NOAA 1999b), and is also a part of the Chesapeake Bay Program strategy.

Circulation Enhancement

Coastal developments, bridges, jetties, causeways, breakwaters, and flood control structures often lead to altered circulation patterns in an estuary. Reduced ocean exchange leads to lack of flushing, longer residence times, and more time for eutrophication processes to proceed. An extreme example is finger canals (Figure 9-4) that support extensive residential development, a pattern that has now been prohibited in most states. Finger canals lead to dead-ends and stagnant water, ripe for eutrophication and other pollution hazards. In many cases, circulation can be enhanced by tidal pumping promoted through new connectivity via culverts, pipes, and bridges. Small differences in tidal amplitude at multiple outlets, coupled with frictional resistance that depends on the flow direction, can create a net circulation through a looped system. Hence, where reduced flushing and increased residence time can contribute significantly to eutrophication problems, engineering alternatives may exist for mitigation through hydraulic controls. General guidance on flushing characteristics of estuaries and implications for mixing, residence times, and eutrophication potential can be found in references such as Officer (1976), McDowell and O'Connor (1977), Fischer et al. (1979), and Kjerfve (1988).

In Situ Biological Treatment Options

Eutrophication and other effects of nutrient over-enrichment in a waterbody are affected in part by the grazing activity of animals—both zooplankton and benthic filter feeders—on phytoplankton. The abundance of zooplankton that feed on phytoplankton is in part regulated by the abundance of zooplanktivorous fish, and these in turn are regulated in part by the abundance of higher predators (Carpenter et al. 1985). To some extent, nutrient problems in lakes can be managed by managing the populations of predatory fish, with the effect cascading down to zooplanktivorous fish, zooplankton, and then phytoplankton. The same principles apply to estuaries and coastal waters (Ingrid et al. 1996). However,

FIGURE 9-4 Finger canals in the Port Charlotte, Florida area, in 1974. Systems of canals with dead-ends have reduced circulation and flushing and can lead to accumulation of pollutants and nutrients, with resultant eutrophication problems (photo by W. Huber).

whereas lakes can be considered relatively closed systems with regard to fish populations, fish readily migrate between estuaries and coastal waters. While fishery practices in coastal areas may have impacts on nutrient enrichment in coastal areas, it would be exceedingly difficult to manage this through manipulations of fishery populations.

Benthic filter feeders such as oysters, mussels, and many species of clams can have a major influence on phytoplankton populations in coastal waters (Lucas et al. 1997; Meeuwig et al. 1998). In fact, it has been suggested that eutrophication of Chesapeake Bay is due in part to loss of oyster populations there: when oyster populations were high in the bay, they may have filtered the water as frequently as once per day on average, which would have been a significant control on phytoplankton abundance. Currently, oysters are believed to filter the water of Chesapeake Bay on average only once per year (Newell 1988). The data behind this speculation are sparse, but the logic is sound, and such grazing activity currently is a major factor regulating primary production in San Francisco Bay (Lucas et al. 1998) and in many small estuaries in Prince Edward Island, Canada (Meeuwig et al. 1998). Manipulation of benthic filter feeding organisms deserves further study as a possible mechanism for partially managing nutrient loads in estuaries.

Control of Bloom Incidence through Nutrient Reductions

Options for control of harmful algal blooms (HABs) include prevention, control, and mitigation (Boesch et al. 1997) (Box 9-2). HAB algae, just like all plants, require certain major and minor nutrients for their nutrition. These can be supplied either naturally or through human activities, such as pollution. As described in Chapter 4, a strong case has been made in several areas of the world that increases in pollution are linked to increases in the frequency and abundance of red tides (e.g., Smayda 1990; Okaichi 1997). It follows that a reduction in pollution can sometimes lead to a decrease in HAB frequency or magnitude. It should be emphasized, however, that it is exceedingly difficult to predict with any certainty what the effect of pollution control strategies will be on HAB incidence, except in situations where the pollution loading is massive (e.g., in Tolo Harbour or the Inland Sea of Japan [Box 4-2]) where it is now clear that increasing pollution was associated with increasing algal biomass, and therefore with more red tides/HABs. Given the high pollution loads to many estuaries and coastal waters, there is little doubt that these inputs contribute to some of the harmful blooms that occur. What is not clear, however, is the nature of that linkage—how much, and in what specific ways, the pollution must change before the number of HABs will decrease to "acceptable" levels. What is needed from a management perspective is the development of quantitative relationships between nutrient loading parameters and HAB incidence, such as the relationship between nitrogen:phosphorus ratios and dinoflagellate abundance described by Hodgkiss and Ho (1997) for Tolo Harbour, Hong Kong. However, the validity of such a relationship needs to be evaluated more thoroughly and if found to be robust, expanded to include other watersheds and hydrographic systems before it can be used to justify major policy decisions on water quality options in any particular region.

In general, the argument can be made that to reduce HAB incidence in an area, strict pollution control regulations should be instituted. However, a reduction in pollution loading will not lead to a complete absence of red tides/HABs—it is likely to reduce red tides in general, but some toxic species that thrive in relatively clean waters may find the new conditions suitable for growth. Given these uncertainties, it is difficult to jus-

BOX 9-2
Can Harmful Algal Blooms be Controlled with a Natural Parasite?

Could naturally occurring parasites be tapped as a biological tool to help control harmful algal blooms? New evidence is emerging that this might be possible (Delgado 1999). During one red tide covered by a paralytic shellfish poison-producing dinoflagellate called *Alexandreum catenella* in Catalonia, Spain, scientists noticed that a decline in the bloom corresponded with the presence of unknown round cysts. Further work showed that these cysts infected the *A. catenella* cells. They named the new organisms "diablillo parasites" (a Spanish word related to devil) and conducted further research to understand how they attacked the algae. The new diablillo parasite develops rapidly, so hundreds of new parasites per infected host can grow in about the 48 hours it takes for the host to reach maturity. Additional work continues to see if the parasite could be used to control harmful algal blooms in the natural environment. One feature of the parasite is that it only infects motile cells, and not the quiet cell stages of the hosts, and this could limit its ability to totally destroy problem algae populations in natural settings. Scientists continue to work to culture the parasite on non-toxic and widespread dinoflagellate hosts, which might facilitate its use in different geographic areas.

tify major pollution reduction programs solely on the basis of an expected reduction in HABs. Instead, reductions of nutrient inputs into coastal waters should be rigorously pursued as a key element of general estuarine and coastal management. The potential benefits of reductions of nutrient loadings in terms of decreased frequency and severity of HABs should be one of several considerations driving pollution policy decisions in estuarine and coastal management programs, but it should not be the sole justification.

Marketable Permits

Prior to 1990, marketable permits played only a minor role in environmental policy design, but the 1990 Clean Air Act Amendments promoted the use of marketable permits for sulfur dioxide emissions and provided an impetus for increased attention to this policy instrument. There are now a number of trading programs in place. As noted earlier, EPA advocated the use of a tradable permit system for reductions in NO_x emissions. In addition, several watersheds are experimenting with trading programs as a means of meeting water quality goals at lower cost. For example, the State of Connecticut estimates that trading will reduce the state's cost of meeting its target level of nitrogen removal by more than $200 million (CDEP 1998).

Although tradable permit systems have the potential to reduce the costs of achieving water quality goals, they have drawbacks that could limit their use. These include the administrative burden of operating a trading market (if it is publicly run) and the difficulty of establishing appropriate trading ratios when the environmental impact of a given discharge level varies by site.

NEXT STEPS FOR SOURCE REDUCTION

Much research has been conducted seeking agricultural, forest, and urban management practices that reduce the potential for nonpoint nutrient export. Yet we have not been successful at implementing cost-effective remedial measures in certain critical areas such as animal waste management. In most cases we know how to minimize nutrient export and input to coastal waters (the science is there), so that the major barriers to implementation now involve overcoming economic constraints, societal pressures, and political forces. New mechanisms to encourage implementation of BMPs, and remedial strategies in general, are necessary.

A critical component for facilitating widespread BMP implementation is by funding of cost-share programs and development of alliances among stakeholders. Stakeholder alliances encourage collaborative rather than adversarial relationships among affected groups.

There is a considerable scientific basis for reduction of nutrient releases from agriculture sources (e.g., enzyme adjustments for poultry and hogs to promote efficiency of phosphorus uptake; genetically-engineered corn to reduce unavailable phosphorus content). But there is a tendency on the part of farmers to over feed and urban dwellers to over fertilize—that is, to provide more nutrient supplements than are scientifically justified. Agricultural practices that reduce nutrient export must continually be communicated to end users in an effort to overcome the intuitive but false premise that "more is better." Many approaches can help people manage fertilizer application. The most fundamental principle is to supply only the nitrogen and phosphorus needed to meet the needs of the current crop, and to apply them in synchrony with crop use.

Manure generated from confined animal feeding operations has a significant potential to discharge nutrients to receiving waters. Many options are available to mitigate this source. Entrepreneurial activities should be encouraged to take advantage of management practices that require more than just field-scale activities (e.g., transport of manure from one location to another).

Methods for managing nitrogen and phosphorus transport may differ because nitrogen losses can occur from any location in a watershed via subsurface pathways, while phosphorus loss occurs most often in areas prone to surface runoff. Hence, remedial strategies for nitrogen may be applied to the whole watershed, whereas the most effective phosphorus strategy would be a combination of stringent measures at the most vulnerable sites to minimize loss of phosphorus in surface runoff with simple measures over the whole watershed to avoid excessive nutrient buildup, and thereby limit losses in subsurface flow. Extreme events must also be considered when designing phosphorus management strategies because one or two extremely high flows may be responsible for the bulk of annual phosphorus export.

There is a serious lack of post-implementation monitoring to assess the effectiveness and long-term viability of BMPs. Monitoring programs should be established to determine the long-term effectiveness of BMPs on nonpoint nitrogen and phosphorus reduction. A database of effective measures and design parameters should be maintained by appropriate federal agencies (e.g., EPA for urban areas and U.S. Department of Agriculture [USDA] for agricultural areas). The cost of implementing control measures is an important planning consideration, and databases of BMP effectiveness should also include relevant cost data. Economic data is also needed to help determine the economic benefits of management strategies and to see implementation costs in relation to the relative costs of different problems caused by nutrient over-enrichment.

The Clean Air Act may be as important as the Clean Water Act in protecting the nation's coastal waters from nitrogen pollution. Air pollution policy for nitrogen control is driven mainly by concerns other than nutrient over-enrichment (e.g., smog and human health, ozone, acid rain, global warming), but addressing these concerns can yield some nutrient-related benefits. Thus the effects of nitrogen on coastal waters should be considered in the formulation of air pollution policy. While eutrophication reduction is an additional benefit of air pollution control, policymakers will need to recognize that year-round emission control is necessary to affect eutrophication, not just summer controls as are used to combat smog. Also, the full range of nitrogen emissions need attention, not just NO_x, and thus ammonia-based scrubbers are inappropriate.

Although biologically-based secondary treatment of municipal wastewater is practiced at many U.S. cities and has some indirect nutrient benefits, point source discharges from publicly owned treatment works can still constitute a significant source of nutrients to coastal waters. In general, nutrient reduction strategies should address the least cost solutions first. Advanced waste treatment options of point sources for additional nutrient removal are often cheaper (on the basis of dollars per kilogram of nitrogen or phosphorus removed) than is control of nonpoint sources and should be examined carefully when planning strategies for nutrient reduction.

The larger the tributary area to the coastal waters, the more important is source control in the control of coastal nutrient over-enrichment. However, managers have options beyond source control. For coastal waters with smaller tributary areas, mitigation strategies such as enhancement of coastal wetlands might be a possibility. Most nutrient management schemes rely on a combination of measures. In all cases, maintenance of natural systems, including water column biota and shellfish, is important. Economic incentives, such as tradable permits, have potential to be used to facilitate the design of comprehensive cost-effective management strategies.

References

Aber, J.D. and C.T. Driscoll. 1997. Effects of land use, climate variation, and nitrogen deposition on nitrogen cycling and carbon storage in northern hardwood forests. *Global Biogeochemical Cycles* 11:639-648.

Aber, J.D., K.J. Nadelhoffer, P. Steudler, and J.M. Melillo. 1989. Nitrogen saturation in northern forest ecosystems. *BioScience* 39:378-386.

Aber, J.D., S.V. Ollinger, and C.T. Driscoll. 1997. Modeling nitrogen saturation in forest ecosystems in response to land use and atmospheric deposition. *Ecological Modeling* 101:61-78.

Adamowicz, W., J. Louviere, and M. Williams. 1994. Combining revealed and stated preference methods for valuing environmental amenities. *Journal of Environmental Economics and Management* 26:271-292.

Adamowicz, W., J. Swait, P. Boxall, J. Louviere, and M. Williams. 1997. Perceptions versus objective measures of environmental quality in combined revealed and stated preference models of environmental valuation. *Journal of Environmental Economics and Management* 32: 65-84.

Alberini, A., W. Harrington, and V. McConnell. 1995. Determinants in participation in accelerated vehicle retirement programs. *RAND Journal of Econ*. 26(1):93-111.

Alberini, A., W. Harrington, and V. McConnell. 1996. Estimating an emissions supply function from accelerated vehicle retirement programs. *RESTAT* 78(2):251-265.

Alley, W.M. and P.E. Smith. 1982. *Multi-Event Urban Runoff Quality Model*. U.S. Geological Survey Open File Report 82-344, Gulf Coast Hydroscience Center, NSTL Station, MS.

Alpine, A. E. and J. Cloern. 1992. Trophic interactions and direct physical effects control phytoplankton biomass and production in an estuary. *Limnology and Oceanography* 37:946-955.

Ambrose, R.B., Jr., T.O. Barnwell, Jr., S.C. McCutcheon, and J.R. Williams. 1996. Computer models for water quality analysis. In: *Handbook of Water Resources*, Mays, L.W. (ed.). McGraw-Hill, New York.

Ambrose, R.B., Jr., J.P. Connolly, E. Southerland, T.O. Barnwell, Jr., and J.L. Schnoor. 1988. Waste allocation simulation models. *J. Water Pollution Control Federation* 60(9):1646-1655.

Ambrose, R.B., Jr.; T.A. Wool; and J.L. Martin. 1993. *The Water Quality Analysis Simulation Program, WASP5: Model Documentation.* Environmental Protection Agency, Office of Research and Development Environmental Research Laboratory, Athens, GA.

American Society of Civil Engineers (ASCE) and Environmental Protection Agency (EPA). (1999). *National Stormwater Best Management Practices (BMP) Database*, [Online]. Available: http://www.bmpdatabase.org [2000, January 7].

American Society of Civil Engineers (ASCE) and Water Environment Federation (WEF). 1998. *Urban Runoff Quality Management.* ASCE-WEF, Reston, VA, 259pp.

Anderson, D.M. 1995. *The Ecology and Oceanography of Harmful Algal Blooms (ECOHAB): A National Research Agenda.* Woods Hole Oceanographic Institution, Woods Hole, MA, 66pp.

Arnold, J.G. and J.R. Williams. 1994. *SWRRB: A Watershed Scale Model for Soil and Water Resource Management.* Grassland Soil and Water Research Laboratory, U.S. Department of Agriculture, Agricultural Research Service, Temple, TX.

Arnold, J.G., J.R. Williams, R. Srinivasan, and K.W. King. 1995. *SWAT: Soil and Water Assessment Tool.* Grassland Soil and Water Research Laboratory, U.S. Department of Agriculture, Agricultural Research Service, Temple, TX.

Arrow K., R. Solow, P. Portney, E. Leamer, R. Radner, and H. Schuman. 1993. *Report of NOAA Panel on Contingent Valuation.* 58 Fed. Reg. 46.

Association of National Estuary Programs (ANEP). 1997. *Key Management Issues.* Background Paper and Summary of a National Estuary Program Workshop held in San Francisco.

Association of National Estuary Programs (ANEP). 1998. *Preserving Our Heritage, Securing Our Future: A Report to the Citizens of the Nation.* Washington, DC.

Atkinson, A.B. and J.E. Stiglitz. 1980. *Lectures in Public Economics.* McGraw-Hill, New York.

Bacon, E.F. and H.S. Greening. 1998. Holding the line: Tampa Bay's cooperative approach to trading. In: *Watershed Management: Moving From Theory to Implementation.* Proceedings from the Water Environment Federation Conference May 3-6, 1998. Denver, CO, 137-144pp.

Banner, A.H. 1974. Kaneohe Bay, Hawaii: urban pollution and a coral reef ecosystem. *Proc. 2nd Inter. Coral Reef Symp.* 2:685-702.

Baretta, J.W., W. Ebenhoh, and P. Ruardij. 1995. The European regional seas ecosystem model, a complex marine ecosystem model. *Neth. J. Sea Res.* 33(3-4):233-246.

Barnwell, T.O., Jr., D.C. Bouchard, R.B. Ambrose, Jr., and D.W. Disney. 1995. *Environmental Software Available at the EPA's Center for Exposure Assessment Modeling.* Proceedings from the 1995 Environmental Conference, Technical Association of the Pulp and Paper Industry, Atlanta, GA, 347-361pp.

Bashkin, V.N. 1997. The critical load concept for emission abatement strategies in Europe: a review. *Environmental Conservation* 24:5-13.

Baumol, W.J. and W.E. Oates. 1988. *The Theory of Environmental Policy* (2nd ed.). Cambridge University Press, New York.

Beasley, D.B. and L.F. Huggins. 1981. *ANSWERS Users Manual.* Environmental Protection Agency, Region V, Chicago, IL.

Bedient, P.B. and W.C. Huber. 1992. *Hydrology and Floodplain Analysis* (2nd ed.). Addision-Wesley Publishing Co., Reading, MA.

Behrendt, H., P. Huber, M. Kornmilch, D. Opitz, O. Schmoll, G. Scholz, and R. Uebe. 1999. *Nutrient Emissions into River Basins of Germany.* German Federal Environmental Agency.

Bell, P.R.F. 1992. Eutrophication and coral reefs: some examples in the Great Barrier Reef lagoon. *Water Research* 26:553-568.

Belval, D.L. and L.A. Sprague. 1999. *Monitoring Nutrients in the Major Rivers Draining to Chesapeake Bay.* U.S. Geological Survey, Water-Resources Investigations Report 99-4238.

Berry, M.W., B.C. Hazen, R.L. MacIntyre, and R.O. Flamm. 1996. Lucas: a system for modeling land-use change. *IEEE Computational Science & Engineering* 3:24-35.

Besanko, D. 1987. Performance versus design standards in the regulation of pollution. *Journal of Public Economics* 34(1):19-44.

Bicknell, R.R., J.C. Imhoff, J.L. Kittle, Jr., A.S. Donigian, Jr., and R.C. Johanson. 1993. *Hydrological Simulation Program–Fortran: User's Manual for Release 10.* Environmental Protection Agency, Office of Research and Development, Athens, GA.

Bierman, V.J., Jr., S.C. Hinz, D. Zhu, W.J. Wiseman, Jr., N.N. Rabalais, and R.E. Turner. 1994. A preliminary mass balance of primary productivity and dissolved oxygen in the Mississippi River Plume/Inner Gulf Shelf region. *Estuaries* 17(4):886-899.

Billen, G., C. Lancelot, and M. Meybeck. 1991. Nitrogen, phosphorus, and silicon retention along the aquatic continuum from land to ocean. In: *Ocean Margin Process in Global Change,* Mantoura, R.F.C., J.M. Martin, and R. Wollast (eds.), Wiley.

Birkeland, C. 1982. Terrestrial runoff as a cause of outbreaks of *Acanthaster plancii* (Echinodermata:Asteroidea). *Mar. Biol.* 69:175-185.

Blumberg, A.F. and G.L.Mellor. 1987. A description of a three-dimensional coastal ocean circulation model. In: *Three-Dimensional Coastal Ocean Models, Coastal and Estuarine Sciences 4,* Heaps, N.S. (ed.). American Geophysical Union, Washington DC, 1-16pp.

Bock, B.R. 1984. Efficient use of nitrogen in cropping systems. In: *Nitrogen in Crop Production,* Hauck, R.D. (ed.). Am. Soc. Agron., Madison, WI, 273-294pp.

Bockstael, N.E., K.E. McConnell, and I.E. Strand. 1989. Measuring the benefits of improvements in water quality: the Chesapeake Bay. *Marine Resource Economics* 6:1-18.

Boesch, D.F., E.M. Anderson, R.A. Horner, S.E. Shumway, P.A. Tester and T.E. Whitledge. 1997. *Harmful Algal Blooms in Coastal Waters: Options for Prevention, Control, and Mitigation.* NOAA Coastal Ocean Program Decision Analysis Series No. 10. NOAA Coastal Ocean Office, Silver Spring, MD.

Boesch, D.F., R.B. Brinsfield, and R.E. Magnien. 2000. Chesapeake Bay eutrophication: scientific understanding, ecosystem restoration, and challenges for agriculture. *Journal of Environmental Quality* in press.

Bohm, P. and C.S. Russell. 1985. Comparative analysis of alternative policy instruments. In: *Handbook of Natural Resource and Energy Economics,* Kneese, A.V. and J.L. Sweeny (eds.). North Holland, NY.

Bosch, D.J., Z.L. Cook, and K.O. Fuglie. 1995. Voluntary versus mandatory agricultural policies to protect water quality: adoption of nitrogen testing in Nebraska. *Review of Agricultural Economics* 17(1):13-24.

Bottcher, A.B. and T. Tremwell. 1995. Best management practices for water quality improvement. *Ecol. Engr.* 5:341-356.

Bouchard, D.C., R.B. Ambrose, Jr., T.O. Barnwell, Jr., and D.W. Disney. 1994. Environmental software available at the environmental protection agency's center for exposure assessment modeling. In: *Computer Assisted Risk Management,* Beroggi, G.E.G. and W.A. Wallace (eds.). Kluwer Academic Publishers, New York.

Bouwman, A.F. and H. Booij. 1998. Global use and trade of feedstuffs and consequences for the nitrogen cycle. *Nutrient Cycling in Agroecosystems* 52:261-267.

Bouwman, A.F., D.S. Lee, A.H. Asman, F.J. Dentener, K.W. van der Hoek, and J.G.J. Olivier. 1997. A global high-resolution emission inventory for ammonia. *Global Biogeochemical Cycles* 11:561-587.

Boyle, K.J., S.R. Lawson, H.J. Michael, and R. Bouchard. 1998. *Lakefront Property Owners' Economic Demand for Water Clarity in Maine Lakes.* Maine Agricultural and Forest Experiment Station, University of Maine.

Boynton, W.R., J.H. Garber, R. Summers, and W.M. Kemp. 1995. Inputs, transformations, and transport to nitrogen and phosphorus in Chesapeake Bay and selected tributaries. *Estuaries* 18:285-314.

Boynton, W.R. and W.M. Kemp. 1985. Nutrient regeneration and oxygen consumption by sediments along an estuarine salinity gradient. *Marine Ecology Progress Series* 23:45-55.

Boynton, W.R., W.M. Kemp, and C.W. Keefe. 1982. A comparative analysis of nutrients and other factors influencing estuarine phytoplankton production. In: *Estuarine Comparisons*, Kennedy, V.S. (ed.), Academic Press, San Diego.

Bradley, P.M. and J.T. Morris. 1990. The influence of oxygen and sulfide concentrations on nitrogen uptake kinetics in *Spartina alterniflora*. *Ecology* 71:282-287.

Bredemeier, M., K. Blanck, Y.J. Xu, A. Tieteam, A.W. Boxman, B.A. Emmett, F. Moldan, P. Gundersen, P. Schleppi, and R.F. Wright. 1998. Input-output budgets at the NITREX sites. *For. Ecol. Manag.* 101:57-64.

Bretherton, F.P., R.E. Davis, and C.B. Fandry. 1976. A technique for objective analysis and design of oceanographic experiments applied to MODE-73. *Deep-Sea Res.* 23:559-582.

Bricker, S.B., C.G. Clement, D.E. Pirhalla, S.P. Orlando, and D.G.G. Farrow. 1999. *National Estuarine Eutrophication Assessment: Effects of Nutrient Enrichment in the Nation's Estuaries.* Special Projects Office and the National Centers for Coastal Ocean Science, National Ocean Service, National Oceanic and Atmospheric Administration. Silver Spring, MD.

Briggs, J.C. 1974. *Marine Zoogeography.* McGraw-Hill Book Company, New York.

Brockman, U.H., R.W.P.M. Laane, and H. Postma. 1990. Cycling of nutrient elements in the North Sea. *Neth. J. Sea Res.* 26:239-264.

Broecker, W.A. 1974. *Chemical Oceanography.* Harcourt, Brace, Jovanovich, New York.

Brunet, P.C., G. Pinay, F. Gazelle, and L. Roques. 1994. Role of the floodplain and riparian zone in suspended matter and nitrogen retention in the Adour River, southwest France. *Regulated Rivers Research and Management* 9:55-63.

Brush, S., M. Jennings, P.J. Young, and H. McWreath. 1994. NPDES monitoring–Dallas-Ft. Worth, Texas region. In: *Stormwater NPDES Related Monitoring Needs: a Proceedings of an Engineering Foundation Conference*, Torro, H.C. (ed.). American Society of Civil Engineers, New York, 115-143pp.

Buchak, E.M. and J.E. Edinger. 1984a. *Generalized, Longitudinal-Vertical Hydrodynamics and Transport: Development, Programming and Applications*, Document No. 84-18-R. U.S. Army Corps of Engineers, WES, Vicksburg, MS.

Buchak, E. M. and J.E. Edinger. 1984b. *Simulation of a Density Underflow into Wellington Reservoir using Longitudinal-Vertical Numerical Hydrodynamics*, Document No. 84-18-R. U.S. Army Corps of Engineers, WES, Vicksburg, MS.

Burkholder, J.A. 1997. Written testimony submitted to the U.S. House of Representatives Committee on Government Reform and Oversight. September 25, 1997

Burkholder, J.A. and H.B. Glasgow, Jr. 1997. *Pfiesteria piscicidia* and other Pfiesteria-dinoflagellates behaviors, impacts, and environmental controls. *Limnology and Oceanography* 42(5):1052-1075.

Burtraw, D. 1996. Cost savings SANS allowance trades: evaluating the SO_2 emission trading program to date. *Resources for the Future* 95(30):28.

Burtrow, D. 1996. The SO_2 emissions trading programme: cost savings without allowance trades. *Contemporary Econ. Policy* 14:79-94.

Buzzelli, C.P. 1998. Dynamic simulation of littoral zone habitats in lower Chesapeake Bay: ecosystem characterization related to model development. *Estuaries* 21:659-672.

Buzzelli, C.P. and R. Wetzel. 1996. Modeling the lower Chesapeake Bay littoral zone and fringing wetlands: ecosystem processes and habitat linkages: model sensitivity analysis, validation, and estimates of ecosystem processes. *Annual Report to EPA: Region III*. Annapolis, MD, 40pp.

Buzzelli, C.P., R. Wetzel, and M. Meyers. 1995. Modeling the lower Chesapeake Bay littoral zone and fringing wetlands: ecosystem processes and habitat linkages: simulation model development and description. *Annual Report to EPA: Region III*. Annapolis, MD, 24pp.

Buzzelli, C.P., R. Wetzel, and M. Meyers. 1999. Dynamic simulation of littoral zone habitats in lower Chesapeake Bay: seagrass habitat primary production and water quality relationships. *Estuaries* 21:673-689.

Cadée, G.C. 1990. Increase of Phaeocystis blooms in the westernmost inlet of the Wadden Sea, the Marsdiep, since 1973. *Water Pollution Research Report* 12:105-112.

Caffrey, J.M., B.E. Cole, J.E. Cloern, J.R. Rudek, A.C. Tyler, and A.D. Jassby. 1994. *Studies of the San Francisco Bay, California, Estuarine Ecosystem: Pilot Regional Monitoring Results, 1993*. U.S. Geological Survey Open-File Report 94-82.

Calvert, J. G., J. B. Heywood, R. F. Sawyer, and J.H. Seinfeld. 1993. Achieving acceptable air quality: some reflections on controlling vehicle emissions. *Science* 261:37-45.

Cameron, T.A. 1992. Combining contingent valuation and travel cost data for the valuation of nonmarket goods. *Land Economics* 68:302-317.

Camp, Dresser, and McKee, Inc. 1992. *Point/Nonpoint Source Pollution Loading Assessment: Phase 1*. Sarasota Bay National Estuary Program, Sarasota, FL.

Camp, Dresser, and McKee, Inc.; Larry Walker Associates; Uribe and Associates; and Resources Planning Associates. 1993. *California Storm Water Best Management Practice Handbooks*. Public Works Agency, County of Alameda, CA.

Campbell, J.L., J.W. Hornbeck, W.H. McDowell, D.C. Buso, J.B. Shaley, and G.E. Likens. 2000. Dissolved organic nitrogen budgets for upland, forested ecosystems in New England. *Biogeochemistry* 49(2):123-142.

Caperon, J.S., A. Cattell, and G. Krasnick. 1971. Phytoplankton kinetics in a subtropical estuary: eutrophication. *Limnol. Oceanogr.* 16:599-607.

Caraco, N., J.J. Cole, and G.E. Likens. 1989. Evidence for sulfate-controlled phosphorus release from sediments of aquatic systems. *Nature* 341:316-318.

Caraco, N., J.J. Cole, and G.E. Likens. 1990. A comparison of phosphorus immobilization in sediments of freshwater and coastal marine systems. *Biogeochemistry* 9:277-290.

Carpenter, S.R., N.F. Caraco, D.L. Correll, R.W. Howarth, A.N. Sharpley, and V.H. Smith. 1998. Nonpoint pollution of surface waters with phosphorus and nitrogen. *Ecological Applications* 8:559-568.

Carpenter, S.R. and J.F. Kitchell (eds.). 1993. *The Trophic Cascade in Lakes*. Cambridge University Press, New York.

Carpenter, S.R., J.F. Kitchell, and J.R. Hodgson. 1985. Cascading trophic interaction and lake productivity. *Bioscience* 35:634-639.

Carpentier, C.L., D.J. Bosch, and S.S. Batie. 1998. Using spatial information to reduce costs of controlling agricultural nonpoint source pollution. *Agricultural and Resource Economics Review* 27(1):72-84.

Carraro, C. and F. Léveque (eds.). 1999. *Voluntary Approaches in Environmental Policy*. Kluwer Academic Publishers, Dordrecht.

Carraro, C. and D. Siniscalco. 1994. Environmental policy reconsidered: the role of technological innovation. *European Economic Review* 38(3-4):545-554.

Carson, R. and R. Mitchell. 1993. The value of clean water: the public's willingness to pay for boatable, fishable, and swimmable quality water. *Water Resources Research* 29(7):2445-2454.

Carter, M.R., W.J. Parton, I.C. Rowland, J.E. Schultz, and G.R. Steed. 1993. Simulation of soil organic carbon and nitrogen changes in cereal and pasture systems of southern Australia. *Aust. J. Soil Res.* 31:481-491.

Cerco, C.F. and T. Cole. 1993. Three-dimensional eutrophication model of Chesapeake Bay. *J. Envir. Engrg.* 119(6):1006-1025.

Cerco, C.F. and T. Cole. 1995. *User's Guide to the CE-QUAL-ICM.* Technical Report EL-95-1. U.S. Army Engineer Waterways Experiment Station, Vicksburg, MS.

Charpy-Roubaud, C. and A. Sournia. 1990. The comparative estimation of phytoplanktonic, microphytobenthic, and macrophytobenthic primary production in the oceans. *Mar. Microb. Food Webs* 4:31-57.

Chen, C., D.A. Wiesenberg, and L. Xie. 1997. Influences of river discharge on biological production in the inner shelf: a coupled biological and physical model of the Louisiana-Texas Shelf. *J. Mar. Sci.* 55:293-320.

Chernick, H. and A. Reschovsky. 1997. Who pays the gasoline tax? *National Tax Journal* 50(2):233-259.

Chesapeake Bay Program. 1995. In: *The State of the Chesapeake Bay 1995*, Magnien, R., D. Boward, and S. Bieber (eds.). U.S. Government Printing Office, Washington, DC, 45pp.

Chesapeake Bay Program. 1998. *Federal Agencies' Chesapeake Ecosystem Unified Plan.* U.S. Government Printing Office, Washington, DC.

Chiappone, M. and K.M. Sullivan. 1997. Rapid assessment of reefs in the Florida Keys: results from a synoptic survey. *Proc. 8th Int. Coral Reef Symp.* 2:1509-1514.

Christensen, P.B. (ed.). 1998. *The Danish Marine Environment: Has Action Improved its State?* Ministry of Environment and Energy, Danish Environmental Protection Agency.

Cleveland, C.C., A.R. Townsend, D.S. Schimel, H. Fisher, R.W. Howarth, L.O. Hedin, S.S. Perakis, E.F. Latty, J.C. von Fischer, A. Elseroad, and M.F. Wasson. 1999. Global patterns of terrestrial biological nitrogen (N_2) fixation in natural ecosystems. *Global Biogeochemical Cycles* 13:623-645.

Cloern, J.E. 1982. Does the benthos control phytoplankton biomass in South San Francisco Bay? *Mar. Ecology Progress Series* 9:191-202.

Cloern, J.E. 1987. Turbidity as a control on phytoplankton biomass and productivity in estuaries. *Cont. Shelf Res.* 7:1367-1382.

Cloern, J.E. 1991. Annual variations in river flow and primary production in the south San Francisco Bay estuary. In: *Estuaries and Coasts: Spatial and Temporal Intercomparisons,* Elliot, M. and D. Ducrotoy (eds.). Olsen and Olsen, 91-96pp.

Cloern, J.E. 1996. Phytoplankton bloom dynamics in coastal ecosystems: a review with some general lessons from sustained investigation of San Francisco Bay, California. *Reviews of Geophysics* 34(2):127-168.

Cloern, J.E. 1999. The relative importance of light and nutrient limitation of phytoplankton growth: a simple index of coastal ecosystems sensitivity to nutrient enrichment. *Aquatic Ecology* 33:3-16.

Cloern, J.E., A.E. Alpine, B.E. Cole, R.L.J. Wong, J.F. Arthur, and M.D. Ball. 1983. River discharge controls phytoplankton dynamics in the northern San Francisco Bay estuary. *Estuarine, Coastal and Shelf Science* 16:415- 429.

Coale, K.H., K.S. Johnson, S.E. Fitzwater, R.M. Gordon, S. Tanner, F.P. Chave, I. Ferioli, C. Sakamoto, P. Rogers, F. Millero, P. Steinberg, P. Nightingale, D. Cooper, W.P. Cochlan, M.R. Landy, J. Constantinou, G. Rollwaten, A. Travina, and R. Kudela. 1996. A massive phytoplankton bloom induced by an ecosystem-scale iron fertilization experiment in the equatorial Pacific Ocean. *Nature* 383:495-501.

Colwell, R.R. 1983. *Vibrios in the Environment.* Wiley, New York.

Colwell, R.R. 1996. Global climate and infectious disease: the Cholera paradigm. *Science* 274:2025-2031.

Committee on Environment and Natural Resources (CENR). 1997. *Integrating the Nation's Environmental Monitoring and Research Networks and Programs: A Proposed Framework.* Office of Science and Technology Policy, Washington, DC.

Committee on Environment and Natural Resources (CENR). 1998. *CENR Strategic Planning,* [Online]. Available: www.whitehouse.gov/WH/EOP/OSTP/NSCT [1999, November 15].

Conley, D.J., C.L. Schelske, and E.F. Stoermer. 1993. Modification of the biogeochemical cycle of silica with eutrophication. *Marine Ecology Progress Series* 101:179-192.

Connecticut Department of Environmental Protection (CDEP). 1998. *Nitrogen Trading Plan to Facilitate Hypoxia Control in Long Island Sound.* Hartford, CT.

Connecticut Department of Environmental Protection (CDEP) and New York State Department of Environmental Conservation (NYSDEC). 1998. *A Total Maximum Daily Load Analysis to Achieve Water Quality Standards for Dissolved Oxygen in Long Island Sound.* Hartford, CT.

Cooke, T., D. Drury, R. Katznelson, C. Lee, P. Mangarella, and K. Whitman. 1994. Storm water NPDES monitoring in Santa Clara Valley. In: *Stormwater NPDES Related Monitoring Needs: A Proceedings of an Engineering Foundation Conference,* Torro, H.C. (ed.). American Society of Civil Engineers, New York, 144-171pp.

Cooper, A.B. 1990. Nitrate depletion in the riparian zone and stream channel of a small headwater catchment. *Hydrobiologia* 202:13-26.

Cooper, J.C. and R.W. Keim. 1996. Incentive payments to encourage farmer adoption of water quality protection practices. *American Journal of Agricultural Economics* 78(1):54-64.

Cooper, S.R. and G.S. Brush. 1991. Long-term history of Chesapeake Bay anoxia. *Science* 254:992-996.

Corredor, J.E., R.W. Howarth, R.R. Twilley, and J.M. Morrell. 1999. Nitrogen cycling and anthropogenic impact in the tropical interamerican seas. *Biogeochemistry* 46(1/3):163-178.

Correll D.L. and D.E. Weller. 1989. Factors limiting processes in freshwater wetlands: an agricultural primary stream riparian forest. In: *Freshwater Wetlands and Wildlife,* Sharitz, R. and J. Gibbons (eds.). U.S. Department of Energy, 9-23pp.

Correll, D.L., T.E. Jordan, and D.E. Weller. 1992. Nutrient flux in a landscape: effects of coastal land use and terrestrial community mosaic on nutrient transport to coastal waters. *Estuaries* 15:431-442.

Council for Agricultural Science and Technology. 1999. *Gulf of Mexico Hypoxia: Land and Sea Interactions.* Ames, IA.

Cowardin, L., V. Carter, F. Golet, and E. LaRoe. 1979. *Classification of Wetlands and Deepwater Habitats of the United States.* U.S. Fish and Wildlife Service Publication FWS/OBS-79/31, Washington, DC, 103pp.

Crutchfield, S.R., D. Letson, and A.S. Malik. 1994. Feasibility of point-nonpoint source trading for managing agricultural pollutant loadings to coastal waters. *Water Resources Research* 30(10):2825-2836.

Cunningham, C. and K. Walker. 1996. Enhancing public access to the coast through the CZMA. *Current: The Journal of Marine Education* 14(1):8-12.

D'Avanzo, C. and J.N. Kremer. 1994. Diel oxygen dynamics and anoxic events in an eutrophic estuary of Waquoit Bay, Massachusetts. *Estuaries* 17:131-139.

D'Elia, C.F., W. Buddemeier, and S.V. Smith. 1991. Workshop on coral bleaching, coral reef ecosystems, and global change. *Report of Proceedings.* Maryland SeaGrant Publication UM-SG-TS-91-03.

Dale, B., T.A. Thorsen, and A. Fjellsa. 1999. Dinoflagellate cysts as indicators of cultural eutrophication in the Oslofjord, Norway. *Estuarine, Coastal and Shelf Science* 48(3):371-382.

Davies, T. and J. Mazurek. 1996. *Industry Incentives for Environmental Improvement: Evaluation of U.S. Federal Initiatives.* Global Environmental Management Initiative, Resources for the Future, Washington, DC.

Debo, T.N. and A.J. Reese. 1995. *Municipal Storm Water Management.* Lewis Publishers, Boca Raton, FL.

Delgado, M. 1999. A new "diablillo parasite" in the toxic dinoflagellate *Alexandrium catenella* as a possibility to control harmful algal blooms. *Harmful Algal News* 19:1-3.

Deliman, P.N., R.H. Glick, and C.E. Ruiz. 1999. *Review of Watershed Water Quality Models.* U.S. Army Corps of Engineers, Waterways Experiment Station, Vicksburg, MS.

Dennison, W.C., R.J. Orth, K.A. Moore, J.C. Stevenson, V. Carter, S. Kollar, P.W. Bergstrom, and R.A. Batiuk. 1993. Assessing water quality with submerged aquatic vegetation. *Bioscience* 43:86-94.

Devito, K.J., P.J. Dillon, and B.D. Lazerte. 1989. Phosphorus and nitrogen retention in five Precambrian shield wetlands. *Biogeochemistry* 8:185-204.

DeVries, J.J. and T.V. Hromadka. 1993. Computer models for surface water. In: *Handbook of Hydrology,* Maidment, D.A. (ed.). McGraw-Hill, New York.

Diamond, P.A. and J.A. Hausman. 1994. Contingent valuation: Is some number better than no number? *Journal of Economic Perpsectives* 8(4):45-64.

Diaz, R. and R. Rosenberg. 1995. Marine benthic hypoxia: a review of its ecological effects and the behavioral responses of benthic macrofauna. *Oceanography and Marine Biology Annual Review* 33:245-303.

Diaz, R.J. and A. Solow. 1999. *Ecological and Economic Consequences of Hypoxia.* Topic 2 Report for the Integrated Assessment of Hypoxia in the Gulf of Mexico. NOAA Coastal Ocean Program Decision Analysis Series No. 16. NOAA Coastal Ocean Program, Silver Spring, MD, 45pp.

DiToro, D.M., S.A. Lowe, and J.J. Fitzpatrick. 2000. An evaluation of the kinetics of a modern eutrophication model: part I: results. *J. Env. Eng.* in press.

DiToro, D.M. and M.J. Small. 1984. Probability model of stream quality due to runoff. *J. Environmental Engineering* 110(3):607-628.

Dodds, W.K., J.M. Blari, G.M. Henebry, J.K. Koelliker, R. Ramundo, and C.M. Tate. 1996. Nitrogen transport from tall grass prairie watersheds. *J. Environ. Qual.* 25:973-981.

Doering, P.H., C.A. Oviatt, L.L. Beatty, V.F. Banzon, R. Rice, S.P. Kelly, B.K. Sullivan, and J.B. Frithsen. 1989. Structure and function in a model coastal ecosystem: silicon, the benthos, and eutrophication. *Mar. Ecol. Prog. Ser.* 52:287-299.

Doering, O.C., F. Diez-Hermelo, C. Howard, R. Heimlich, F. Hitzhusen, R. Kazmierczak, J. Lee, L. Libby, W. Milon, T. Prato, and M. Ribaudo. 1999. *Evaluation of the Economic Costs and Benefits of Methods for Reducing Nutrient Loads to the Gulf of Mexico.* Topic 6 Report for the Integrated Assessment of Hypoxia in the Gulf of Mexico. NOAA Coastal Ocean Program Decision Analysis Series No. 20. NOAA Coastal Ocean Program, Silver Spring, MD, 115pp.

Donigian, A.S., Jr. and W.C. Huber. 1991. *Modeling of Nonpoint Source Water Quality in Urban and Non-Urban Areas.* Environmental Protection Agency, Athens, GA.

Donigian, A.S., Jr., W.C. Huber, and T.O. Barnwell, Jr. 1995. Modeling of nonpoint source water quality in urban and non-urban areas. In: *Nonpoint Pollution and Urban Stormwater Management,* Novotny, V. (ed.). Technomic Publishing Co., Inc., Lancaster, PA, 293-345pp.

Doremus, C. 1982. Geochemical control of dinitrogen fixation in the open ocean. *Biol. Oceanogr.* 1:429-436.

Driscoll, E.D. and W. James. 1987. Evaluation of alternatives. In: *Pollution Control Planning*, James, W. (ed.). Proc. Ontario Ministry of the Environmental Technology Transfer Workshop. CHI, Guelph, Ontario.

Driscoll, E.D, P.E. Shelley, and E.W. Strecker. 1989. *Pollutant Loadings and Impacts from Highway Stormwater Runoff: Design Procedure*. Office of Engineering and Highway Operations Research and Development, Federal Highway Administration, McLean, VA.

Driver, N.E. and G.D. Tasker. 1990. *Techniques for Estimation of Storm-Runoff Loads, Volumes, and Selected Constituent Concentrations in Urban Watersheds in the United States*. U.S. Geological Survey, Reston, VA.

Duarte, C.M. 1995. Submerged aquatic vegetation in relation to different nutrient regimes. *Ophelia* 41:87-112.

Durka, W., E.D. Schultze, G. Gebauer, and S. Voerkelius. 1994. Effects of forest decline on uptake and leaching of deposited nitrate determined from ^{15}N and ^{18}O measurements. *Nature* 372:765-767.

Dustan, P. 1977. Vitality of coral populations off Key Largo, FL.: recruitment and mortality. *Environ. Geol.* 2:51-58.

Dyer, K. 1973. *Estuaries: A Physical Introduction*. Wiley-Interscience, 140pp.

Edinger, J.E. and E.M. Buchak. 1983. *Developments in LARM2: A Longitudinal-Vertical, Time-Varying Hydrodynamic Reservoir Model*. Technical Report E-83-1. U.S. Army Corps of Engineers, Waterways Experiment Station, Vicksburg, MS.

Edmondson, W.T. 1970. Phosphorus, nitrogen, and algae in Lake Washington after diversion of sewage. *Science* 169:690-691.

Ekholm, P. 1994. Bioavailability of phosphorus in agriculturally loaded rivers in southern Finland. *Hydrobiologia* 287:179-194.

Electric Power Research Institute (EPRI). 1998. *Watershed Analysis Risk Management Framework: A Decision Support System for Watershed Approach and Total Maximum Daily Load Calculation*, Report TR-110709. Electric Power Research Institute, Palo Alto, CA.

Elmgren, R. and U. Larsson. 1997. Himmerfjarden: forandringar i ett naringsbelastat kustekyosystem i Ostersjon. *Rapport* 4565. Naturvardsverket Forlag.

Emmett, B.A. and B. Reynolds. 1996. Nitrogen critical loads for spruce plantations in Wales: is there too much nitrogen? *Forestry* 69:200-214.

Emmett, B.A., D. Boxman, M. Bredemeier, P. Gundersen, O.J. Kjønaas, F. Moldan, P. Schleppi, A. Tietema, and R.F. Wright. 1998. Predicting the effects of atmospheric deposition in conifer stands: evidence from the NITREX ecosystem-scale experiments. *Ecosystems* 1:352-360.

Englin, J. and T.A. Cameron. 1996. Augmenting travel cost models with contingent behavior data: Poisson regression analyses with individual panel data. *Environmental and Resource Economics* 7:133-147.

Environmental and Hydraulics Laboratories. 1986. *CE-QUAL-W2: A Numerical Two-Dimensional Model of Hydrodynamics and Water Quality, User's Manual*, Instruction Report E-86-5. U.S. Army Corps of Engineers' Waterways Experiment Station, Vicksburg, MS.

Environmental Protection Agency (EPA). 1983. *Results of the Nationwide Urban Runoff Program*. Environmental Protection Agency, Office of Water, U.S. Govt. Printing Office, Washington, DC.

Environmental Protection Agency (EPA). 1987. *Quality Criteria for Water*. Office of Water Regulations and Standards, Washington, DC.

Environmental Protection Agency (EPA). 1988. *Lake and Reservoir Restoration Guidance Manual*. Criteria and Standards Division Nonpoint Sources Branch, Washington, DC.

Environmental Protection Agency (EPA). 1989. *Saving Bays and Estuaries: A Primer for Establishing and Managing Estuary Programs.* Office of Marine and Estuarine Protection, Washington, DC, 58pp.

Environmental Protection Agency. 1990. *Technical Guidance Manual for Performing Wasteload Allocations.* Book III. Estuaries. Part I: Estuaries and waste load allocation models. Office of Water, Washington, DC.

Environmental Protection Agency (EPA). 1991a. *Technical Support Document for Water Quality Based Toxics Control.* Environmental Protection Agency, Office of Water, U.S. Govt. Printing Office, Washington, DC.

Environmental Protection Agency (EPA). (1991b). *Guidance for Water Quality-Based Decisions: The TMDL Process,* [Online]. Available: http://www.epa.gov/OWOW/tmdl/decisions/ [1999, June 29].

Environmental Protection Agency (EPA). (1991c). *CAFE Info. Sheet 1.* Certification Division, Motor Vehicle Emission Laboratory, [Online]. Available: http://www.epa.gov/orcdizux/cert/factshts/cafe1.pdf [1999, November 11].

Environmental Protection Agency (EPA). 1992. *Evaluation of Wet Weather Design Standards for Controlling Pollution from Combined Sewer Overflows.* 230-R-92-006, Office of Policy, Planning and Evaluation. Washington, DC.

Environmental Protection Agency (EPA). 1993a. *Guidance Specifying Management Measures for Sources of Nonpoint Pollution in Coastal Waters.* EPA-840-B-93-001c. Office of Water, Washington, DC.

Environmental Protection Agency (EPA). 1993b. *Combined Sewer Overflow Control Manual.* EPA/625/R-93/007. Office of Research and Development. Washington, DC.

Environmental Protection Agency (EPA). (1994). *Milestones in Auto Emissions Control,* [Online]. Available: http://www.epa.gov/r10earth/offices/air/emission.html [1999, November 11].

Environmental Protection Agency (EPA). 1995. *QUAL-2E: The Enhanced Stream Water Quality Model. Model Documentation and User's Manual.* Athens, GA.

Environmental Protection Agency (EPA). 1996. *Level III Ecoregions of the Continental United States.* National Health and Environmental Effects Research Laboratory, Corvallis, OR.

Environmental Protection Agency (EPA). 1997. *Compendium of Tools for Watershed Assessment and TMDL Development.* Washington, DC.

Environmental Protection Agency (EPA). 1998a. *Tier 2 Report to Congress.* U.S. Government Printing Office, Washington, DC.

Environmental Protection Agency (EPA). 1998b. *The Regional NO_x SIP Call and Reduced Atmospheric Deposition of Nitrogen: Benefits to Selected Estuaries.* Office of Water, Washington, DC.

Environmental Protection Agency (EPA). (1998c). *National Strategy for the Development of Regional Criteria.* Office of Water, Washington, DC, [Online]. Available: http://www.epa.gov/OST/standards/nutrient.html [1999, June 30].

Environmental Protection Agency (EPA). 1998d. *Condition of the Mid-Atlantic Estuaries.* Office of Research and Development, Washington, DC.

Environmental Protection Agency (EPA). 1998e. *Water Quality Criteria and Standards Plan: Priorities for the Future.* Office of Water, Washington, DC.

Environmental Protection Agency (EPA). (1999a). *Total Maximum Daily Load (TMDL) Program.* Office of Water, Washington, DC, [Online]. Available: http://www.epa.gov/OWOW/tmdl [1999, December 20].

Environmental Protection Agency (EPA). 1999b. New animal waste laws spring up across the nation. *Nonpoint Source News-Notes* 57.

Environmental Protection Agency (EPA). 1999c. Deposition of air pollutants to the great waters. *Third Report to Congress.* U.S. Government Printing Office, Washington, DC.

Ertl, D.S., K.A. Young, and V. Raboy. 1998. Plant genetic approaches to phosphorus management in agricultural production. *J. Environ. Qual.* 27:299-304.

Espey, M. 1998. Gasoline demand revisited: an international meta-analysis of elasticities. *Energy Economics* 20(3):273-295.

Esseks, J.D. and S.E. Kraft. 1988. Why eligible landowners did not participate in the first four sign-ups of the conservation reserve program. *Journal of Soil and Water Conservation* 43(3):251-258.

Evans, R., L.C. Cuffman-Neff, and R. Nehring. 1996. Increases in agricultural productivity, 1948-1993. In: *Updates on Agricultural Resources and Environmental Indicators* No. 6. U.S. Department of Agriculture-Economic Research Service. U.S. Govt. Printing Office, Washington, DC, 85 pp.

Fagoonee, I.H., B. Wilson, M.P. Hassell, and J.R. Turner. 1999. The dynamics of zooxanthellae populations: a long-term study in the field. *Science* 283:843-845.

Fail, J.L., Jr., M.N. Hamzah, B.L. Haines, and R.L. Todd. 1986. Above and below ground biomass, production, and element accumulation in riparian forests of an agricultural watershed. In: *Watershed Research Perspectives*, Correll, D.L. (ed.). Smithsonian Institution, Washington, DC, 193-224pp.

Fischer, H. 1976. Mixing and dispersion in estuaries. *Annual Review of Fluid Mechanics* 8:107-133.

Fischer, H.B., E.J. List, R.C.Y. Koh, J. Imberger and N.H. Brooks. 1979. *Mixing in Inland and Coastal Waters*. Academic Press, Burlington, MA.

Fisher, H.B. and M. Oppenheimer. 1991. Atmospheric nitrate deposition and the Chesapeake Bay estuary. *Ambio* 20:102.

Fixen, P.E. 1998. Soil test levels in North America. *Better Crops* 82:16-18.

Flett, R.J., D.W. Schindler, R.D. Hamilton, and N.E.R. Campbell. 1980. Nitrogen fixation in Canadian Precambrian Shield lakes. *Can. J. Fish. Aquat. Sci.* 37:494-505.

Fogg, G.E. 1987. Marine planktonic cyanobacteria. In: *The Cyanobacteria*, Fay, P. and C.V. Baalen (eds.). Elsevier.

Fong, P., R.M. Donohoe, and J.B. Zedler. 1993. Competition with macroalgae and benthic cyanobacterial mats limits phytoplankton abundance in experimental microcosms. *Mar. Ecol. Prog. Ser.* 100:97-102.

Food and Agriculture Organization (FAO). (1999). *FAOSTAT Agriculture Data*, [Online]. Available: http://apps.fao.org/cgi-bin/nph-dp.pl?subset=agriculture [2000, February 24].

Freeman, A.M., III. 1993. *The Measurement of Environmental and Resource Values: Theory and Methods*. Resources for the Future, Washington, DC.

Freeman, A.M., III. 1995. The benefits of water quality improvements for marine recreation: a review of the empirical evidence. *Marine Resource Economics* 10(4):385-406.

Friederichs, C.T. and O.S. Madsen. 1992. Nonlinear diffusion of the tidal signal in frictionally dominated embayments. *Journal of Geophysical Research* 97:5637-5650.

Frithsen, J.B., A.A. Keller, and M.E.Q. Pilsen. 1985. *Effects of Inorganic Nutrient Additions in Coastal Areas: A Mesocosm Experiment Data Report*, Volume 1. Marine Ecosystem Research Laboratory, Report 3, University of Rhode Island, Kingston.

Frolking, S.E., A.R. Mosier, D.S. Ojima, C. Li, W.J. Parton, C.S. Potter, E. Priesack, R. Stenger, C. Haberbosch, P. Dorsch, H. Flessa, and K.A. Smith. 1998. Comparison of N_2O emissions from soils at three temperate agricultural sites: simulations of year-round measurements by four models. *Nutrient Cycling in Agroecosystems* 52:77-105.

Fullerton, D., S.P. McDermott, and J.P. Caulkins. 1997. Sulfur dioxide compliance of a regulated utility. *J. Environ. Econom. Management* 34(1):32-53.

Galloway, J.N., W.H. Schlesinger, C. Levy, A. Michaels, and J.L. Schnoor. 1995. Nitrogen fixation: anthropogenic enhancement environmental response. *Global Biogeochem. Cycles* 9:235-252.

Gardner, G. 1998. Recycling organic wastes. In: *State of the World*, Brown, L., C. Flavin, and H. French (eds.). W.W. Norton, New York, 96-112pp.

Gburek, W.J. and A.N. Sharpley. 1998. Hydrologic controls on phosphorus loss from upland agricultural watersheds. *J. Environ. Qual.* 27:267-277.

Geraci, J.R., D.M. Anderson, R.J. Timperi, D.J. St. Aubin, G.A. Early, J.H. Perscott, and C.A. Mayo. 1990. Humpback whales (*Megaptera novaeangliae*) fatally poisoned by dinaflagellate toxin. *Can. J. Fish. Aquat. Sci.* 46:1895-1898.

Geyer, W.R., J. Morris, F. Prahl, and D. Jay. 1999. Interaction between physical processes and ecosystem structure: a comparative approach. In: *Estuarine Synthesis: The Next Decade*, Hobbie, J. (ed.). Island Press.

Gijsman, A.J., A. Oberson, H. Tiessen, and D.K. Friesen. 1996. Limited applicability of the century model to highly weathered tropical soils. *Agron. J.* 88:894-903.

Gilbert, P.M., D.J. Conley, T.R. Fisher, L.W. Harding, and T.C. Malone. 1995. Dynamics of the 1990 winter/spring bloom in Chesapeake Bay. *Marine Ecology Progress Series* 122:27-43.

Gilbertson, C.B., F.A. Norstadt, A.C. Mathers, R.F. Holt, A.P. Barnett, T.M. McCalla, C.A. Onstad, and R.A. Young. 1979. *Animal Waste Utilization on Cropland and Pastureland.* U.S. EPA Rep. No. EPA 600/2-79-059 and USDA Utilization Research Report No. URR6. U.S. Government Printing Office, Washington, DC, 135pp.

Gilliam, J.W., R.W. Skaggs, and S.B. Weed. 1979. Drainage control to diminish nitrate loss from agricultural fields. *J. Environ. Qual.* 8:137-179.

Gilmanov, T.G., W.J. Parton, and D.S. Ojima. 1997. Testing the CENTURY ecosystem level model on data sets from eight grassland sites in the former USSR representing a wide climatic/soil gradient. *Ecol. Model.* 96:191-210.

Giraldez, C. and G. Fox. 1995. An economic analysis of groundwater contamination from agricultural emissions in southern Ontario. *Canadian Journal of Agricultural Economics* 43(3):387-402.

Glynn, P.W. 1997. Bioerosion and coral reef growth: a dynamic balance. In: *Life and Death of Coral Reefs*, Birkeland, C. (ed.). Chapman and Hall, New York, 68-95.

Goldberg, P. 1998. The effects of the corporate average fuel efficiency standards in the United States. *J. Industrial Econ.* 46(1):1-33.

Goldstein, A.L. and G.J. Ritter. 1993. A performance-based regulatory program for phosphorus control to prevent the accelerated eutrophication of Lake Okeechobee, Florida. *Water Sci. Tech.* 28:13-26.

Goodman, J.L., K.A. Moore, and W.C. Dennison. 1995. Photosynthetic responses of eelgrass (*Zostera marina* L.) to light and sediment sulfied in a shallow barrier island lagoon. *Aquat. Bot.* 50:37-47.

Goolsby, D.A., W.A. Battaglin, G.B. Lawrence, R.S. Artz, B.T. Aulenbach, R.P. Hooper, D.R. Kenney, and G.J. Stensland. 1999. Flux and *Sources of Nutrients* in the Mississippi-Atchafalaya River *Basin: Topic 3 Report for the Integrated Assessment of Hypoxia in the Gulf of Mexico. NOAA Coastal Ocean Program Decision Analysis Series* No. 17. NOAA Coastal Ocean Office, Silver Spring, MD, 130pp.

Gosner, K.L. 1971. *Guide to Identification of Marine and Estuarine Invertebrates, Cape Hatteras to the Bay of Fundy.* Wiley-Interscience, New York.

Granéli, E., K. Wallstrom, U. Larsson, W. Granéli, and R. Elmgren. 1990. Nutrient limitation of primary production in the Baltic Sea area. *Ambio* 19:142-151.

Grattan, L.M., D. Oldach, T.M. Perl, M.H. Lowitt, D.L. Matuszak, C. Dickson, C. Parrott, R.C. Shoemaker, C.L. Kaufman, M.P. Wasserman, J.R. Hebel, P. Charache, and J.G. Morris, Jr. 1998. Learning and memory difficulties after environmental exposure to waterways containing toxin-producing *Pfiesteria* or *Pfiesteria*-like dinoflagellates. *Lancet* 352:532-539.

Greening, H.S. 1999. Progress towards goals of the Tampa Bay CCMP. In: *Baywide Environmental Monitoring Report 1993-1998, Tampa Bay, Florida,* Pribble, J.R., A.J. Janicki, and H.S. Greening (eds.). Technical Publication #07-99, Tampa Bay Estuary Program, St. Petersburg, FL, 1-11pp.

Greve, W. and T.R. Parsons. 1977. Photosynthesis and fish production: hypothetical effects of climate change and pollution. *Helgolander Wiss. Meeresunters* 30:666-672.

Groffman, P.M., A.J. Gold, and R.C. Simmons. 1992. Nitrate dynamics in riparian forests: microbial studies. *J. Environ. Qual.* 21:666-71.

Gundersen, P. and V. Bashkin. 1994. Nitrogen cycling. In: *Biogeochemistry of Small Catchments: A Tool for Environmental Research,* Moldan, N. and J. Cerny (eds.). Wiley and Sons, Chichester.

Hager, S.W. 1994. *Dissolved Nutrient and Suspended Particulate Matter Data for the San Francisco Bay Estuary, California, October 1991 through November 1993.* U.S. Geological Survey Open-File Report 94-471.

Hahn, R.W. 1994. United States environmental policy: past, present, and future. *Natural Resources Journal* 34(2):305-348.

Hahn, R.W. and M.S. Borick. 1996. Why energy transitions matter: a case study of methanol. *J. Regulatory Econ.* 9(2):133-155.

Hamrick, J.M. 1996. *A User's Manual for the Environmental Fluid Dynamics Computer Code (EFDC),* Special Report 331. The College of William and Mary, Virginia Institute of Marine Science, Gloucester Point, VA.

Hanemann, W.M. 1994. Valuing the environment through contingent valuation. *Journal of Economic Perspectives* 8(4):19-43.

Hansen, D.V. and M. Rattray. 1966. New dimensions in estuary classification. *Limnol. Oceanogr.* 11:319-326.

Hansson, S. and L.G. Rudstam. 1990. Eutrophication and Baltic Sea fish communities. *Ambio* 19:123-125.

Harper, H.H., M.P. Wanielista, D.M. Baker, B.M. Fries and E.H. Livingston. 1988. *Treatment Efficiencies for Residential Stormwater Runoff in a Hardwood Wetland.* Annual International Symposium on Lake and Watershed Management, St. Louis, MO.

Harrington, W. 1997. Fuel economy and motor vehicle emissions. *J. Environ. Econom. Management* 33(3):240-252.

Haughton, J. and S. Sarkar. 1996. Gasoline tax as a corrective tax: estimates for the United States, 1970-1991. *Energy Journal* 17(2):103-126.

Hauhs, M., K. Rost-Siebert, G. Ragen, T.Paces, and B. Vigerus. 1989. Summary of European data: the role of nitrogen in the acidification of soils and surface waters. *Miljorapport* 10:92.

Hawkins, C.P., J.L. Kershner, P.A. Bisson, M.D. Bryant, and L.M. Decker. 1993. A hierarchical approach to classifying stream habitat features. *Fisheries* 18:3-12.

Haycock, N.E., G. Pinay, and C. Walker. 1993. Nitrogen retention in river corridors: European perspective. *Ambio* 22:340-346.

Hayden, B. and R. Dolan. 1976. Coastal marine fauna and marine climates of the Americas. *J. Biogeography* 3:71-81.

Hearn, C.J. 1998. Application of the Stommel model to shallow Mediterranean estuaries and their characterization. *Journal of Geophysical Research* 103:10,391-10,405.

Heatwole, C.D., P.L. Diebel, and J.M. Halstead. 1990. Management and policy effects on potential groundwater contamination from dairy waste. *Water Resources Bulletin* 26(1):25-34.

Hecky, P.E. 1998. Low nitrogen:phosphorus ratios and the nitrogen fix: why watershed nitrogen removal will not improve the Baltic. In: *Effects of Nitrogen in the Aquatic Environment*, Hellström, T. (ed.). Swedish National Committee for IAWQ, Royal Swedish Academy of Sciences, Stockholm.

Hecky, P.E. and P. Kilham. 1988. Nutrient limitation of phytoplankton in freshwater and marine environments: a review of recent evidence on the effects of enrichment. *Limnol. Oceanogr.* 33:796-822.

Hedin, L.O., J.J. Armesto, and A.H. Johnson. 1995. Patterns of nutrient loss from unpolluted old-growth temperate forests: evaluation of biogeochemical theory. *Ecology* 76:493-509.

Hedley, M.J. and A.N. Sharpley. 1998. Strategies for global nutrient cycling. In: *Long-Term Nutrient Needs for New Zealand's Primary Industries: Global Supply, Production Requirements and Environmental Constraints*, Currie, L. (ed.). The Fertilizer and Lime Research Centre, Massey University, Palmerston North, New Zealand, 70-95pp.

Hein, M., M.F. Pedersen, and K. Sand-Jensen. 1995. Size-dependent nitrogen uptake in micro- and macroalgae. *Mar. Ecol. Prog. Ser.* 118:247-253.

Helfand, G.E. and B.W. House. 1994. Regulating nonpoint source pollution under heterogeneous conditions. *American Journal of Agricultural Economics* 77(5):1024-1032.

Hellström, T. 1996. An empirical study of nitrogen dynamics in lakes. *Water Env. Res.* 68:55-65.

Hellström T. 1998. Why nitrogen is not limiting production in the seas around Sweden. In: *Effects of Nitrogen in the Aquatic Environment*, Hellström, T. (ed.). Swedish National Committee for IAWQ, Royal Swedish Academy of Sciences, Stockholm.

Herring, H.J., M. Inoue, G.L. Mellor, C.N.K. Mooers, P.P. Niiler, L.Y. Oey, R.C. Patchen, F.M. Vukovich, and W.J. Wiseman, Jr. 1999. *Coastal Ocean Modeling Program for the Gulf of Mexico*. U.S. Dept. of Interior, Minerals Management Service, Environmental Division, Technical Analysis Group.

Hetling, L.J., N.A. Jaworski, and D.J. Garetson. 1996. Comparison of input loadings and riverine export fluxes in large watersheds. In: *Third International Conference on Diffuse Pollution*. International Association of Water Quality, Edinburgh, Scotland.

Hinga, K.R., H. Jeon, and N.F. Lewis. 1995. *Marine Eutrophication Review*. NOAA Coastal Ocean Program Decision Analysis Series No. 4. NOAA Coastal Ocean Office, Silver Spring, MD, 156pp.

Hobbie, J.E. and G.E. Likens. 1973. Output of phosphorus, organic carbon, and fine particulate carbon from Hubbard Brook watersheds. *Limnol. Oceanog.* 18:734-742.

Hodgkiss, I.J. and K.C. Ho. 1997. Are changes in nitrogen:phosphorus ratios in coastal waters the key to increased red tide blooms? *Hydrobiologia* 852:141-147.

Holland, E.A., B.H. Braswell, J.F. Lamarque, A. Townsend, J. Sulzman, J.F. Muller, F. Dentener, G. Brasseur, H. Levy II, J.E. Penner, and G.J. Roelofs. 1997. Variations in the predicted spatial distribution of atmospheric nitrogen deposition and their impact on carbon uptake by terrestrial ecosystems. *J. Geophys. Res.* 102 (15):849-866.

Holland, E.A., F.J. Dentener, B.H. Braswell, and J.M. Sulzman. 1999. Contemporary and pre-industrial global reactive nitrogen budgets. *Biogeochemistry* 46(1/3):7-43.

Hopkinson, C.S. and J. Vallino. 1995. The nature of watershed perturbations and their influence on estuarine metabolism. *Estuaries* 18:598-621.

Horne, A.J. 1977. Nitrogen fixation: a review of this phenomenon as a polluting process. *Prog. Water Technol.* 8:359-372.

Horner, RR., J.J. Skupien, E.H. Livingston, and H.E. Shaver. 1994. *Fundamentals of Urban Runoff Management: Technical and Institutional Issues*. Terrene Institute, Washington, DC.

House, W.A. and H. Casey. 1988. Transport of phosphorus in rivers. In: *Phosphorus Cycles in Terrestrial and Aquatic Ecosystems: Europe*, Tiessen, H. (ed.). SCOPE, UNEP, University of Saskatchewan, Saskatoon, Canada, 253-282pp.

Howarth, R.W. 1988. Nutrient limitation of net primary production in marine ecosystems. *Ann. Rev. Ecol. and Syst.* 19:89-110.

Howarth, R.W. 1996. *Nitrogen Cycling in the North Atlantic Ocean and its Watersheds: A report from the International SCOPE Nitrogen Project*. Kluwer, Dordrecht, The Netherlands.

Howarth, R.W. 1998. An assessment of human influences on inputs of nitrogen to the estuaries and continental shelves of the North Atlantic Ocean. *Nutrient Cycling in Agroecosystems* 52:213-223.

Howarth, R.W., G. Billen, D. Swaney, A. Townsend, N. Jaworski, K. Lajtha, J.A. Downing, R. Elmgren, N. Caraco, T. Jordan, F. Berendse, J. Freney, V. Kudeyarov, P. Murdoch, and Z. Zhao-Liang. 1996. Regional nitrogen budgets and riverine nitrogen and phosphorus fluxes for the drainages to the North Atlantic Ocean: natural and human influences. *Biogeochemistry* 35:75-79.

Howarth, R.W., F. Chan, and R. Marino. 1999. Do top-down and bottom-up controls interact to exclude nitrogen-fixing cyanobacteria from the plankton of estuaries: an exploration with a simulation model. *Biogeochemistry* 46:203-231.

Howarth, R.W. and J.J. Cole. 1985. Molybdenum availability, nitrogen limitation, and phytoplankton growth in natural waters. *Science* 229:653-655.

Howarth, R.W., H.S. Jensen, R. Marino, and H. Postma. 1995. Transport to and processing of phosphorus in near-shore and oceanic waters. In: *Phosphorus in the Global Environment*, Tiessen, H. (ed.), Wiley, New York.

Howarth, R.W. and R. Marino. 1990. Nitrogen-fixing cyanobacteria in the plankton of lakes and estuaries: a reply to the comment by Smith. *Limnol. Oceanogr.* 35:1859-1863.

Howarth, R.W. and R. Marino. 1998. A mechanistic approach to understanding why so many estuaries and brackish waters are nitrogen limited. In: *Effects of Nitrogen in the Aquatic Environment*, Hellström, T. (ed.), KVA Report 1, Royal Swedish Academy of Sciences, Stockholm.

Howarth, R.W., R. Marino, and J.J. Cole. 1988a. Nitrogen fixation in freshwater, estuarine, and marine ecosystems: biogeochemical controls. *Limnol. Oceanogr.* 33:688-701.

Howarth, R.W., R. Marino, J. Lane, and J.J. Cole. 1988b. Nitrogen fixation in freshwater, estuarine, and marine ecosystems: rates and importance. *Limnol. Oceanogr.* 33:669-687.

Howarth, R.W., D.P. Swaney, T.J. Butler, and R. Marino. 2000. Climatic control on eutrophication of the Hudson River estuary. *Ecosystems* in press.

Howell, P. and D. Simpson. 1994. Abundance of marine resources to dissolved oxygen in Long Island Sound. *Estuaries* 17:394-402.

Huang, J.H., T.C. Haab, and J.C. Whitehead. 1997. Willingness to pay for quality improvements: Should revealed and stated preference data be combined? *Journal of Environmental Economics and Management* 34(3):240-255.

Hubbard, D.K. 1997. Reefs as dynamic systems. In: *Life and Death of Coral Reefs*, Birkeland, C. (ed.). Chapman and Hall, New York, 43-67pp.

Hubbard, T.N. 1997. Using inspection and maintenance programs to regulate vehicle emissions. *Contemporary Economic Policy* 15(2):52-62.

Hubbard, T.N. 1998. An empirical examination of moral hazard in the vehicle inspection market. *Rand Journal of Economics* 29(2):406-426.

Huber, A.L. 1986. Nitrogen fixation by *Nodularia spumigina* Mertins (Cyanobacteriaceae): field studies and the contribution of blooms to the nitrogen budget of the Peel-Harvey estuary, western Australia. *Hydrobiol.* 131:193-203.

Huber, W.C. and R.E. Dickinson. 1988. *Storm Water Management Model Version 4, User's Manual*. Environmental Protection Agency, Athens, GA.

Huggins, W.H. 1963. *An Algebra for Signal Representation: Representation and Analysis of Signals, Part XVII: Reprints on Signal Theory*. Dept. of Electrical Eng., Johns Hopkins University, Baltimore, MD.

Hughes, R.M. 1985. Use of watershed characteristics to select control streams for estimating effects of metal mining wastes on extensively disturbed streams. *Environmental Management* 9:253-262.

Hughes, R.M. and D.P. Larsen. 1988. Ecoregions: an approach to surface water protection. *Journal of the Water Pollution Control Federation* 60:486-493.

Hughes, R.M., D.P. Larsen, and J.M. Omernik. 1986. Regional reference sites: a method for assessing stream potentials. *Environmental Management* 10:629-635.

Hughes, R.M., E. Rexstad, and C.E. Bond. 1987. The relationship of aquatic ecoregions, river basins, and physiographic provinces to the ichthyogeographic regions of Oregon. *Copeia* 1987:423-432.

Hunsacker, C.T. and D.A. Levine. 1995. Hierarchical approaches to the study of water quality in rivers. *BioSci.* 45(3):193-203.

Huq, A., E.B. Small, P.A. West, M.I. Huq, R. Rahman, and R.R. Colwell. 1983. Ecological relationships between Vibrio cholerae and planktonic crustacean copepods. *Appl. Environ. Microbiol.* 45:275-283.

Hydrologic Engineering Center (HEC). 1977. *STORM: Storage, Treatment, Overflow Runoff Model User's Manual*. CPD-7, U.S. Army Corps of Engineers, Davis, CA.

Hydroqual, Inc. 1995. *A Water Quality Model for Massachusetts and Cape Cod Bays: Calibration of the Bays Eutrophication Model (BEM)*. Mahwah, NJ, 346pp.

Hydroqual, Inc. 1998. *A Water Quality Model for Jamaica Bay: Calibration of the Jamaica Bay Eutrophication Model (JEM)*. Mahwah, NJ, 405pp.

Ingrid, G., T. Andersen, and O. Vadstein. 1996. Pelagic food webs and eutrophication of coastal waters: impact of grazers on algal communities. *Mar. Poll. Bull.* 33:22-35.

Inoue, M., W.J. Day, Jr., and D. Justic. 1996. *Ecological Linkages Among Wetland, Bay, and Shelf Systems: Development of Linked Hydrodynamic and Ecological Spatial Models*. Louisiana Sea Grant Program, LSU, 68pp.

Ippen, A. and D. Harlemann. 1961. *Technical Bulletin* 5. Committee for Tidal Hydraulics, U.S. Army Corps of Engineers.

Isserman, K. 1990. Share of agriculture in nitrogen and phosphorus emissions into the surface waters of Western Europe against the background of their eutrophication. *Fert. Res.* 26:253-269.

Jacobs, T.C. and J.W. Gilliam. 1985. Riparian losses of nitrate from agricultural drainage waters. *J. Environ. Qual.* 14:472-78.

Jaffe, A.B. and K. Palmer. 1997. Environmental regulation and innovation: a panel data study. *RESTAT* 79(4):610-619.

Jansson, M., R. Andersson, H. Berggren, and L. Leonardson. 1994. Wetlands and lakes as nitrogen traps. *Ambio* 23:320-325.

Jaworski, N.A. 1981. Sources of nutrient and the scale of eutrophication problems in estuaries. In: *Estuaries and Nutrients*, Neilson, B.J. and L.E. Cronin (eds.). Humana, NY.

Jaworski, N.A. 1990. Retrospective study of the water quality issues of the upper Potomac estuary. *Aquat. Sci.* 3:11-40.

Jaworski, N.A., R.W. Howarth, and L.J. Hetling. 1997. Atmospheric deposition of nitrogen oxides onto the landscape contributes to coastal eutrophication in the northeast United States. *Environ. Sci. Technol.* 31:1995-2004.

Jay, D.A. and J.D. Smith. 1988. Circulation in and classification of shallow, stratified estuaries. In: *Physical Processes in Estuaries*, Dronkers, J. and W. van Leussen (eds.). Springer-Verlag, 21-41pp.

Jay, D.A., W.R. Geyer, and D.R. Montgomery. 1999. An ecological perspective on estuarine classification. In: *Estuarine Synthesis: The Next Decade*, Hobbie, J. (ed.). Island Press.

Jensen, H.S., K.J. McGlathery, R. Marino, and R.W. Howarth. 1998. Forms and availability of sediment phosphorus in carbonate sand of Bermuda seagrass beds. *Limnol. Oceanogr.* 43:799-810.

Johansson, J. and H. Greening. 2000. Seagrass restoration in Tampa Bay: a resource-based approach to estuarine management. In: *Subtropical and Tropical Seagrass Management Ecology*, Bortone, S. (ed.). CRC Press, Boca Raton, FL.

Johnson, B.H., K.W. Kim, R.E. Heath, B.B. Hsieh, and H.L. Butler. 1993. Validation of three-dimensional hydrodynamic model of Chesapeake Bay. *J. Hydraulic Engineering* 119 (1):2-20.

Johnson, D.W. 1992. Nitrogen retention in forest soils. *J. Envir. Qual.* 21:1-12.

Johnson, S.L., R.M. Adams, and G.M. Perry. 1991. The on-farm costs of reducing groundwater pollution. *American Journal of Agricultural Economics* 73(4):1063-1073.

Johnston, C.A., N.E. Detenbeck, and G.J. Niemi. 1990. The cumulative effect of wetlands on stream water quality and quantity: a landscape approach. *Biogeochemistry* 10:105-141.

Johnston, D.W. and S.E. Lindberg (eds.). 1992. *Atmospheric Deposition and Forest Nutrient Cycling: A Synthesis of the Integrated Forest Study*. Springer-Verlag, New York.

Jones, G.J., S.I. Blackburn, and N.S. Parker. 1994. A toxic bloom of Nodularia spumigena Mertens in Orielton Lagoon, Tasmania. *Aust. J. Mar. Freshwater Res.* 45:787-800.

Jordan, T.E., D.L. Correll, and D.E. Weller. 1993. Nutrient interception by a riparian forest receiving inputs from cropland. *J. of Envir. Qual.* 22:467-473.

Jørgensen, B.B. and K. Richardson. 1996. *Eutrophication in Coastal Marine Systems*. American Geophysical Union, Washington, DC.

Joskow, P.L., R. Schmalensee, and E.M. Bailey. 1998. The market for sulfur dioxide emissions. *American Economic Review* 88(4):669-685.

Joye, S. and J. Hollibaugh. 1995. Influence of sulfide inhibition of nitrification on nitrogen regeneration in sediments. *Science* 270:623-625.

Just, R.E. and N. Bockstael (eds.). 1991. *Commodity and Resource Policies in Agricultural Systems*. Springer-Verlag, Berlin.

Justic, D., N.N. Rabalais, R.E. Turner, and Q. Dortch. 1995. Changes in nutrient structure of river-dominated coastal waters: stoichiometric nutrient balance and its consequences. *Estuarine, Coastal and Shelf Science* 40:339-356.

Justic, D., N.N. Rabalais, R.E. Turner, and W.J. Wiseman, Jr. 1993. Seasonal coupling between riverborne nutrients, net productivity, and hypoxia. *Marine Pollution Bulletin* 26:184-189.

Kahn, M.E. 1996. New evidence on trends in vehicle emissions. *Rand J. Econ.* 27(1):183-196.

Kahn, J.R. and W.M. Kemp. 1985. Economic losses associated with the degradation of an ecosystem: the case of submerged aquatic vegetation in the Chesapeake Bay. *Journal of Environmental Economics and Management* 12(3):246-163.

Kaneko, T. and R.R. Colwell. 1975. Adsorption of Vibrio parahaemolyticus onto chitin and copepods. *Appl. Microbiol.* 29:269-274.

Kaoru, Y., V.K. Smith, and J.L. Liu. 1995. Using random utility models to estimate the recreational value of estuarine resources. *American Journal of Agricultural Economics* 77:141-151.

Karr, J. R. and E. Chu. 1997. *Biological Monitoring and Assessment: Using Multimetric Indexes Effectively*. EPA 235-R97-001. Office of Research and Development, Washington, DC.

Kazimi, C. 1997. Evaluating the environmental impact of alternative-fuel vehicles. *J. Environ. Econom. Management* 33(2):163-185.

Kellogg, R.L. and C.H. Lander. 1999. Trends in the potential for nutrient loadings from confined livestock operations. In: *The State of North America's Private Land*. USDA-NRCS, U.S. Government Printing Office, Washington, DC, 5pp.

Kelly, C.A., J.M. Rudd, R.H. Hesslein, D.W. Schindler, P.J. Dillon, C.T. Driscoll, S.A. Gherinie, and R.E. Hecky. 1987. Prediction of biological acid neutralization in acid-sensitive lakes. *Biogeochemistry* 3:129-140.

Ketchum, B. 1951. *The Dispersion and Fate of Pollution Discharged into Tidal Waters, and the Viability of Enteric Bacteria in the Area*. Woods Hole Oceanographic Institution, Woods Hole, MA, 16pp.

Khazzoom, J.D. 1995. An econometric model of the regulated emissions for fuel efficient new vehicles. *J. Environ. Econom. Management* 28(2):190-204.

Kinsey, D.W. and P. Davies. 1979. Effects of elevated nitrogen and phosphorus on coral reef growth. *Limnol. Oceanogr.* 24:935-940.

Kinsman, B. 1957. Proper and improper use of statistics in Geophysics. *Tellus* IX:408-418.

Kirchner, W.B. and P.J. Dillon. 1975. An empirical method of estimating the retention of phosphorus in lakes. *Wat. Resour. Res.* 11:1881-1182.

Kjerfve, B. 1988. *Hydrodynamics of Estuaries: Estuarine Physics, Vol. 1.* CRC Press, Inc., Boca Raton, FL.

Klaassen, G. 1996. *Acid Rain and Environmental Degradation: The Economics of Emission Trading*. Edward Elgar Publishing Co., Cheltenhan, U.K.

Kladivko, E.J., G.E. Van Scoyoc, E.J. Monke, K.M. Oates, and W. Pask. 1991. Pesticide and nutrient movement into subsurface tile drains on a silt loam soil in Indiana. *J. Environ. Qual.* 20:264- 270.

Kling, C.L. 1994. Emissions trading versus rigid regulations in the control of vehicle emissions. *Land Econ.* 70(2):174-188.

Knisel, W.G. 1993. *GLEAMS: Groundwater Loading Effects of Agricultural Management Systems*. U.S. Department of Agriculture, Agricultural Research Service, Southeast Watershed Laboratory, Tifton, GA.

Knisel, W.G. (ed.). 1980. *CREAMS: A Field Scale Model for Chemicals, Runoff and Erosion from Agricultural Management Systems*, Conservation Research Report No. 26. U.S. Department of Agriculture, Agricultural Research Service, Washington, DC.

Knuuttila, S., O.P. Pietilainen, and L. Kauppi. 1994. Nutrient balances and phytoplankton dynamics in two agriculturally loaded shallow lakes. *Hydrobiologia* 275/276:359-369.

Koch, M.S., I.A. Mendelssohn, and K.L. McKee. 1990. Mechanism for the hydrogen sulfide-induced growth limitation in wetland macrophytes. *Limnology and Oceanography* 35:399-408.

Koebel, J.W., Jr. 1995. An historical perspective on the Kissimmee River restoration project. *Restoration Ecology* 3(3):149-159.

Koebel, J.W., Jr., B.L. Jones, and D.A. Arrington. 1999. Restoration of the Kissimmee River, Florida: water quality impacts from canal backfilling. *Environmental Monitoring and Assessment* 57:85-107.

Konyar, K. and C.T. Osborn. 1990. A national-level economic analysis of conservation reserve program participation: a discrete choice approach. *Journal of Agricultural Economic Research* 42(2):5-12.

Kros, H. and P. Warfvinge. 1995. Evaluation of model behavior with respect to the biogeochemistry at the soling spruce site. *Ecol. Model.* 83:255-262.

Krug, A. 1993. Drainage history and land use pattern of a Swedish river systems–their importance for understanding nitrogen and phosphorus load. *Hydrobiologia* 251:285-296.

Krupnick, A.J. 1993. Vehicle emissions, urban smog, and clean air policy. In: *The Environment of Oil: Studies in Industrial Organization, vol. 17*, Gilbert, R.J. (ed.). Kluwer Academic, Norwell, MA and Dordrecht, 143-177pp.

Krupnick, A.J. and M.A. Walls. 1992. The cost-effectiveness of methanol for reducing motor vehicle emissions and urban ozone. *J. of Policy Analysis and Management* 11(3):373-396.

Laffont, J. and J. Tirole. 1994. Environmental policy, compliance and innovation. *European Economic Review* 38(3-4):555-562.

Laffont, J. and J. Tirole. 1996. Pollution permits and environmental innovation. *Journal of Public Economics* 62(1-2):127-140.

Lahlou, M., L. Shoemaker, S. Choudhury, R. Elmer, A. Hu, H. Manguerra, and A. Parker. 1998. *BASINS Version 2.0 User's Manual.* Environmental Protection Agency, Office of Water, Washington, DC.

Lajtha, K., B. Seely, and I. Valiela. 1995. Retention and leaching losses of atmospherically-derived nitrogen in the aggrading coastal watershed of Waquoit Bay, MA. *Biogeochemistry* 28:33-54.

Lam, C.W.Y. and K.C. Ho. 1989. Red tides in Tolo Harbor, Hong Kong. In: *Red Tides: Biology Environmental Science and Toxicology*, Okaichi, T., D.M., Anderson, and T. Nemoto (ed.). Elsevier, New York, 49-52pp.

Lancelot, C., M.D. Keller, V. Rousseau, W.O. Smith, Jr., and S. Mathot. 1998. Autoecology of the marine haptophyte *Phaeocystis sp.* In: *Physiological Ecology of Harmful Algal Blooms*, Anderson, D.M., A.D. Cembella, and G.M. Hallegraeff (eds.). Springer-Verlag, Berlin, Heidelberg, 209-224pp.

Lander, C.H., D. Moffitt, and K. Alt. 1997. Nutrients available from livestock manure relative to crop growth requirements. In: *Resource Assessment and Strategic Planning Working Paper* 98:1. U.S. Department of Agriculture, Natural Resources Conservation Service, Washington, DC.

Lander, C.H., D. Moffitt, and K. Alt. 1998. *Nutrients Available from Livestock Manure Relative to Crop Growth Requirements.* Resource Assessment and Strategic Planning Working Paper 98-1. U.S. Department of Agriculture, Natural Resources Conservation Service, Washington, DC.

Lanyon, L.E. 1999. Nutrient management: regional issues affecting the Chesapeake Bay. In: *Agricultural Phosphorus in the Chesapeake Bay Watershed: Current Status and Future Trends*, Sharpley, A. (ed.). Scientific and Technical Advisory Committee of the Chesapeake Bay Program, Annapolis, MD.

Lanyon, L.E. and P.B. Thompson. 1996. Changing emphasis of farm production. In: *Animal Agriculture and the Environment: Nutrients, Pathogens, and Community Relations*, Salis, M. and J. Popow (eds.). Northeast Regional Agricultural Engineering Service, Ithaca, NY, 15-23pp.

Lapointe, B.E. 1997. Nutrient thresholds for bottom-up control of macroalgal blooms on coral reefs in Jamaica and southeast Florida. *Limnol. Oceanogr.* 42(5, part 2):1119-1131.

Lapointe, B.E. 1999. Simultaneous top-down and bottom-up forces control macroalgal blooms on coral reefs. *Limnol. Oceanogr.* 44(6):1586-1592.

Lapointe, B.E. and M.W. Clark. 1992. Nutrient inputs from the watershed and coastal eutrophication in the Florida Keys. *Estuaries* 15:465-476.

Lapointe, B.E., M.M. Littler, and D.S. Littler. 1993. Modification of benthic community structure by natural eutrophication: the Belize Barrier Reef. *Proc. 7th Inter. Coral Reef Symp. Guam* 1:323-334.

Lapointe, B.E. and W.R. Matzie. 1996. Effects of stormwater nutrient discharges on eutrophication processes in nearshore waters of the Florida Keys. *Estuaries* 19:422-435.

Lapointe, B.E. and J.D. O'Connell. 1989. Nutrient-enhanced productivity of *Cladophora prolifera* in Harrington Sound, Bermuda: eutrophication of a confined, phosphorus-limited marine ecosystem. *Est. Coast Shelf Sci.* 28:347-360.

Lapointe, B.E., J.D. O'Connell, and G.S. Garrett. 1990. Nutrient couplings between on-site sewage disposal systems, groundwaters, and nearshore surface waters of the Florida Keys. *Biogeochemistry* 10:289-307.

Larned, S.T. and J. Stimson. 1996. Nitrogen-limited growth in the coral reef chlorophyte *Dictyosphaeria cavernosa*, and the effect of exposure to sediment derived nitrogen on growth. *Mar. Ecol. Prog. Ser.* 145:95-108.

Larsen, D.P., J.M. Omernik, R.M. Hughes, C.M. Rohm, T.R. Whittier, A.J. Kinney, A.L. Gallant, and D.R. Dudley. 1986. The correspondence between spatial patterns in fish assemblages in Ohio streams and aquatic ecoregions. *Environ. Manage.* 10:815-828.

Legates, D.R and G.J. McCabe, Jr. 1999. Evaluating the use of "goodness-of-fit" measures in hydrologic and hydroclimatic model validation. *Water Resources Research* 35(1):233-241.

Letson, D. 1992. Point/nonpoint source pollution reduction trading: an interpretive survey. *Natural Res. J.* 32(2):219-232.

Lewis, W.M., J.M. Melack, W.H. McDowell, M. McClain, and J.E. Richey. 1999. Nitrogen yields from undisturbed watersheds in the Americas. *Biogeochemistry* 46:149-162.

Lindahl, G. and K. Wallstrom. 1985. Nitrogen fixation (acetylene reduction) in planktonic cyanobacteria in Oregrundsgrepen, southwest Bothnian Sea. *Arch. Hydrobiol.* 104:193-204.

Lipton, D.W. 1998. *Pfiesteria's Economic Impact on Seafood Industry Sales and Recreational Fishing.* Paper presented at Economics of Policy Options for Nutrient Management and Dinoflagellates Conference, Center for Agriculture and Natural Resource Policy, Department of Agricultural and Resource Economics, University of Maryland, November 1998.

Lipton, D.W. and K.F. Wellman. 1995. *Economic Valuation of Natural Resources: A Handbook for Coastal Resource Policymakers.* NOAA Coastal Ocean Program, Decision Analysis Series No. 5, U.S. Department of Commerce.

Litke, D.W. 1999. *Review of Phosphorus Control Measures in the United States and their Effects on Water Quality.* Water-Resources Investigations Report 99-4007, U.S. Geological Survey, Denver, CO.

Livernois, J. and L. Karp. 1994. Using automatic tax changes to control pollution emissions. *Journal of Environmental Economics and Management* 27(1):38-48.

Lohr, L. and T.A. Park. 1995. Voluntary versus mandatory agricultural policies to protect water quality: adoption of nitrogen testing in Nebraska. *Review of Agricultural Economics* 17(1):13-24.

Lovett, G.M. and S.E. Lindberg. 1993. Atmospheric deposition and canopy interactions of nitrogen in forests. *Can. J. For. Res.* 23:1603-1616.

Lowrance, R.R., L.S. Altier, J.D. Newbold, R.R. Schnabel, P.M. Groffman, J.M. Denver, J.W. Gilliam, J.L. Robinson, R.B. Brinsfield, K.W. Staver, W. Lucas, and A.H. Todd. 1995. *Water Quality Functions of Riparian Forest Buffer Systems in the Chesapeake Bay Watershed.* Nutrient Subcommittee Chesapeake Bay Program, Environmental Protection Agency, Annapolis, MD.

Lowrance, R.R., R.A. Leonard, and J.M. Sheridan. 1985. Managing riparian ecosystems to control nonpoint pollution. *J. Soil and Water Conserv.* 40:87-91.

Lowrance, R.R., R.L. Todd, and L.E. Asmussen. 1984a. Nutrient cycling in an agricultural watershed: stream flow and artificial drainage. *J. Environ. Qual.* 13:27-32.

Lowrance, R.R., R.L. Todd, J. Fail, O. Hendrickson, Jr, R. Leonard, and L. Asmussen. 1984b. Riparian forests as nutrient filters in agricultural watersheds. *BioScience* 34:374-377.

Lucas, L.V., J.E. Cloern, J.R. Koseff, S.G. Monismith, and J.K. Thompson. 1998. Does the Sverdrup critical depth model explain bloom dynamics in estuaries? *J. Mar. Res.* 56:375-415.

Lynch, D.R. and A.M. Davies. 1995. Quantitative skill assessment for coastal ocean models. In: *Coastal and Estuarine Studies* 47. American Geophysical Union, Washington, DC, 510pp.

Lynne, G.D., P. Conroy, and F.J. Prochaska. 1981. Economic valuation of marsh areas for marine production processes. *Journal of Environmental Economics and Management* 8:175-186.

Magdoff, F.R., D. Ross, and J. Amadon. 1984. A soil test for nitrogen availability for corn. *Soil Sci. Soc. Am. J.* 48:1301-1304.

Magill, A.H., J.D. Aber, J.J. Hendricks, R.D. Bowden, J.M. Melillo, and P.A. Steuder. 1997. Biogeochemical response of forest ecosystems to simulated chronic nitrogen deposition. *Ecological Applications* 7:402-415.

Magnien, R., D. Boward, and S. Bieber (eds.). 1995. *The State of the Chesapeake 1995.* Environmental Protection Agency, Annapolis, MD.

Malik, A.S., D. Letson, and S.R. Crutchfield. 1993. Point/nonpoint source trading of pollution abatement: choosing the right trading ratio. *Amer. J. Agr. Econ.* 75(4):959-967.

Malone, T.C. 1977. Environmental regulation of phytoplankton productivity in the lower Hudson Estuary. *Estuar. Coast. Mar. Sci.* 5:151-171.

Malone, T.C., D.J. Conley, T.F., Fisher, T. F., P.M. Gilbert, L.W. Harding, and K.G. Sellner. 1996. Scales of nutrient-limited phytoplankton productivity in Chesapeake Bay. *Estuaries* 19:371-385.

Mapp, H.P., D.J. Bernardo, G.S. Sabbagh, S. Gelata, and K.B. Watkins. 1994. Economic and environmental impacts of limiting nitrogen use to protect water quality: a stochastic regional analysis. *American Journal of Agricultural Economics* 76(4):889-903.

Marcus, N.H. and F. Boero. 1998. Minireview: the importance of benthic-pelagic coupling and the forgotten role of life cycles in coastal aquatic systems. *Limnology and Oceanography* 43:763-768.

Marino, R., R.W. Howarth, J. Shamess, and E.E. Prepas. 1990. Molybdenum and sulfate as controls on the abundance of nitrogen-fixing cyanobacteria in saline lakes in Alberta. *Limnol. Oceanogr.* 35:245-259.

Markager, S. and K. Sand-Jensen. 1990. Heterotrophic growth of *Ulva lactuca* (Chlorohyceae). *J. Phycol.* 26:670-673.

Martin, J.H., K.H. Coale, K.S. Johnson, S.E. Fitzwater, R.M. Gordon, S.J. Tanner, C.N. Hunter, V.A. Elrod, J.L. Nowicki, T.L. Coley, R.T. Barber, S. Lindley, A.J. Watson, K. Van Scoy, C.S. Law, M.I. Liddicoat, R. Ling, T. Stanton, J. Stockel, C. Collins, A. Anderson, R. Bidigare, M. Ondrusek, M. Latasa, F.J. Millero, K. Lee, W. Yao, J.Z. Zhang, G. Friederich, C. Sakamoto, F. Chavez, K. Buck, Z. Kolber, R. Greene, P. Falkowski, S.W. Chisholm, F. Hoge, R. Swift, J. Yungel, S. Turner, P. Nightingale, A. Hatton, P. Liss, and N.W. Tindale. 1994. Testing the iron hypothesis in ecosystems of the equatorial Pacific Ocean. *Nature* 371:123-129.

Marubini, F. and P. S. Davies. 1996. Nitrate increases zooxanthellae density and reduces skeletogenesis in corals. *Mar. Biol.* 127:319-328.

Maryland Department of Natural Resources. (2000). *The Cambridge Consensus: Forum on Land-Based Pollution and Toxic Dinoflagellates in Chesapeake Bay*, [Online]. Available: http://www.dnr.state.md.us/pfiesteria/ccc.html [2000, January 31].

Maryland General Assembly. 1998. *Water Quality Improvement Act of 1998.* Annapolis, MD.

Matson, P.A., W.J. Parton, A.G. Power, and M.J. Swift. 1997. Agricultural intensification and ecosystem properties. *Science* 277:504-509.

Mattson, W.J., Jr. 1980. Herbivory in relation to plant nitrogen content. *Ann. Rev. Ecology and Systematics* 11:119-161.

McCollum, R.E. 1991. Buildup and decline in soil phosphorus: 30-year trends on a Typic Umprabuult. *Agronomy Journal* 83(1):77-85.

McCutcheon, S.C., J.L. Martin and T.O. Barnwell, Jr. 1993. Water quality. *Handbook of Hydrology*, Maidment, D.L. (ed.). McGraw-Hill, New York.

McDowell, D.M. and B.A. O'Connor. 1977. *Hydraulic Behavior of Estuaries.* John Wiley and Sons, New York.

McGlathery, K.J., D. Krause-Jense, S. Rysgaard, and P.B. Christensen. 1997. Patterns of ammonium uptake within dense mats of the filamentous macroalga Chaetomorpha linum. *Aquat. Bot.* 59:99-115.

McGlathery, K.J., R. Marino, and R.W. Howarth. 1994. Variable rates of phosphate uptake by shallow marine carbonate sediments: mechanisms and ecological significance. *Biogeochemistry* 25:127-146.

McLean-Meyinsse, P.E., J. Hui, and R. Joseph, Jr. 1994. An empirical analysis of Louisiana small farmers' involvement in the conservation reserve program. *Journal of Agricultural and Applied Economics* 26(2):379-385.

McSweeny, W.T. and J.S. Shortle. 1989. Reducing nutrient application in intensive livestock areas: policy implications of alternative producer behavior. *Northeast Journal of Agricultural and Resource Economics* 18(1):1-11.

Meade, R.H. 1988. Movement and storage of sediment in river systems. In: *Physical and Chemical Weathering in Geochemical Cycles*, Lerman, A. and M. Meybeck (eds.). Kluwer, Dordrecht, 165-179pp.

Meals, D.W. 1990. LaPlatte River watershed water quality monitoring and analysis program: comprehensive final report. *Program Report No. 12.* Vermont Water Resour. Res. Center, Univ. of Vermont, Burlington.

Meeuwig, J.J., J.B. Rasmussen, and R.H. Peters. 1998. Turbid waters and clarifying mussels: their moderation of empirical chlorine:nutrient relations in estuaries in Prince Edward Island, Canada. *Mar. Ecol. Prog. Ser.* 171:139-150.

Meisinger, J.J., V.A. Bandel, J.S. Angle, B.E. O'Keefe, and C.M. Reynolds. 1992. Preside dress soil nitrate test evaluation in Maryland. *Soil Sci. Soc. Am. J.* 56:1527-1532.

Messinger, F. and A. Arakawa. 1976. *Numerical Methods Used in Atmospheric Models.* Global Atmospheric Research Programme, WMO-ICSU Joint Organizing Committee, GARP Publication.

Michaelis, L. 1995. The abatement of air pollution from motor vehicles: the role of alternative fuels. *J. Transport. Econ. Policy* 29(1):71-84.

Millhouser, W.C., J. McDonough, J.P. Tolson, and D. Slade. 1998. *Managing Coastal Resources.* National Oceanic and Atmospheric Administration, NOAA's State of the Coast Report, Silver Spring, MD.

Mills, W.B., D.B. Porcella, M.J. Ungs, S.A. Gherini, K.V. Summers, M. Lingfung, M. Rupp, G.L. Bowie, and D.A. Haith. 1985. *Water Quality Assessment: A Screening Procedure for Toxic and Conventional Pollutants.* Environmental Protection Agency, Athens, GA.

Mitsch, W.J. and J.G. Gosselink. 1986. *Wetlands.* van Nostrand Reinhold, New York.

Moffa, P.E. (ed.). 1990. *Control and Treatment of Combined Sewer Overflows.* van Nostrand Reinhold, New York.

Moffat, A.S. 1998. Global nitrogen overload problem grows critical. *Science* 279:988-989.

Moffett, J.W., L.E. Brand, P.L.Croot, and K.A. Barbeau. 1997. Copper speciation and cyanobacterial distribution in harbors subject to anthropogenic copper inputs. *Limnol. Oceanogr.* 42(5):789-799.

Montgomery, D. and J. Buffington. 1993. *Channel Classification, Prediction of Channel Response, and Assessment of Channel Condition.* Washington State Dept. of Natural Resources Report TFW-SH10-93-002 for SHAMW Committee of the Washington State Timber/Fish/Wildlife Agreement, 84pp.

Moore, P.A., Jr. and D.M. Miller. 1994. Decreasing phosphorus solubility in poultry litter with aluminum, calcium, and iron amendments. *J. Environ. Qual.* 23:325-330.

Moore, P.A., Jr., T.C. Daniel, D.R. Edwards, and D.M. Miller. 1995. Effect of chemical amendments on ammonia volatilization from poultry litter. *J. Environ. Qual.* 24:293-300.

Moore, M.V., M.L. Pace, J.R. Mather, P.S. Murdoch, R.W. Howarth, C.L. Folt, C.Y. Chen, H.F. Hemond, P.A. Flebbe, and C.T. Driscoll. 1997. Potential effects of climate change on freshwater ecosystems of the New England/mid-Atlantic region. *Hydrol. Proc.* 11:925-947.

Morse, D., H.H. Head, C.J. Wilcox, H.H. van Horn, C.D. Hissem, and B. Harris, Jr. 1992. Effects of concentration of dietary phosphorus on amount and route of excretion. *J. Dairy Sci.* 75:3039-3045.

Moxey, A. and B. White. 1994. Efficient compliance with agricultural nitrate pollution standards. *Journal of Agricultural Economics* 45(1):27-37.

Muscatine, L. and J. Porter. 1977. Reef corals: mutualistic symbiosis adapted to nutrient-poor environments. *BioScience* 27:454-460.

Myers, V.B. and R.I. Iverson. 1981. Phosphorus and nitrogen limited phytoplankton productivity in northeastern Gulf of Mexico coastal estuaries. In: *Estuaries and Nutrients*, Nielson, B.J. and L.E. Cronin (eds.), Humana, NY.

Naiman, R.J., P.A. Bisson, R.G. Lee, and M.G. Turner. 1998. Watershed management In: *River Ecology and Management: Lessons from the Pacific Coastal Ecoregion*. Springer, NY, 642-661pp.

Napier, T.L. and E.J. Johnson. 1998. Impacts of voluntary conservation initiatives in the Darby Creek Watershed of Ohio. *Journal of Soil and Water Conservation* 53(1):78-84.

National Ocean Research Leadership Council (NORLC). 1999. *An Integrated Ocean Observing System: A Strategy for Implementing the First Steps of a U.S. Plan.* National Oceanographic Partnership Program Office, Consortium for Oceanographic Research and Education, Washington, DC.

National Oceanic and Atmospheric Administration (NOAA). 1998. *Population: Distribution, Density, and Growth*, Culliton, T.J. (ed.). NOAA's State of the Coast Report, Silver Spring, MD.

National Oceanic and Atmospheric Administration (NOAA). (1999a). *Applying Science and Education to Improve the Management of Estuaries,* [Online]. Available: http://wave.nos.noaa.gov/ocrm/nerr [2000, February 17].

National Oceanic and Atmospheric Administration (NOAA). (1999b). *Atmospheric Loadings to Coastal Ecosystems,* [Online]. Available: www.arl.noaa.gov/research/programs/coast1/html [1999, November 11].

National Oceanic and Atmospheric Administration (NOAA). 1999c. *Hypoxia in the Gulf of Mexico: Progress Towards the Completion of an Integrated Assessment.* National Ocean Service, NOAA, Silver Spring, MD.

National Research Council (NRC). 1986. *Acid Deposition: Long-Term Trends.* National Academy Press, Washington, DC.

National Research Council (NRC). 1989. *Nutrient Requirements of Dairy Cattle,* 6th Ed. National Academy Press, Washington, DC.

National Research Council (NRC). 1990. *Managing Troubled Waters: The Role of Marine Environmental Monitoring*. National Academy Press, Washington, DC, 125pp.

National Research Council (NRC). 1991a. *Rethinking the Ozone Problem in Urban and Regional Air Pollution*. National Academy Press, Washington, DC.

National Research Council (NRC). 1991b. *Restoration of Aquatic Ecosystems: Science, Technology, and Public Policy*. National Academy Press, Washington, DC.

National Research Council (NRC). 1992. *Automotive Fuel Economy: How Far Should We Go?* National Academy Press, Washington, DC.

National Research Council (NRC). 1993a. *Managing Wastewater in Coastal Urban Areas*. National Academy Press, Washington, DC.

National Research Council (NRC). 1993b. *Soil and Water Quality: An Agenda for Agriculture*. National Academy Press, Washington, DC.

National Research Council (NRC). 1994. *Review of EPA's Environmental Monitoring and Assessment Program: Forest and Estuaries Components*. National Academy Press, Washington, DC.

National Research Council (NRC). 1997a. *Valuing Ground Water: Economic Concepts and Approaches*. National Academy Press, Washington, DC.

National Research Council (NRC). 1997b. *Striking a Balance: Improving Stewardship of Marine Areas*. National Academy Press, Washington, DC.

National Research Council (NRC). 1999a. *New Strategies for America's Watersheds*. National Academy Press, Washington, DC.

National Research Council (NRC). 1999b. *From Monsoons to Microbes: Understanding the Ocean's Role in Human Health*. National Academy Press, Washington, DC.

National Science and Technology Council (NSTC). (1998). *National Science and Technology Council 1998 Annual Report*, [Online]. Available: www.whitehouse.gov/WH/EOP/OSTP/NSTC [1999, November 15].

National Water Quality Evaluation Project. 1988. Status of Agricultural Nonpoint Source Projects. North Carolina State Univ., Raleigh.

Nelson, D. 1985. Minimizing nitrogen losses in non-irrigated eastern areas. In: *Plant Nutrient Use and the Environment*. Fertilizer Institute, Washington, DC, 173-209pp.

Newell, R.I.E. 1988. Ecological changes in Chesapeake Bay: are they the result of over harvesting the American oyster, *Crassotrea virginica*? In: *Proceedings of Understanding the Estuary: Advances in Chesapeake Bay Research*, March 1988. Chesapeake Bay Consortium, Baltimore, MD, 29-31pp.

Nichols, D.J., T.C. Daniel, P.A. Moore, Jr., D.R. Edwards, and D.H. Pote. 1997. Runoff of estrogen hormone 17-estradiol from poultry litter applied to pasture. *J. Environ. Qual.* 26:1002-1006.

Nihoul, J.C.J. 1998. Optimum complexity in ecohydrodynamic modeling: an ecosystem dynamics standpoint. *J. Mar. Sys.* 16:3-5.

Nixon, S.W. 1981. Remineralization and nutrient cycling in coastal marine ecosystems. In: *Estuaries and Nutrients*, Neilson, B.J. and L.E. Cronin (eds.). Humana Press, Clifton, NJ, 111-138pp.

Nixon, S.W. 1988. Physical energy inputs and the comparative ecology of lake and marine ecosystems. *Limnol. Oceanogr.* 33:1005-1025.

Nixon, S.W. 1992. Quantifying the relationship between nitrogen input and the productivity of marine ecosystems. *Proceedings Advances in Marine Technology Conference* 5:57-83.

Nixon, S.W. 1995. Coastal marine eutrophication: a definition, social causes, and future concerns. *Ophelia* 41:199-219.

Nixon, S.W. 1997. Prehistoric nutrient inputs and productivity in Narragansett Bay. *Estuaries* 20(2):253-261.

Nixon, S.W. 1998. Physical energy inputs and the comparative ecology of lake and marine ecosystems. *Limnol. Oceanogr.* 33(4):1005-1025.

Nixon, S.W., J.W. Ammerman, L.P. Atkinson, V.M. Berounsky, G. Billen, W.C. Boicourt, W.R. Boyton, T.M. Church, D.M. DiToro, R. Elmgren, J.H. Garber, A.E. Giblin, R.A. Jahnke, N.J.P. Owens, M.E.Q. Pilson, and S.P. Seitzinger. 1996. The fate of nitrogen and phosphorus at the land-sea margin of the North Atlantic Ocean. *Biogeochemistry* 35:141-180.

Nixon, S.W., S.L. Granger, and B.L. Nowicki. 1995. An assessment of the annual mass balance of carbon, nitrogen, and phosphorus in Narragansett Bay. *Biogeochemistry* 31:15-61.

Nixon, S.W., J.R Kelly, B.N. Furnas, C.A. Oviatt, and S.S. Hale. 1980. Phosphorus regeneration and the metabolism of coastal marine bottom communities. In: *Marine Benthic Dynamics*, Tenore, K.R. and B.C. Coull (eds.), University of South Carolina Press.

Nixon, S.W., C.A. Oviatt, J. Frithsen, and B. Sullivan. 1986. Nutrients and the productivity of estuarine and coastal marine ecosystems. *J. Limnol. Soc. South Africa* 12(1/2):43-71.

Nixon, S.W. and M.E.Q. Pilson. 1983. Nitrogen in estuarine and coastal marine ecosystems. In: *Nitrogen in the Marine Environment*, Carpenter, E.J. and D.G. Capone (eds.). Academic Press, San Diego.

Norko, A. and E. Bonsdorff. 1996. Rapid zoobenthic community responses to accumulations of drift algae. *Mar. Ecol. Prog. Ser.* 131:143-157.

Novotny, V. (ed.). 1995. *Nonpoint Pollution and Urban Stormwater Management*. Technometric Publishing Co., Lancaster, PA.

Novotny, V., K.R. Imhoff, M. Olthof, and P.A. Krenkel. 1989. *Karl Imhoff's Handbook of Urban Drainage and Wastewater Disposal*. John Wiley and Sons, New York.

Novotny, V. and H. Olem. 1994. *Water Quality: Prevention, Identification, and Management of Diffuse Pollution*. van Nostrand Reinhold, New York.

Nowlin, W.D., Jr. 1999. A strategy for long-term ocean observations. *Bull. Am. Met. Soc.* 80(4):621-627.

Nunes Vaz, R.A. and C.W. Lennon. 1991. Modulation of estuarine stratification and mass transport at tidal frequencies. In: *Progress in Tidal Hydrodynamics*, Parker, B.B. (ed.). John Wiley and Sons, New York, 505-520pp.

Odum, H.T., B.J. Copeland, and E. McMahon. 1974. *Coastal Ecological Systems of the United States*. The Conservation Foundation, Washington, DC, 4 vols.

Oey, L.Y. 1984. On the steady salinity distribution and circulation in partially mixed and well mixed estuaries. *J. Phys. Oceanogr.* 14:629-645.

Officer, C.B. 1976. *Physical Oceanography of Estuaries and Associate Coastal Waters*. Wiley-Interscience, New York.

Officer, C.B., R. Biggs, J. Taft, L.E. Cronin, M.A. Tyler, and W.R. Boynton. 1984. Chesapeake Bay anoxia: origin, development, and significance. *Science* 223:22-27.

Okaichi, T. 1997. Red tides in the Seto Inland Sea. In: *Sustainable Development in the Seto Inland Sea, Japan: From the Viewpoint of Fisheries*, Okaichi, T. and T. Yanagi (eds.). Terra Scientific Publishing Company, Tokyo, 251-304pp.

Omernik, J.M. 1977. *Nonpoint Source–Stream Nutrient Level Relationships: A Nationwide Study*. EPA-600/3-77-105. Corvallis, OR.

Omernik, J.M. 1987. Ecoregions of the conterminous United States. *Annals of the Association of American Geographers* 77(1):118-125.

Oregon State University. (1999). *Watershed Analysis Tools on the Web*, [Online]. Available: http://biosys.bre.orst.edu/watersheds.htm [1999, November 16].

Oviatt, C.A., P. Doering, B. Nowicki, L. Reed, J. Cole, and J. Frithsen. 1995. An ecosystem level experiment on nutrient limitation in temperate coastal marine environments. *Mar. Ecol. Prog. Ser.* 116:171-179.

Paerl, H.W. 1985. Enhancement of marine primary production by nitrogen-enriched acid rain. *Nature* 316:747-749.

Paerl, H.W. 1990. Physiological ecology and regulation of nitrogen fixation in natural waters. In: *Advances in Microbial Ecology*. Marshall, K.C. (ed.), Plenum.

Paerl, H.W. 1997. Coastal eutrophication and harmful algal blooms: the importance of atmospheric and groundwater as "new" nitrogen and other nutrient sources. *Limnol. Oceanogr.* 42:1154-1165.

Paerl, H.W. and R. Whitall. 1999. Anthropogenically-derived atmospheric nitroegn deposition, marine eutrophication and harmful algal bloom expansion: Is there a link? *Ambio* 28:307-311.

Paerl, H.W. and J.P. Zehr. 2000. Nitrogen fixation. In: *Marine Microbial Ecology*, Kirchman, D.F. (ed.). Academic Press, San Diego.

Pan, J.H. and I. Hodge. 1994. Land use permits as an alternative to fertilizer and leaching taxes for the control of nitrate pollution. *Journal of Agricultural Economics* 45(1):102-112.

Pardo, L. and C. Driscoll. 1993. A critical review of mass balance methods for calculating critical loads of nitrogen for forested ecosystems. *Environment* 1:145-156.

Parks, P.J. and J.P Schorr. 1997. Sustaining open space benefits in the northeast: an evaluation of the Conservation Reserve Program. *Journal of Environmental Economics and Management* 32(1):85-94.

Parry, R. 1998. Agricultural phosphorus and water quality: a U.S. Environmental Protection Agency perspective. *J. Environ. Qual.* 27:258-261.

Parton, W.J., M. Hartman, D. Ojima, and D. Schimel. 1998. DAYCENT and its land–surface submodel: description and testing. *Global Planet. Change* 19:35-48.

Parton, W.J., J.W.B. Stewart, and C.V. Cole. 1988. Dynamics of carbon, nitrogen, phosphorus, and sulfur in grassland soils: a model. *Biogeochemistry* 5:109-131.

Paustian, K., W.J. Parton, and J. Persson. 1992. Modeling soil organic matter in organic-amended and nitrogen-fertilized long-term plots. *Soil Sci. Soc. Am. J.* 56:476-488.

Pearson, T.H. and R. Rosenberg. 1978. Macrobenthic succession in relation to organic enrichment and pollution of the marine environment. *Oceanogr. Mar. Biol. Ann. Rev.* 163:229-311.

Pedersen, M.F. and J. Borum. 1996. Nutrient control of algal growth in estuarine waters: nutrient limitation and the importance of nitrogen requirements and nitrogen storage among phytoplankton and species of macroalgae. *Mar. Ecol. Prog. Ser.* 142: 261-272.

Peterjohn, W.T. and D.L. Correll. 1984. Nutrient dynamics in an agricultural watershed: observations on the role of a riparian forest. *Ecology* 65:1466-1475.

Pihl, L., S.P. Baden, and R.J. Diaz. 1991. Effects of periodic hypoxia on distribution of demersal fish and crustaceans. *Marine Biology* 108:349-360.

Pihl, L., S.P. Baden, R.J. Diaz, L.C. Schaffner. 1992. Hypoxia-induced structural changes in the diet of bottom-feeding fish and crustacea. *Marine Biology* 112:349-361.

Pinay, G., L. Roques, and A. Fabre. 1993. Spatial and temporal patterns of denitrification in a riparian forest. *J. Appl. Ecol.* 30:581-91.

Postma, H. 1985. Eutrophication of Dutch coastal waters. *Neth. J. Zool.* 35:348-359.

Pote, D.H., T.C. Daniel, A.N. Sharpley, P.A. Moore, D.R. Edwards, and D.J. Nichols. 1996. Relating extractable phosphorus to phosphorus losses in runoff. *Soil Sci. Soc. Am. J.* 60:855-859.

Prandle, D. 1986. Generalized theory of estuarine dynamics. In: *Physics of Shallow Estuaries and Bays*, van de Kreek, J. (ed.). Springer-Verlag, Berlin, 42-57pp.

Preston, S.D., R.A. Smith, G.E. Schwarz, R.B. Alexander, and J.W. Brakebill. 1998. *Spatially Referenced Regression Modeling of Nutrient Loading in the Chesapeake Bay Watershed*. Proceedings of the First Federal Interagency Hydrologic Modeling Conference, Las Vegas, NV.

Prince, R. and S.R. Milliman. 1989. Firm incentives to promote technological change in pollution control. *Journal of Environmental Economics and Management* 17(3):247-265.

Pritchard, D. 1952. Estuarine hydrography. *Advances in Geophysics* 1:243-280.

Pritchard, D. 1967. What is an estuary: physical viewpoint. In: *Estuaries*, Lauff, G. (ed.). Publ. 83, AAAS, Washington, DC, 3-5pp.

Probert, M.E., B.A. Keating, J.P. Thompson, and W.J. Parton. 1995. Modeling water, nitrogen, and crop yield for a long-term fallow management experiment. *Australian Journal of Experimental Agriculture* 35:941-950.

Prospero, J.M., K. Barrett, T. Church, F. Dentner, R.A. Duce, J.N. Galloway, H. Levy, J. Moddy, and P. Quinn. 1996. Atmospheric deposition of nutrient to the North Atlantic basin. *Biogeochemistry* 35:75-139.

Pryor, D., R. Bierbaum, and J. Melillo. 1998. Environmental monitoring and research initiative: a priority activity for the Committee on Environmental and Natural Resources. *Environmental Monitoring and Assessment* 51:3-14.

Rabalais, N.N., R.E. Turner, D. Justic, Q. Dortch, and W.J. Wiseman, Jr. 1999. *Characterization of Hypoxia: Topic 1 Report for the Integrated Assessment of Hypoxia in the Gulf of Mexico.* NOAA Coastal Ocean Program Decision Analysis Series No. 15. NOAA Coastal Ocean Program, Silver Spring, MD, 167pp.

Rabalais, N.N., R.E. Turner, D. Justic, Q. Dortch, W.J. Wiseman, Jr., and B.K. Sen Gupta. 1996. Nutrient changes in the Mississippi River and system responses on the adjacent continental shelf. *Estuaries* 19:386-407.

Rabalais, N.N., R.E. Turner, W.J. Wiseman, Jr., and D.F. Boesch. 1991. A brief summary of hypoxia on the northern Gulf of Mexico continental shelf: 1985-1988. In: *Modern and Ancient Continental Shelf Anoxia*, Tyson, R.V. and T.H. Pearson (eds.). Geological Society Special Publication No. 58, The Geological Society, London, 470pp.

Radach, G., J. Berg, and E. Hagmeier. 1990. Long-term changes of the annual cycles of meteorological, hydrographic, nutrient and phytoplankton time series at Helgoland and at LV ELBE 1 in the German Bight. *Continental Shelf Research* 10:305-328.

Radach, G. and H.J. Lenhart. 1995. Nutrient dynamics in the North Sea: fluxes and budgets in the water column derived from the European regional seas ecosystem model. *Neth. J. Sea Res.* 33:301-335.

Ralph, S.C., G.C. Poole, L.L. Conquest, and R.J. Naiman. 1994. Stream channel condition and in-stream habitat in logged and unlogged basins of western Washington. *Canadian Journal of Fisheries and Aquatic Sciences* 51:37-51.

Rask, K.N. 1998. Clean Air and renewable fuels: the market for fuel ethanol in the United States from 1984 to 1993. *Energy Economics* 20(3):325-345.

Rast, W. and G.F. Lee. 1978. *Summary Analysis of the North American OECD Eutrophication Project: Nutrient Loading, Lake Response Relationships, and Trophic State Indices.* EPA 600/3-78-008, EPA, Corvallis, OR.

Rauch, W., M. Henze, L. Koncsos, P. Reichert, P. Shanahan, L. Somlyody, and P. Vanrolleghem. 1998. River water quality modeling: state of the art. *Water Science and Technology* 38(11):237-260.

Redfield, A.C. 1958. The biological control of chemical factors in the environment. *American Scientist* 46:205-222.

Reguera, B., J. Blanco, Mª L. Fernández, and T. Wyatt (eds.). *Harmful Algae*. 1998. Xunta de Galicia and Intergovernmental Oceanographic Commission of UNESCO.

Reichelderfer, K.H. 1990. National agroenvironmental incentives programs: the U.S. experience. In: *Agriculture and Water Quality: International Perspectives*, Braden, J.B. and S.B. Lovejoy (eds.). Lynne Rienner Publishers, Boulder, CO and London.

Ribaudo, M. 1998. Lessons learned about the performance of USDA agricultural nonpoint source pollution programs. *Journal of Soil and Water Conservation* 53(1):4-40.

Ribaudo, M. and D. Hellerstein. 1992. *Estimating Water Quality Benefits: Theoretical and Methodological Issues.* U.S. Department of Agriculture, Economic Research Service.

Richards, R.P. and D.B. Baker. 1993. Trends in nutrient and suspended sediment concentrations in Lake Erie tributaries, 1975-1990. *J. Great Lakes Res.* 19:200-211.

Riegman, R., A.A.M. Noordeloos, and G.C.Cadée. 1992. Phaeocystis blooms and eutrophication of the continental coastal zones of the North Sea. *Mar. Biol.* 112:479-484.

Ries, T. 1993. The Tampa Bay experience. In: *Proceedings and Conclusions of Workshops on Submerged Aquatic Vegetation and Photosynthetically Active Radiation*, Morriss, L. and D. Tomasko (eds.). Special publication SJ93-SP13. St. Johns River Water Management District, Palatka, FL, 19-24pp.

Rigler, F.H. and R.H. Peters. 1995. Science and limnology, vol. 6. In: *Excellence in Ecology*, Kinne, O. (ed.). Ecology Institute, Oldendorf/Luhe, Germany.

Ritter, M.C. and P.A. Montagna. 1999. Seasonal hypoxia and models of benthic response in a Texas bay. *Estuaries* 22:7-20.

Robblee, M.B., T.R. Barber, P.R. Carlson, M.J. Durako, J.W. Fourqurean, L.K. Muehlstein, D. Porter, L.A. Yarbro, R.T. Zieman, and J.C. Zieman. 1991. Mass mortality of the tropical seagrass (*Thalassia testudinum*) in Florida Bay (USA). *Mar. Prog. Ser.* 71:297-299.

Roesner, L.A., J.A. Aldrich, and R.E. Dickinson. 1988. *Storm Water Management Model Version 4, User's Manual: EXTRAN Addendum*. Environmental Protection Agency, Athens, GA.

Rohm, C.M., J.W. Giese, and C.C. Bennett. 1987. Evaluation of an aquatic ecoregion classification of streams in Arkansas. *Freshwater Ecology* 4:127-140.

Roman, M., A.L. Gauzens, W.K. Rhinehart, and J. White. 1993. Effects of low oxygen water on Chesapeake Bay zooplankton. *Limnol. Oceanogr.* 38:1603-1614.

Rosen, S. 1974. Hedonic prices and implicit markets: product differentiation in perfect competition. *Journal of Political Economy* 82(1):34-55.

Rosenberg, R. 1985. Eutrophication: the future marine coastal nuisance. *Marine Pollution Bulletin* 16:227-231.

Rosenberg, R., R. Elmgren, S. Fleischer, P. Jonsson, G. Persson, and H. Dahlin. 1990. Marine eutrophication case studies in Sweden. *Ambio* 19(3):102-108.

Rosenberg, R., O. Lindhal., and H. Blanck. 1988. Silent spring in the sea. *Ambio* 17:289-290.

Rosgen, D.L. 1994. A classification of natural rivers. *Catena* 22:169-199.

Ryden, J.C., J.K. Syers, and R.F. Harris. 1973. Phosphorus in runoff and streams. *Adv. Agron.* 25:1-45.

Ryther, J.H., and W.M. Dunstan. 1971. Nitrogen, phosphorus, and eutrophication in the coastal marine environment. *Science* 171:1008-1012.

Sand-Jensen, K. 1977. Effect of epiphytes on eelgrass photosynthesis. *Aquat. Bot.* 3:55-63.

Sand-Jensen, K. and J. Borum. 1991. Interactions among phytoplankton, periphyton, and macrophytes in temperate freshwaters and estuaries. *Aquatic Botany* 41:137-175.

Sandquist, R.L. and C. Paulson. 1998. The Cherry Creek watershed: a case study of water quality trading. *Watershed-Based Effluent Trading Demonstration Projects: Results Achieved and Lessons Learned.* Water Environment Research Foundation Workshop Proceedings, Orlando.

Sarasota Bay National Estuary Program. 1995. *Sarasota Bay: The Voyage to Paradise Reclaimed: The Comprehensive Conservation and Management Plan for Sarasota Bay*, Anderson, M., C. Ciccolella, and P. Roat (eds.). Sarasota Bay National Estuary Program, Sarasota, FL.

Savchuck, O. and F. Wulff. 1999. Modeling regional and large-scale response of Baltic Sea ecosystems to nutrient load reductions. *Hydrobiologia* 393:35-43.

Schaffner, L., P. Jonsson, R.J. Diaz, R. Rosenberg, and P. Gapcynski. 1992. Benthic communities and bioturbation history of estuarine and coastal systems: effects of hypoxia and anoxia. *Science of the Total Environment* Supplement:1001-1016.

Schelske, C.L. 1988. Historical trends in Lake Michigan silica concentrations. *Int. Rev. Hydrobiol.* 73:559-591.

Schepers, J.S., T.M. Blackmer, and D.D. Francis. 1992. Predicting nitrogen fertilizer needs for corn in humid regions: using chlorophyll meters. In: *Tennessee Valley Authority Bulletin Y-226*, Bock, B.R. and K.R. Kelley (eds.). Proc. SSSA Symp., Denver, CO, 105-114pp.

Schindler, D.W. 1977. Evolution of phosphorus limitation in lakes. *Science* 195:260-262.

Schlesinger, W.H. 1997. *Biogeochemistry: An Analysis of Global Change*. Academic Press, San Diego.

Scholin, C.A., F. Gulland, G.J. Doucette, S. Benson, M. Busman, F.P. Chavez, J. Cordaro, R. DeLong, A. De Vogelaere, J. Harvey, M. Haulena, K. Lefebvre, T. Lipscomb, S. Loscutoff, L.J. Lowenstine, R. Marin III, P.E. Miller, W.A. McLellan, P.D.R. Moeller, C.L. Powell, T. Rowles, P. Silvagni, M. Silver, T. Spraker, V. Trainer, and F.M. Van Dolah. 2000. Mortality of sea lions along the central California coast linked to a toxic diatom bloom. *Nature* 403:80-84.

Schreiber, J.D., P.D. Duffy, and D.C. McClurkin. 1976. Dissolved nutrient losses in storm runoff from five southern pine watersheds. *J. Environ. Qual.* 5:201-205.

Schueler, T.R. 1987. *Controlling Urban Runoff: A Practical Manual for Planning and Designing Urban Best Management Practices*. Metropolitan Washington Council of Governments, Washington, DC.

Schueler, T.R. 1992. *Design of Stormwater Wetland Systems: Guidelines for Creating Diverse and Effective Stormwater Wetlands in the Mid-Atlantic Region*. Metropolitan Washington Council of Governments, Washington, DC.

Schueler, T.R. 1996. Crafting better urban watershed protection plans. *Watershed Protection Techniques* 2(2):1-9.

Schulze, E.D., W. de Vries, M. Hauhs, K. Rosen, L. Rasmussen, O.C. Tann, and J. Nilsson. 1989. Critical loads for nitrogen deposition in forest ecosystems. *Water, Air, and Soil Pollution* 48:451-456.

Segerson, K. and N. Li. 1999. Voluntary approaches to environmental protection. *The International Yearbook of Environmental and Resource Economics 1999/2000: A Survey of Current Issues*, Folmer, H. and T. Tietenberg (eds.). Edward Edgar, Cheltenham, UK.

Segerson, K. and T. Miceli. 1998. Voluntary environmental agreements: good or bad news for environmental protection? *Journal of Environmental Economics and Management* 36(2):109-130.

Seitzinger, S.P. 1988. Denitrification in freshwater in coastal marine ecosystems: ecological and geochemical significance. *Limnol. Oceanogr.* 33:702-724.

Seitzinger, S.P. and A.E. Giblin. 1996. Estimating denitrification in North Atlantic continental shelf sediments. *Biogeochem.* 35:235-260.

Sfriso, A., B. Pavoni, A. Marcomini, and A.A. Orio. 1992. Macroalgae, nutrient cycles, and pollutants in the Lagoon of Venice. *Estuaries* 15:517-528.

Shanahan, P., M. Henze, L. Koncsos, W. Rauch, P. Reichert, L. Somlyody, and P. Vanrolleghem. 1998. River water quality modeling: problems of the art. *Water Science and Technology* 38(11):245-252.

Sharpley, A.N. 1993. Assessing phosphorus bioavailability in agricultural soils and runoff. *Fertilizer Research* 36:259-272.

Sharpley, A.N. 1995. Dependence of runoff phosphorus on soil phosphorus. *J. Environ. Qual.* 24:920-926.

Sharpley, A.N. (ed.). 2000. *Agriculture and Phosphorus Management: The Chesapeake Bay*. Lewis Publishers, Boca Raton, FL.

Sharpley, A.N., S.C. Chapra, R. Wedepohl, J.T. Sims, T.C. Daniel, and K.R. Reddy. 1994. Managing agricultural phosphorus for protection of surface waters: issues and options. *J. Envir. Qual.* 23:437-451.

Sharpley, A.N., W.J. Gburek, and G. Folmar. 1998. Integrated phosphorus and nitrogen management in animal feeding operations for water quality protection. In: *Animal Feeding Operations and Ground Water: Issues, Impacts, and Solutions*, Masters, R.W. and D. Goldman (eds.). National Ground Water Association, Westerville, OH, 72-95pp.

Sharpley, A.N., M.J. Hedley, E. Sibbesen, A. Hillbricht-Ilkowsk, W.A House, and L. Ryszkowski. 1995. Phosphorus transfers from terrestrial to aquatic ecosystems. In: *Phosphorus in the Global Environment*, Tiessen, H. (ed.), Wiley, 173-242pp.

Sharpley, A.N. and E.I. Lord. 1998. The loss of nitrogen and phosphorus in agricultural runoff: processes and management. In: *Fertilization for Sustainable Plant Production and Soil Fertility,* Van Cleemput, O. and G. Hofman (eds.). 11th World Fertilizer Congress of International Scientific Centre of Fertilizers, Gent, Belgium, 548-565 pp.

Sharpley, A.N., J.J. Meisinger, A. Breeuwsma, J.T. Sims, T.C. Daniel, and J.S. Schepers. 1998. Impacts of animal manure management on ground and surface water quality. In: *Animal Waste Utilization: Effective Use of Manure as a Soil Resource,* Hatfield, J.L. and B.A. Stewart (eds.). Ann Arbor Press, Chelsea, MI, 173-242pp.

Sharpley, A.N., J.J. Meisinger, J. Power, and D. Suarez. 1992. Root extraction of nutrients associated with long-term soil management. *Adv. Soil Sci.* 19:151-217.

Sharpley, A.N. and S. Rekolainen. 1997. Phosphorus in agriculture and its environmental implications. In: *Phosphorus Loss from Soil to Water,* Tunney, H., O.T. Carton, P.C. Brookes, and A.E. Johnston. (eds.). CAB International Press, Cambridge, England, 1-54pp.

Sharpley, A.N. and S.J. Smith. 1991. Effect of cover crops on surface water quality. In: *Cover Crops for Clean Water,* Hargrove, W.L. (ed.). Soil and Water Conserv. Soc., Ankeny, IA, 41-50pp.

Sharpley, A.N. and S.J. Smith. 1994. Wheat tillage and water quality in the Southern Plains. *Soil Tillage Research* 30:33-38.

Sharpley, A.N., S.J. Smith, J.A. Zollweg, and G.A. Coleman. 1996. Gully treatment and water quality in the Southern Plains. *J. Soil Water Conserv.* 51:512-517.

Sharpley, A.N. and J.K. Syers. 1979. Phosphorus inputs into a stream draining an agricultural watershed: amounts and relative significance of runoff types. *Water, Air, and Soil Pollution* 11:417-428.

Sharpley, A.N. and H. Tunney. 2000. Phosphorus research strategies to meet agricultural and environmental challenges of the 21st century. *J. Environ. Qual.* 29:176-181.

Shaver, R. and W.T. Howard. 1995. Are we feeding too much phosphorus? *Hoards Dairyman* 140:280.

Shaw, S. and C. Fredine. 1956. *Wetlands of the United States, their Extent, and their Value for Waterfowl and Other Wildlife.* U.S. Department of Interior, Fish and Wildlife Service, Circular 39, Washington, DC, 67pp.

Shepherd, M.A. and E.I. Lord. 1996. Nitrate leaching from a sandy soil: the effect of previous crop and post-harvest soil management in an arable rotation. *J. Ag. Sci.* 127:215-229.

Short, F.T., D.M. Burdick, and J.E. Kaldy. 1995. Mesocosm experiments to quantify the effects of eutrophication on eelgrass, *Zostera marina. Limnol. Oceanogr.* 40:740-749.

Short, F.T., W.C. Dennison, and D.G. Capone. 1990. Phosphorus-limited growth of the tropical seagrass Syringodium filiforme in carbonate sediments. *Mar. Ecol. Prog. Ser.* 62:169-174.

Shreve, B.R., P.A. Moore, Jr., T.C. Daniel, D.R. Edwards, and D.M. Miller. 1995. Reduction of phosphorus in runoff from field-applied poultry litter using chemical amendments. *J. Environ. Qual.* 24:106-111.

Shumway, S.E. 1990. A review of the effects of algal blooms on shellfish and aquaculture. *J. World Aquacult. Soc.* 21:65-104.

Silk, J. and F.L. Joutz. 1997. Short and long run elasticities in U.S. residential electricity demand: a co-integration approach. *Energy Economics* 19(4):493-513.

Simpson, J. and J.R. Hunter. 1974. Fronts in the Irish Sea. *Nature* 250:404-406.

Simpson, T.W. 1998. *A Citizen's Guide to the Water Quality Improvement Act of 1998.* Univ. of Maryland, College of Agriculture and Natural Resources, Maryland Cooperative Extension, College Park, MD.

Sims, J.T., R.R. Simard, B.C. Joern. 1998. Phosphorus losses in agricultural drainage: historical perspective and current research. *J. Environ. Qual.* 27:277-293.

Singh, V.P. (ed.). 1995. *Computer Models of Watershed Hydrology.* Water Resources Publications, Highlands Ranch, CO.

Singleton, F.L., R.W. Atwell, M.S. Janghi, and R.R. Colwell. 1982. Influence of salinity and organic nutrient concentration on survival and growth of *Vibrio cholerae* in aquatic microcosms. *Appl. Environ. Microbiol.* 43:1080-1085.

Sirois, D.L. and S.W. Fredrick. 1978. Phytoplankton and primary production in the lower Hudson River estuary. *Est. Coast. Mar. Sci.* 7:413-423.

Skaggs, R.W., and J.W. Gilliam. 1981. Effect of drainage system design and operation on nitrate transport. *Transactions Am. Soc. Agric. Eng.* 24(4):929-934 and 940.

Skeffington, R.A. 1999. The use of critical loads in environmental policy making: a critical appraisal. *Env. Sci. Tech.* 33:245.

Smayda, T.J. 1990. Novel and nuisance phytoplankton blooms in the sea: evidence for global epidemic. In: *Toxic Marine Phytoplankton,* Graneli, E., B. Sundstrom, R. Edler, and D.M. Anderson (eds.). Elsevier, New York, 29-40pp.

Smayda, T.J. 1974. Bioassay of the growth potential of the surface water of lower Narragansett Bay over an annual cycle using the diatom *Thalasiosira pseudonana* (oceanic clone, 13-1). *Limnol. Oceanogr.* 19:889-901.

Smayda, T.J. 1989. Primary production and the global epidemic of phytoplankton blooms in the sea: a linkage? In: *Novel Phytoplankton Blooms: Causes and Impacts of Recurrent Brown Tides and Other Unusual Blooms,* Cosper, E.M., E.J. Carpenter, and V.M. Bricelj (eds.), Springer-Verlag, Berlin.

Smith, V.K. 1996. *Estimating Economic Values for Nature: Methods for Nonmarket Valuation,* Elgar, E. (ed.). Cheltenham, UK.

Smith, R.A., R.B. Alexander, and M.G. Wolman. 1987. Water-quality trends in the nation's rivers. *Science* 235:1607-1615.

Smith, R.A., G.E. Schwarz, and R.B. Alexander. 1997. Regional interpretation of water-quality monitoring data. *Water Resources Research* 33(12):2781-2798.

Smith, S.V. 1984. Phosphorus versus nitrogen limitation in the marine environment. *Limnol. Oceanogr.* 29:1149-1160.

Smith, S.V. and M.J. Atkinson. 1984. Phosphorus limitation of net production in a confined aquatic ecosystem. *Nature* 207:626-627.

Somlyody, L., M. Henze, L. Koncsos, W. Rauch, P. Reichert, P. Shanahan, and P. Vanrolleghem. 1998. River water quality modeling: future of the art. *Water Science and Technology* 38(11):253-260.

South Florida Water Management District (SFWMD). 1998. *A Closer Look at Kissimmee River Research.* South Florida Water Management District, West Palm Beach, FL.

Sparrow, L.A., A.N. Sharpley, and D.J. Reuter. 2000. Safeguarding soil and water quality. In: Opportunities for the 21st Century: Expanding the Horizons for Soil, Plant, and Water Analysis. *Communications in Soil Science and Plant Analysis* 31(11-14): in press.

Stranlund, J.K. 1995. Public mechanisms to support compliance to an environmental norm. *Journal of Environmental Economics and Management* 28(2):205-222.

Sterner, R.W., J.J. Elser, and D.O. Hessen. 1992. Stoichiometric relationships among producers, consumers, and nutrient cycling in pelagic ecosystems. *Biogeochemistry* 17:49-67.

Stommel, H. and H. Farmer. 1952. On the nature of estuarine circulation. *Woods Hole Oceanographic Institute Reference Notes* 52-51, 52-63 and 52-88.

Strecker, E.W. 1996. The use of wetlands for stormwater pollution control. *Infrastructure* 1(3):48-66.

Strecker, E.W., J. Kersnar, E.D. Driscoll, and R. Horner. 1992. *The Use of Wetlands for Controlling Stormwater Pollution*. Terrene Institute, Washington, DC, 66pp.

Strecker, E.W., M.M. Quigley and B.R. Urbonas. 1999. Determining urban stormwater BMP effectiveness. *Proceedings 8th International Conference on Urban Storm Drainage*, Joliffe, I.A. and J.E. Ball (eds.). Institution of Engineers, Canberra, Australia, Volume 2, 1079-1086pp..

Stutler, D., L.A. Roesner and M.F. Schmidt. 1995. Urban runoff and diffuse pollution. In: *Nonpoint Pollution and Urban Stormwater Management*, Novotny, V. (ed.). Technometric Publishing Co., Lancaster, PA.

Sunda, W.G. 1989. Trace metal interactions with marine phytoplankton. *Biological Oceanography* 6:411-442.

Taft, J.L., W.R. Taylor, E.O. Hartwig, and R. Loftus. 1980. Seasonal oxygen depletion in Chesapeake Bay. *Estuaries* 3(4):242-247.

Taylor, A.W., W.M. Edwards, and E.C. Simpson. 1971. Nutrients in streams draining woodland and farmland near Coshocton, Ohio. *Water Resour. Res.* 7:81-90.

Thayer, G.W., D.A. Wolfe, and R.B. Williams. 1975. The impact of man on seagrass systems. *American Scientist* 63:288-296.

Theis, T.L. and P.J. McCabe. 1978. Phosphorus dynamics in hypereutrophic lake sediments. *Water Res.* 12:677-685.

Thybo-Christesen, M., M.B. Rasmussen, and T.H. Blackburn. 1993. Nutrient fluxes and growth of *Cladophora sericea* in a shallow Danish bay. *Mar. Ecol. Prog. Ser.* 23:179-191.

Tietenberg, T. 1985a. Emissions trading: an exercise in reforming pollution policy. In: *Resources for the Future*. Washington, DC.

Tietenberg, T. 1985b. The valuation of environmental risks using hedonic wage models. In: *Horizontal Equity, Uncertainty, and Economic Well-Being: NBER Studies in Income and Wealth Series* 50:385-91, Martin, D. and T. Smeeding (eds.). University of Chicago Press.

Tiktak, A. and H.J.M. van Grinsven. 1995. Review of sixteen forest-soil-atmosphere models. *Ecol. Model.* 83:35-53.

Tobey, J.A. and H. Smets. 1996. The polluter pays principle in the context of agriculture and the environment. *World Economy* 19(1):63-87.

Tomascik, T. and F. Sander. 1985. Effects of eutrophication on reef-building corals: growth rate of the reef-building coral *Montastrea annularis*. *Marine Biology* 87:143-155.

Tomasko, D.A. and B.E. Lapointe. 1991. Productivity and biomass of *Thalassia testudinum* as related to water column nutrient availability and epiphyte levels: field observations and experimental studies. *Mar. Ecol. Prog. Ser.* 75:9-17.

Townsend, A. 1999. *Nitrogen Cycling in the Temperate and Tropical Americas: A Report from the International SCOPE Nitrogen Project*. Kluwer, Dordrecht, The Netherlands.

Tukey, J.W. 1977. *Exploratory Data Analysis*. Addison-Wesley Pub. Co., Reading, MA, 688pp.

Turner, M.G., V.H. Dale, and R.H. Gardner. 1989a. Predicting across scales: theory development and testing. *Landscape Ecology* 3:245-252.

Turner, M.G., R.V. O'Neill, R.H. Gardner, and B.T. Milne. 1989b. Effects of changing spatial scale on the analysis of landscape pattern. *Landscape Ecology* 3:153-162.

Turner, R.E., N. Qureshi, N.N Rabalais, Q. Dortch, D. Justic, R.F. Shaw, and J. Cope. 1998. Fluctuating silicate:nitrate ratios and coastal plankton food webs. *Proc. Natl. Acad. Sci.* 95:13,048-13,051.

Turner, R.E. and N.N. Rabalais. 1991. Changes in Mississippi River water quality this century: implications for coastal food webs. *BioScience* 41:140-148.

Turner, R.E. and N.N. Rabalais. 1994. Coastal eutrophication near the Mississippi River delta. *Nature* 368:619-621.

Tyrrell, T. 1999. The relative influences of nitrogen and phosphorus on oceanic primary production. *Nature* 400:525-531.

U.S. Bureau of the Census. 1998. *Statistical Abstract of the United States: 1998* (118th edition). Washington, DC.

U.S. Congress. 1991. *Catching our Breath: Next Steps for Reducing Urban Ozone*. Office of Technology Assessment, Government Printing Office, Washington, DC.

U.S. Department of Agriculture (USDA). 1989. *Fact Book of Agriculture*. Misc. Publ. No. 1063, Office of Public Affairs, Washington, DC, 17pp.

U.S. Department of Agriculture (USDA) and Agricultural Stabilization and Conservation Service (ASCS). 1992. *Conestoga Headwaters Project, Pennsylvania: Rural Clean Water Program, 10-Year Report 1981-1991*. USDA-ASCS, Washington, DC.

U.S. Department of Agriculture (USDA) and Environmental Protection Agency (EPA). 1998a. *Clean Water Action Plan: Restoring and Protecting America's Waters*. EPA-840-R-98-001. 89 pp.

U.S. Department of Agriculture (USDA) and Environmental Protection Agency (EPA). (1998b). *Draft Unified National Strategy for Animal Feeding Operations*, [Online]. Available: http://www.nhq.nrcs.usda.gov/cleanwater/afo/index.html [1999, June 30].

U.S. Department of Agriculture (USDA) and Environmental Protection Agency (EPA). 1999. *Unified National Strategy for Animal Feeding Operations*. U.S. Government Printing Office, Washington, DC.

U.S. Department of the Interior (DOI), Centers for Disease Control and Prevention, U.S. Food and Drug Administration (FDA), U.S. Department of Agriculture (USDA), Environmental Protection Agency (EPA), National Oceanic and Atmospheric Administration (NOAA), and National Institute for Environmental Health Sciences. 1997. *National Harmful Algal Bloom Research and Monitoring Strategy: An Initial Focus on Pfiesteria, Fish Lesions, Fish Kills, and Public Health*. Washington, DC, 26pp.

U.S. Geological Survey (USGS). (1999a). *United States NWIS-W Data Retrieval*, [Online]. Available: http://waterdata.usgs.gov/nwis-w/us/ [2000, February 17].

U.S. Geological Survey (USGS). (1999b). *Archives of Models and Modeling Tools*, [Online]. Available: http://smig.usgs.gov/SMIG/model_archives.html [1999, November 16].

Urbonas, B. and P. Stahre. 1993. *Stormwater Best Management Practices and Detention for Water Quality, Drainage and Combined Sewer Overflow Management. Prentice Hall, Englewood Cliffs, NJ.*

Valiela, I., J. McClelland, J. Hauxwell, P.J. Behr, D. Hersh, and K. Foreman. 1997. Macroalgal blooms in shallow estuaries: controls and ecophysiological and ecosystem consequences. *Limnol. Oceanogr.* 42:1105-1118.

Valigura, R.A., R.B. Alexander, D.A. Brock, M.S. Castro, T.P. Meyers, H.W. Paerl, P.E. Stacey, and D. Stanley (eds.). 2000. *An Assessment of Nitrogen Inputs to Coastal Areas with an Atmospheric Perspective.* AGU Coastal Estuaries Series.

Vallino, J.J. and C.S. Hopkinson. 1998. Estimation of dispersion and characteristic mixing times in Plum Island Sound Estuary. *Estuarine, Coastal, and Shelf Science* 46:333-350.

Vallis, I., W.J. Parton, B.A. Keating, and A.W. Wood. 1996. Simulation of the effects of trash and nitrogen fertilizer management on soil organic matter levels and yields of sugarcane. *Soil Till. Res.* 38:115-132.

van Breement, N., P.A. Burroughs, E.J. Velthorst, H.F. van Dobben, T. de Wit, T.B. Ridder, and H.F.R. Reijnders. 1982. Soil acidification from atmospheric ammonium sulphate in forest canopy throughfall. *Nature* 299:548-550.

van Raaphost, W., P. Ruadij, and A.G. Brinkman. 1988. The assessment of benthic phosphorus regeneration in an estuarine ecosystem model. *Neth. J. Sea Res.* 22:23-26.

Varela, R.A., A. Cruzado, and J.E. Gabaldon. 1995. Modeling primary production in the North Sea using the European regional seas ecosystem model. *Neth. J. Sea Res.* 33(3/4):337-361.

Viessman, Jr., W. and M.J. Hammer. 1998. *Water Supply and Pollution Control*, 6th Ed. Addison-Wesley, Reading, MA.

Vince, S., and I. Valiela. 1973. The effects of ammonium and phosphate enrichment on chlorophyll a, a pigment ratio, and species composition of phytoplankton of Vineyard Sound. *Mar. Biol.* 19:69-73.

Vitousek, P.M., J. Aber, S.E. Bayley, R.W. Howarth, G.E. Likens, P.A. Matson, D.W. Schindler, W.H. Schlesinger, and G.D. Tilman. 1997. Human alteration of the global nitrogen cycle: causes and consequences. *Ecol. Issues* 1:1-15.

Vitousek, P.M. and R.W. Howarth. 1991. Nitrogen limitation on land and sea: how can it occur? *Biogeochemistry* 13:87-115.

Vitousek, P.M. and W.A. Reiners. 1975. Ecosystem succession and nutrient retention: a hypothesis. *Bioscience* 25:376-381.

Vollenweider, R.A. 1976. Advances in defining critical loading levels for phosphorus in lake eutrophication. *Mem. First Ital. Idrobiol.* 33:53-83.

Vought, L., J. Dahl, C. Pedersen, and J. Lacoursiere. 1994. Nutrient retention in riparian ecotones. *Ambio* 23:342-348.

Wadman, W.P., C.M.J. Sluijsmans, and L.C.N. de la Lande-Cremer. 1987. Value of animal manures: changes in perception. In: *Animal Manure on Grassland and Fodder Crops*, van der Meer, H.G. (ed.). Martinus Nijhoff Publishers, Dordrecht, The Netherlands, 1-45pp.

Ward, R.C. 1984. On the response to precipitation of headwater streams in humid areas. *J. Hydrol.* 74:171-189.

Warren, I.R. and H.K. Bach. 1992. MIKE 21: a modelling system for estuaries, coastal waters and seas. Environmental Software 7. Elsevier, New York.

Water Environment Federation (WEF). 1998. *Biological and Chemical Systems for Nutrient Removal.* Alexandria, VA.

Water Pollution Control Federation (WPCF) (now WEF). 1989. Combined sewer overflow pollution abatement. *WPCF Manual of Practice* No. FD-17. Alexandria, VA.

Weersink, A., C. Dutka, and M. Goss. 1998. Crop price risk effects on farmer abatement costs of reducing nitrate levels in groundwater imposed by environmental policy instruments. *Canadian Journal of Agricultural Economics* 46(2):171-190.

Weinraub, J. 1997. Pfiesteria fallout: wholesalers moan, restaurants try to adjust. *The Washington Post*, 1 October, Washington, D.C.

Westphal, P.J., L.E. Lanyon, and E.J. Partenheimer. 1989. Plant nutrient management strategy implications for optimal herd size and performance of a simulated dairy farm. *Agric. Syst.* 31:381-394.

Wetzel, R.L. 1996. *Ecosystem Processes Modeling of Lower Chesapeake Bay Submerged Aquatic Vegetation and Littoral Zones.* Briefing paper to the Chesapeake Bay Program Living Resources Subcommittee, 15p.

Wetzel, R.L. and C. Buzzelli. 1997. Modeling the lower Chesapeake Bay littoral zone and fringing wetlands: ecosystem processes and habitat linkages. *Annual Report to EPA on Region III,* Annapolis, MD, 27p.

Wetzel, R.L. and H. Neckles. 1986. A model of *Zostera marina* L. photosynthesis and growth: simulated effects of selected physical-chemical variables and biological interactions. *Aquatic Botany* 26:307-323.

Whelpdale, D.M., P.W. Summer, and E. Sanhuez. 1997. A global overview of atmospheric acid deposition fluxes. *Environmental Monitoring and Assessment* 48:217-227.

Whitehead, D.C. and N. Raistrick. 1990. Ammonia volatilization from five nitrogen compounds used as fertilizers following surface application to soils. *J. Soil Sci.* 41:387-394.

Whittier, T.R., R.M. Hughes, and D.P. Larsen. 1988. Correspondence between ecoregions and spatial patterns in stream ecosystems in Oregon. *Canadian Journal of Fisheries and Aquatic Sciences* 45:1264-1278.

Wienke, S.M., B.E. Cole, and J.E Cloern. 1993. *Plankton Studies in San Francisco Bay, California: Chlorophyll Distributions and Hydrographic Properties of San Francisco Bay, 1992.* U.S. Geological Survey Open-File Report 93-423.

Williams, M.W., J.S. Baron, N. Caine, R. Sommerfeld, and R.J. Sanford. 1996. Nitrogen saturation in the Rocky Mountains. *Environ. Sci. Tech.* 30:640-646.

Willmott, C.J., S.G. Ackleson, R.E. Davis, J.J. Feddema, K.M. Klink, D.R. Legates, J. O'Donnell, and C.M. Rowe. 1985. Statistics for the evaluation and comparison of models. *J. Geophys. Res.* 90:8995-9005.

Wirl, F. 1997. *The Economics of Conservation Programs.* Kluwer Academic Press, Boston, Dordrecht and London.

Wright, A.C., R.T. Hill, J.A. Johnson, M.C. Roghman, R.R. Colwell, and J.G. Morris, Jr. 1996. Distribution of *Vibrio vulnificus* in the Chesapeake Bay. *Appl. Environ. Microbiol.* 62:717-724.

Wu, Z. and L D. Satter. 1998. Milk production and reproductive performance of dairy cows fed low or normal phosphorous diets. *J. Dairy Sci.* 81 (Supp. 1):326 (Abstract).

Wulff, F., A. Stigebrandt, and L. Rahm. 1990. Nutrient dynamics of the Baltic Sea. *Ambio* 19:126-133.

Yentsch, C.S. and D.A. Phinney. 1989. A bridge between ocean optics and microbial ecology. *Limnol. Oceanogr.* 34:1694-1705.

Yiridoe, E. and A. Weersink. 1998. Marginal abatement costs of reducing groundwater nitrogen pollution from intensive and extensive management choices. *Agricultural and Resource Economics Review* 27(2):169-185.

Young, R.A., C.A. Onstad, D.D. Bosch, and W.P. Anderson. 1994. Agricultural non-point source pollution model. *AGNPS User's Guide,* v. 4.02. North Central Soil Conservation Research Laboratory, Morris, MN.

Zarbock, H.W., A.J. Janicki, and S.S. Janicki. 1996. *Estimates of Total Nitrogen, Total Phosphorus, and Total Suspended Solids to Tampa Bay, Florida.* Tampa Bay National Estuary Program Technical Publication #19-96. St. Petersburg, FL.

Zucker, L.A. and L.C. Brown. 1998. Agricultural drainage: water quality impacts and subsurface drainage studies in the midwest. *Ohio State University Extension Service Bulletin,* No. 871. Columbus, OH.

Appendixes

APPENDIX A

Statement of Task and Committee and Staff Biographies

STATEMENT OF TASK

The committee will recommend actions that can help managers to achieve short-term reductions of eutrophication. The committee will:

- assess how coastal and watershed processes and their interactions affect eutrophication of coastal ecosystems;
- recommend ways to improve coordination and effectiveness of ongoing research, monitoring, and management activities being conducted at the federal, state, and local levels;
- identify means to remove barriers that impede implementation of existing techniques to reduce coastal eutrophication; and
- evaluate the effectiveness of existing strategies for monitoring watersheds, atmospheric deposition, and coastal areas and for managing watersheds.

The committee will also recommend actions that could provide a basis for better watershed management to reduce coastal eutrophication in the future. The committee will:

- delineate potential watershed management approaches for reducing eutrophication and its impacts on coastal ecosystems; and
- identify research needs for better understanding eutrophication

and its effects, particularly focused on reducing the uncertainties in existing models and other methods.

The study committee will evaluate models now used for coastal eutrophication management and will evaluate process-based models as a possible next step for achieving better predictions in the future when watersheds are subjected to changing land-use patterns, global climate variations, and other factors.

COMMITTEE AND STAFF BIOGRAPHIES

Committee Chairman:

Robert Howarth earned a Ph.D. in oceanography from the MIT/Woods Hole Oceanographic Joint Program in 1979, and has been a professor of ecology at Cornell University since 1991. He also co-chairs the International SCOPE Nitrogen Project. His research interests include controls on nitrogen fixation, causes and consequences of eutrophication in estuaries, interaction of biogeochemical cycles, and global and regional analysis of the nitrogen, sulphur, and phosphorus cycles.

Committee Members:

Donald M. Anderson earned a Ph.D. in aquatic sciences from the Massachusetts Institute of Technology in 1977. His research interests include phytoplankton physiological ecology, red tides and other bloom phenomena, ciguatera, dinoflagellate toxins, dinoflagellate resting cysts, and molecular and immunological probes. Dr. Anderson is a Senior Scientist at the Woods Hole Oceanographic Institution, and also serves as the Director of the National Office for Marine Biotoxins and Harmful Algal Blooms.

Thomas Church earned a Ph.D. from the University of California, San Diego in 1970. He is a professor at the University of Delaware. His research interests include the transport of continental emissions to the ocean environment (with emphasis on atmospheric deposition), the chemistry of marine precipitation and deposition of trace elements to the marine environment, sedimentary geochemistry, and marine chemistry.

Holly Greening earned an M.S. from Florida State University in 1980. She has been a senior scientist at the Tampa Bay National Estuary Program since 1991. Her responsibilities there include coordinating state, federal, and university researchers and resource managers investigating the impacts of stormwater runoff. Her research interests include freshwater inflow, atmospheric deposition, and watershed management.

Charles Hopkinson, Jr. earned a Ph.D. in marine science from Louisiana State University in 1979. He is a senior scientist at the Marine Biological Laboratory (Woods Hole, Massachusetts). Dr. Hopkinson's research interests include wetland and aquatic ecology, nutrient cycling in marine and fresh water systems, and ecological modeling.

Wayne Huber earned a Ph.D. in civil engineering from the Massachusetts Institute of Technology in 1968, and has been professor of civil, construction, and environmental engineering at Oregon State University since 1991. His research interests are in the areas of urban hydrology, stormwater management, and transport processes related to water quality.

Nancy Marcus earned her Ph.D. in biological oceanography from Yale University in 1976. She is an oceanographer with expertise in the evolution, ecology and population genetics of marine zooplankton, and developmental responses of organisms to environmental change. Dr. Marcus, a member of the Ocean Studies Board, is a professor at Florida State University.

Robert Naiman earned his Ph.D. from Arizona State University in 1974. His research interests include the ecology of streams and rivers from a watershed perspective, landscape ecology, and the role of large animals in influencing ecosystem dynamics. Dr. Naiman has been a professor in the College of Ocean and Fishery Sciences at the University of Washington since 1988.

Kathleen Segerson earned a Ph.D. from Cornell University in 1984. Her research interests include the use of natural resource economics and law in the compensation of takings, the economic implications of environmental management techniques, and the use of economic incentives in resource policy. Dr. Segerson is a professor of economics at the University of Connecticut.

Andrew Sharpley earned a Ph.D. in soil science from Massey University, New Zealand in 1977. His research investigates the cycling of phosphorus in soil-plant-water systems in relation to soil productivity and water quality and includes the management of fertilizers, animal manures, and crop residues. He focuses on achieving results that are both economically beneficial to farmers and environmentally sound to the general public. Dr. Sharpley is a soil scientist at the U.S. Department of Agriculture,

Agricultural Research Service's, Pasture Systems and Watershed Management Research Laboratory in University Park, Pennsylvania.

William Wiseman, Jr. earned his Ph.D. from The Johns Hopkins University in 1969. His research interests include coastal and estuarine dynamics, and the interactions of physical processes with both biological and geological processes in coastal and estuarine environments. Dr. Wiseman is the director of the Coastal Studies Institute and professor of Oceanography and Coastal Sciences at Louisiana State University.

Staff:

Dan Walker (Study Director), received his Ph.D. in geology from the University of Tennessee in 1990. He is currently a Senior Program Officer with the Ocean Studies Board of the National Research Council. Since joining the Ocean Studies Board in 1995, he has directed a number of studies including *Science for Decisionmaking: Coastal and Marine Geology at the U.S. Geological Survey* (1999), *Global Ocean Sciences: Toward an Integrated Approach* (1998), and *The Global Ocean Observing System: Users, Benefits, and Priorities* (1997). A former member of the both the Kentucky and North Carolina state geologic surveys, Dr. Walker's interests focus on the value of environmental information for policymaking at local, state, and national levels.

Chris Elfring is a Senior Program Officer with the Water Science and Technology Board (WSTB). In her work with the WSTB, she has directed almost two dozen studies including *New Strategies for America's Watersheds* (1999), *A New Era for Irrigation* (1996), *Flood Risk Management and the American River Basin* (1995), *Water Transfers in the West: Efficiency, Equity, and the Environment* (1992), and *Irrigation-Induced Water Quality Problems* (1989). Prior to her work at the Academy, Ms. Elfring worked at Congress's Office of Technology Assessment. She first came to Washington as an AAAS Congressional Fellow (1979-80). Her primary areas of interest include watershed management, water allocation issues, public lands management, the environmental impacts of agriculture, and alternative dispute resolution.

Jodi Bachim received her B.S. in zoology from the University of Wisconsin-Madison in 1998. She is currently a Project Assistant with the Ocean Studies Board.

Appendix B

Acronyms and Abbreviations

AFO	animal feeding operation
AGNPS	Agricultural Nonpoint Source Pollution Model
ANSWERS	Areal, Nonpoint Source Watershed Environment Response Simulation
ASP	amnesic shellfish poisoning
BASINS	Better Assessment Science Integrating Point and Nonpoint Sources
BMP	best management practices
BOD	biological oxygen demand
C	carbon
CASNET	Clean Air Status and Trends Network
CCMP	Comprehensive Conservation and Management Plan
CENR	Executive Office's Committee on Environment and Natural Resources
C-GOOS	Coastal Component of the Global Ocean Observing System
CISNet	Coastal Intensive Site Network
CNSPCP	Coastal Nonpoint Source Pollution Control Program
CREAMS	Chemical, Runoff, and Erosion from Agricultural Management Systems
CRP	Conservation Reserve Program
CSO	combined sewer overflow
CWAP	Clean Water Action Plan

DCP	dissolved concentration potential
DIN	dissolved inorganic nitrogen
DO	dissolved oxygen
DOM	dissolved organic matter
DR3M-QUAL	Distributed Routing, Rainfall, Runoff Model-Quality
DSP	diarrhetic shellfish poisoning
EMAP	Environmental Monitoring and Assessment Program
EMC	event mean concentration
EPA	Environmental Protection Agency
EPRI	Electric Power Research Institute
EQIP	Environmental Quality Incentives Program
EXP	estuarine export potential
GIS	geographical information system
GLEAMS	Groundwater Loading Effects of Agricultural Management Systems
GOMP	Gulf of Mexico Program
GPP	gross primary productivity
HAB	harmful algal bloom
HEC	Hydrologic Engineering Center
HSPF	Hydrologic Simulation Program-FORTRAN
MERL	Marine Ecosystem Research Laboratory
N	nitrogen
NADP	National Atmospheric Deposition Program
NASA	National Aeronautics and Space Administration
NEP	National Estuary Program
NERR	National Estuarine Research Reserve
NERRS	National Estuarine Research Reserve System
NOAA	National Oceanic and Atmospheric Administration
NPDES	EPA's National Pollutant Discharge Elimination System
NRC	National Research Council
NSF	National Science Foundation
NSP	neurotoxic shellfish poisoning
NSTC	National Science and Technology Council
NURP	EPA's Nationwide Urban Runoff Program
P	phosphorus
PSP	paralytic shellfish poisoning

SAV submerged aquatic vegetation
SCOPE Scientific Committee on Problems of the Environment
SCS Soil Conservation Service
SIP State Implementation Plan
SPARROW Spatially Referenced Regressions on Watersheds model
SRP soluble reactive phosphorus
SWAT Soil and Water Assessment Tool
SWEM System-Wide Eutrophication Model
SWMM Storm Water Management Model
SWRRB Simulator for Water Resources in Rural Basins

TMDL total maximum daily load

USDA U.S. Department of Agriculture
USGS U.S. Geological Survey

WQAM Water Quality Assessment Methodology
WQIA Water Quality Improvement Act
WTA willingness to accept
WTP willingness to pay

Appendix C

Programmatic Approaches and Results of a Local Managers Questionnaire

Many federal agencies are associated with programs that provide information relevant to an understanding of coastal eutrophication. Some are devoted solely to this topic, while others treat it only peripherally. Certain programs have been in existence for decades while others remain in the planning stage. Often the programs involve collaboration amongst multiple federal agencies as well as with state, local, and private units. The following brief descriptions include:

1) Representative federal programs currently addressing coastal eutrophication from regulatory, policy, non-regulatory management, educational and incentive program standpoints. The Clean Water Act, Clean Air Act, and Coastal Zone Management Act have been described in Chapter 2 are and not repeated in this appendix;
2) Representative federal monitoring and assessment programs which address coastal conditions, including eutrophication;
3) Management strategies addressing coastal eutrophication, as developed by National Estuary Programs; and
4) Results of a local managers questionnaire conducted by this committee.

The list of programs selected for description is meant to be exemplary, not exhaustive.

REPRESENTATIVE FEDERAL PROGRAMS

Regulatory Programs

The **National Pollutant Discharge Elimination System (NPDES)**, established by the Clean Water Act, requires permits for the discharge of pollutant material into water bodies, with the states ultimately establishing the standard at or below the Environmental Protection Agency (EPA) established maxima. The NPDES permitting program is the primary regulatory program addressing point source discharges, and is often recognized as being responsible for improvements in water quality in specific areas of coastal systems.

Data collected as required by NPDES urban stormwater permits are proving to be very useful to develop area- and land-use specific runoff coefficients, a secondary benefit of the permitting program. To date, sparse literature values, often from areas of the country far from the coastal area of concern, have been the only readily available information for specific loading measurements from a variety of land uses. Runoff coefficients representative of local conditions provide much-needed information for developing adequate nutrient loading budgets and are critical in developing watershed-based nutrient management strategies.

The **Total Maximum Daily Load (TMDL)** program, established under Section 303(d) of the Clean Water Act, focuses on identifying and restoring the Nation's polluted waterbodies.

The goal of a TMDL is the attainment of water quality standards. A TMDL is a written, quantitative assessment of water quality problems and contributing pollutant sources. It can identify the need for point source and nonpoint source controls. Under this provision, States are required to (1) identify and list waterbodies where State water quality standards are not being met following the application of technology-based point source pollution controls; and (2) establish TMDLs for these waters. EPA must review and approve (or disapprove) State lists and TMDLs. If State actions are not adequate, EPA must prepare lists and TMDLs. EPA has revised TMDL program regulations and guidance for public review and comment.

Policy

The **Coastal Nonpoint Source Pollution Control Program (CNSPCP)** (Section 6217 of the Coastal Zone Act Reauthorization Amendments of 1990), administered by National Oceanic and Atmospheric Administration (NOAA), EPA, and the states, addresses nonpoint pollution problems in coastal waters. Section 6217 requires the 29 states and territories with approved Coastal Zone Management Programs to develop CNSPCPs. General requirements for approval of management plans include provisions for protection and management of important resources; policies directed at resource protection, managing development, and simplifying government procedures; and clear guidelines for decisionmakers within the management program.

To help states and territories identify appropriate technologies and tools, EPA issued "Guidance Specifying Management Measures for Sources of Nonpoint Pollution in Coastal Waters", describing the best available, economically achievable approaches used to control nonpoint source pollution from major categories of land management activities that can degrade coastal waters. States may elect to implement alternative measures as long as the alternatives will achieve the same environmental results as those described in the guidance. Mechanisms for ensuring implementation of management measures include permit programs, zoning, enforceable water quality standards, or voluntary approaches like economic incentives if they are backed by appropriate legislation.

The **Clean Water Action Plan** (CWAP) was initiated in 1998 in commemoration of the 25th anniversary of the Clean Water Act (USDA and EPA 1998a). It is a broad effort designed to protect and restore the nation's water resources by highlighting existing activities, starting new activities, and building partnerships among federal, state, tribal, and local decisionmakers, resource managers, and citizens. On the federal side, it is a multi agency effort, initiated by the White House, with the lead agencies being the U.S. Department of Agriculture (USDA); U.S. Department of Commerce, NOAA; U.S. Department of Defense, U.S. Army Corps of Engineers; U.S. Department of the Interior; and EPA. Supporting agencies include the Tennessee Valley Authority, U.S. Department of Energy, U.S. Department of Transportation, and U.S. Department of Justice.

The Action Plan contains 111 key actions designed to reinvigorate efforts to protect rivers, lakes, coastal waters, and wetlands by strengthening leadership at the local level and bringing together federal, state, tribal, and local partners. Many of the actions included in CWAP, if implemented, will directly or indirectly address eutrophication in coastal waters.

Non-Regulatory Management Programs

As of April 1996, 22 sites have been designated in NOAA's **National Estuarine Research Reserve** (NERR) program (Figure C-1); another six reserves are in various stages of development. The program was created with the 1972 passage of the Coastal Zone Management Act. Through linked programs of stewardship, education and research, the NERR

FIGURE C-1 Estuaries and bays of the United States and Puerto Rico that are designated (■) National Estuarine Research Reserves (NERRs) and proposed (●) NERRs, and National Estuary Programs (NEPs), which are the shaded areas.

enhances informed management and scientific understanding of the Nation's estuarine and coastal habitats (NOAA 1999a).

EPA's **National Estuary Program (NEP)** currently includepoint sourcecs 28 estuaries throughout the United States (Figure C-1), with the overall objective to restore and protect "estuaries of national significance". NEP has adopted an "ecosystem or watershed management approach" which encourages cooperation among the various stakeholders and regulators, and promotes practical solutions that attempt to provide the maximum environmental benefit in the most cost-effective manner (ANEP 1998). Although each individual estuary program has developed unique approaches and solutions, four steps are generally followed:

- characterization of the major threats facing an estuary;
- development of a Comprehensive Conservation and Management Plan (CCMP) that sets specific goals for protecting or improving the estuary and allocates responsibility for achieving those goals among NEP partners (regulatory agencies, local governments, and citizens);
- implementation of CCMP by the various NEP partners, emphasizing flexibility in solutions as long as overall goals are met; and
- monitoring to determine progress made toward achieving CCMP goals.

Eutrophication has been identified as a primary management concern in over half of the NEP estuaries, and a variety of nutrient management strategies have been developed through the individual programs (see discussion below).

The 1990 Amendments to the Clean Air Act (Section 112[m]) called for the establishment of the **Great Waters Program**, a joint effort between EPA and NOAA. This program was specifically designed to address air pollution and its impacts on water quality in the Great Lakes, the Chesapeake Bay, Lake Champlain, and other specified coastal waters. Specific elements of the Great Waters Program include a monitoring program of facilities on each of the five Great Lakes, Lake Champlain, Chesapeake Bay, National Estuary Program waters, and National Estuarine Research Reserves, and a research program which seeks to determine sources and deposition rates for air pollutants. Additional research is aimed at improving monitoring as well as determining the fraction that air pollution contributes to the overall pollution of coastal waters. Adverse health and environmental effects are also appraised. This appraisal includes indirect exposure pathways, results of which are then linked to the appropriate provisions of the Clean Water Act and the Safe Drinking Water Act. Finally, a sampling program of living organisms strives to determine the bioavailability of these pollutants. EPA is required to report their findings to Congress on a biennial basis from 1993 onward.

The **Chesapeake Bay Program** (EPA, three states, and the District of Columbia) is the largest locally-based estuarine management program in the United States, and Chesapeake Bay was the nation's first estuary targeted for restoration and protection. In the late 1970s, researchers identified nutrient over-enrichment as one of three areas requiring immediate attention. In 1987, the Chesapeake Bay Agreement set a goal to reduce nitrogen and phosphorus loading to the bay by 40 percent by the year 2000, a goal that was been renewed in 1997, with the acknowledgment that reductions of nitrogen would need to be accelerated to reach the 40 percent reduction goal. Recent assessments indicate that the phosphorus reduction goal has been met, but that despite considerable management effort and progress toward the goal, the nitrogen reduction goal will not be met by 2000. Current efforts to reach goals are highlighted below in this appendix.

The **Gulf of Mexico Program (GOMP)** is patterned after the successful efforts of the Chesapeake Bay Program, with the specific mission to facilitate the protection and restoration of the coastal marine waters of the Gulf of Mexico and its coastal natural habitats through a network of citizens and institutions. To address nutrient enrichment and its effects on hypoxia and harmful algal blooms, GOMP has initiated a Mississippi River/Gulf of Mexico Watershed Nutrient Task Force. It is anticipated that the Task Force will work closely with the federally mandated "HAB and Hypoxia Task Force" (called for by the Harmful Algal Bloom and Hypoxia Research and Control Act of 1998), with GOMP Task Force taking the lead on developing an action plan to implement management and policy recommendations as they are finalized.

Education Programs

Educational and public outreach are important components of several of the larger federal programs, including NOAA's NERRS, EPA's NEPs, and Seagrant programs. In addition, the following program provides funds for implementation of industry-specific education and pollution prevention guidelines.

EPA's **Pollution Prevention Grants Program** provides project grants to States to implement pollution prevention (P2) projects, including educational outreach, training, and technical assistance for businesses, and projects focusing on multimedia pollution prevention as an environmental management priority. States are required to provide at least 50 percent of total project costs.

Incentive Programs

The **Environmental Quality Incentives Program (EQIP)**, administered through USDA and local/state NRCS, was established to provide a single, voluntary conservation program for farmers and ranchers to address significant natural resource needs and objectives. Nationally, it provides technical, financial, and educational assistance, half of it targeted to livestock-related natural resource concerns and the other half to more general conservation practices. EQIP is available to non-federal landowners engaged in livestock operations or agricultural production. Eligible land includes cropland, rangeland, pasture, forest land and other farm and ranch lands. Cost share of up to 75 percent is available for certain conservation practices, with a maximum of $10,000 per person per year and $50,000 over the length of the contract.

While traditionally used to build wastewater treatment facilities, **Clean Water State Revolving Funds** loans, administered through EPA Office of Wastewater and the States, can now be used for other water quality management activities, including 1) agriculture, rural and urban runoff control; 2) estuary improvement projects; 3) wet weather flow control, including stormwater and sewer overflows; 4) alternative wastewater treatment technologies; and 5) nontraditional projects such as landfills and riparian buffers. States lend money to municipalities, communities, citizen groups, Non-Profit Organizations, and private citizens implementing nonpoint source pollution and estuary management activities (Clean Water Act Section 319 and 320). Twenty percent state match is required.

Nonpoint Source Implementation Grants (319 Program), administered by the States with funds from EPA, provides grants to implement nonpoint source projects and programs in accordance with Section 319 of the Clean Water Act. Examples of projects funded through this program include best management practices (BMPs) for animal waste; design and implementation of BMP systems for stream, lake and estuary watersheds; and basinwide landowner education programs. State/local organizations are required to provide 40 percent of the total project cost.

FEDERAL MONITORING AND ASSESSMENT PROGRAMS

Watershed and Airshed

One of the oldest monitoring programs in the United States, the **U.S. Geological Survey (USGS) Streamgaging Network**, was initiated in 1889 to measure stream discharge. The USGS with over 800 different funding partners currently monitor about 7100 stations nationwide. The National Estuarine Eutrophication Assessment notes that "without data from this long-term monitoring program, conclusions drawn from this assessment about linkages between symptoms and nitrogen sources would not have been possible" (Bricker et al. 1999).

The **National Water Quality Assessment (NAWQA)** program was fully initiated by USGS in 1991. NAWQA program is assessing water-quality conditions in more than 50 river basins and aquifers, known as Study Units, which collectively cover more than one-half of the United States. Data collected in each Study Unit include discharge, concentrations of suspended sediment, major ions, nutrients, trace elements, synthetic organic chemicals, and biological conditions. Understanding derived from individual Study Units and regional and national syntheses of information is being compiled with the objective of describing relationships between natural factors, human activities, and water quality conditions and to define those factors that most affect water quality in different parts of the nation. To make the program cost-effective and manageable, intensive assessment activities in each of the study units are being conducted on a rotational schedule, with one-third of the study units being studied at any one time. Topics presently being addressed by national synthesis include nutrients, pesticides, volatile organic chemicals, and trace elements. Plans for an ecological synthesis of streams are in early stages of implementation.

The **National Atmospheric Deposition Program (NADP)/National Trends Network**, coordinated by USGS, was initiated in 1978 and currently includes over 220 sites in the United States. Weekly samples of wet atmospheric deposition are analyzed for common ions, pH, and nitrogen. Data, yearly national isopleth maps, and interpretive reports are available on the Internet. Participants include over 100 partners from public and private entities throughout the states. NADP data have been crucial to the development of national models, including the regional acid deposition model and the Spatially Referenced Regressions on Watersheds (SPARROW) model.

The **Clean Air Status and Trends Network (CASTNET)** (multi-agency, EPA primarily) is a network of 55 sites nationally, sampling weekly (initiated in 1991) for air quality (criteria pollutants, fine particulates, toxics) and wet and dry deposition. Very few stations are located in coastal areas; however, CASTNET data are used in many of the local, state and national atmospheric deposition modeling efforts.

Coastal Waters

EPA's **Environmental Monitoring and Assessment Program (EMAP)** is a research program to develop the tools necessary to monitor and assess the status and trends of national ecological resources. Specific objectives of EMAP are to advance the science of ecological monitoring and risk assessment, guide national monitoring with improved scientific understanding of ecosystem integrity and dynamics, and demonstrate the framework through large regional projects.

NOAA's National Estuarine Research Reserve system has designed a **System-Wide Monitoring Program (SWMP)** to "establish a comprehensive national monitoring program for coastal marine and Great Lakes environmental quality that will provide data and information on the status and trends in the levels and biological effects of natural and anthropogenic stresses needed for our nation to make well-informed decisions concerning the utilization and protection of our coastal resources and environments" (NOAA 1999c). The proposed monitoring network includes three primary elements:

1. A **Nationwide Coastal Environmental Quality Monitoring Network** (20-25 monitoring sites) used primarily for measurements of a set of common parameters that can serve as indicators of long-term trends in environmental quality and ecosystem health. Measurements at these sites will provide the basis for linking state and regional monitoring programs into national-scale assessments, and will serve as sites for linking process research studies with long-term measurements of environmental driving variables.
2. Nested within this network will be the more intensive **Regional Monitoring Programs**, to be developed and implemented primarily through State and local academic institutions. Particular emphasis will be placed on areas considered to be under greatest threat from chemical contamination, nutrient over-enrichment, habitat degradation and other degradative anthropogenic influences.
3. A **National Coastal Monitoring Center** is proposed to provide a national focal point for coordination, data management and archiving, methods development, information dissemination, and development of routine state of the coast reports and other national-scale assessments.

To implement the proposed National Coastal Monitoring Program, SWMP recommends that the Environmental Quality Network sites include:

a. all estuarine regions in the National Estuarine Research Reserves and the National Estuary Program; and
b. additional areas nominated for inclusion by the governor or governors of the states that border a proposed region.

A regional monitoring coordination group will be established for each region to develop and direct a coordinated monitoring program tailored to the needs of its region, and to develop a long-term monitoring plan for its region. NOAA's National Center for Coastal Monitoring and Assessment in the National Ocean Service will support development and improvement of scientific methods and procedures for implementing the National Coastal Monitoring Program.

EPA, NOAA, and the National Aeronautic and Space Administration (NASA) have joined in a partnership to establish pilot sites for the development of a network known as the **Coastal Intensive Site Network (CISNet)**. CISNet is composed of intensive, long-term monitoring and research sites around the U.S. marine and Great Lakes coasts. In this partnership, EPA and NOAA are funding research and monitoring programs at pilot sites that utilize ecological indicators and investigate the ecological effects of environmental stressors. NASA is funding research aimed at developing a remote sensing capability that will augment or enhance *in situ* research and monitoring programs.

To select sites for inclusion in CISNet, criteria were developed by a working group of NOAA and EPA scientists, who then nominated 120 locations (subsequently narrowed to 41) from around the U.S. coasts. An announcement of opportunity for funding monitoring/research activities at one or more of the 41 sites resulted in over 100 proposals, of which 10 were selected for funding.

An **Integrated Coastal Monitoring Program for the Gulf of Mexico** is being developed by the Gulf of Mexico Program Office in cooperation with Gulf State agencies, EPA Regions 4 and 6, EPA's Office of Water and Office of Research and Development (ORD). Goals of the coastal watersheds (estuarine focus) element are to:

- integrate existing federal, state, local and private sector monitoring activities into a statistically-based sampling design that will provide data to support a regional assessment;
- support implementation of the monitoring requirements in the Federal Clean Water Action Plan for gulf coastal watersheds;
- support Gulf state monitoring and modeling assessments in Gulf coastal watersheds and contiguous near coastal waters;
- develop and implement a statistically-based sampling design for monitoring near coastal waters; and
- provide public access to Gulf coastal monitoring data and information.

A proposal to establish a coastal element to the existing **Coastal Componenets of the Global Ocean Observing System (C-GOOS)** is under development in response to a congressional request to propose a plan to achieve a "truly integrated ocean observing system". The purposes of the proposed coastal component are to 1) quantify inputs of energy and materials from land, air, ocean, and human activities; and to 2) detect and predict the effects of these inputs on human populations living in the coastal zone, on coastal ecosystems and living marine resources, and on coastal marine operations (NORLC 1999). Eutrophication issues (including nutrient flux measurements) figured prominently in recommendations from scientists and resource managers participating in a development workshop held in May 1999.

Five key elements are proposed for the C-GOOS system (NORLC 1999):

1. remote sensing to capture the spatial and temporal dimensions of change;
2. in-situ measurements to capture change in time and depth;
3. index sites, pilot projects and test beds to develop the models to link observations to products in the form of predictions and early warnings;
4. real time telemetry and data assimilation for timely access to and applications of environmental data; and
5. an effective data management system that accommodates the disparate coastal observation data systems and sources.

Assessment Programs

The **Long Term Ecological Research** Network (established by the National Science Foundation [NSF] in 1980) is a collaborative effort involving more than 1100 scientists and students investigating ecological processes operating at long time scales and over broad spatial scales. The network promotes synthesis and comparative research across sites and ecosystems, and now consists of 21 sites. Of these current sites, two focus on coastal areas: the Virginia Coast Reserve is a coastal barrier island with a focus on salt marsh ecology; the Plum Island Ecosystem project focuses on linkages between land and coastal waters involving organic carbon and organic nitrogen inputs to estuarine ecosystems from watersheds with various land covers and uses.

In 1997, the Committee on Environmental and Natural Resources (CENR), one of nine committees under the National Science and Technology Council established the CENR Environmental Monitoring Team to develop a **Framework for Integrating the Nation's Environmental Monitoring and Research Networks and Programs.** The team's charge was based on the assessment that current monitoring programs do not provide integrated data across multiple natural resources at the various temporal and spatial scales needed to develop policies based on current scientific understanding of ecosystem processes. The emphasis was on agency cooperation and coordination. The CENR conceptual framework, which was developed by a team representing 13 federal agencies, supports better understanding, evaluation, and forecasting of renewable natural resources at national and regional scales.

Many of the concepts outlined in CENR Monitoring Framework are being incorporated into the outline of the **Coastal Research and Monitoring Strategy**, as an element of the Clean Water Action Plan. A draft outline has been developed (June 1999) by an interagency workgroup consisting of the following organizations (number of representatives on the workgroup is shown in parentheses): EPA (16), NASA (1), the National Institute of Health (3), NOAA (15), NSF (2), the Office of Science and Technology Policy (1), the Smithsonian (1), the U.S. Army Corps of Engineers (2), the U.S. Bureau of Reclamation (1), the U.S. Coast Guard (2), USDA (1), and USGS (5), plus "stakeholder representatives": Association of National Estuary Programs (1), Coastal States Organization (2), National Estuarine Research Reserves (1), Center for Marine Conservation (1), Coastal Alliance (1), Association of State and Interstate Water Pollution Control Administration (1), and Consortium for Oceanographic Research and Education (1). The Strategy will attempt to coordinate existing programs; however, it is not clear to date how and when implementation will take place, or whether a new initiative will be needed to implement recommendations from the workgroup.

National Strategy for the Development of Regional Nutrient Criteria. As part of the Clean Water Action Plan, which calls for expanded efforts to reduce nutrient over-enrichment of waterways, the EPA has begun an effort to accelerate the development of scientific information concerning the levels of nutrients that cause water quality problems and to organize this information by different types of waterbodies (e.g., streams, lakes, coastal waters, wetlands). EPA is to work with states and tribes to adopt criteria (i.e., numeric concentration levels) for nutrients, including nitrogen and phosphorus, as part of enforceable state water quality standards under the Clean Water Act. Draft nutrient criteria guidance for estuaries are scheduled for review by 2002.

The **NOAA Coastal Services Center** has initiated several national projects that directly address access and coordination of data and information concerning eutrophication. These projects include:

- the Coastal Information Directory, which provides a single query point to search a variety of nationwide databases for descriptions of coastal data, information and products;
- the Coastal Management and Geographic Information Systems

Bibliography, which is an international compilation of documented geographic information system and remote sensing applications in the field of coastal management;

- the Coastal Ocean Habitat Project, which aims to produce easily accessible remotely sensed time series imagery for detecting long-term, seasonal and event-specific trends in water turbidity and sea surface temperature for coastal regions;
- NOAA Coastal Services Center Library, which provides information required by coastal resource managers on-line; and
- the Coastal Zone Information Center Collection, which contains all publications produced under the Coastal Zone Management Act of 1972. To date, approximately 5000 of the estimated 12000 pieces have been cataloged and classified, and are available on-line.

MANAGEMENT STRATEGIES ADDRESSING COASTAL EUTROPHICATION

National Estuary Program Strategies

In 1997, 27 of the 28 EPA National Estuary Programs met in an American Assembly format to define key management issues. Eighteen of the 27 NEPs, from every region of the United States, identified the impacts of nutrient overloading as either a high or medium program priority (ANEP 1997). Management actions used to address eutrophication within NEP estuaries vary considerably between programs. Many of the programs reporting actions list regulatory control as the primary technique; however, several have also initiated a mixture of regulatory and nonregulatory (voluntary) approaches. A summary of some of these approaches is presented in Table C-1.

RESULTS OF A MANAGERS QUESTIONNAIRE

Managers and Scientists

To gather information about how federal, state and local programs are currently perceived by those managers in eutrophic coastal areas, the committee talked with local managers and scientists from 18 estuaries and coastal areas throughout the United States in the spring of 1999. These areas were selected from the 48 estuaries identified by NOAA's National Estuarine Eutrophication Assessment as exhibiting the effects or being at high risk from nutrient over-enrichment. Individuals were asked to express their views regarding the tools they currently use for management, whether they consider those tools adequate for the development of

TABLE C-1

Estuary	Approach and Milestones
Albemarle-Pamlico Sounds	Develop basinwide plans, and the Tar-Pamlico Basin Association, a coalition of permitted dischargers, is experimenting with a point/nonpoint source trading strategy.
Barataria-Terrebonne Bays	Develop initiatives to reduce agricultural and sewage pollution and manage stormwater impacts.
Buzzards Bay	Develop a nitrogen loading strategy amending local zoning to reduce future development so as not to exceed critical loading limits.
Indian River Lagoon	Require all domestic wastewater discharges to cease by 1995, develop and implement pollutant loading goals, and implement pollution reduction programs through education and BMPs.
Long Island Sound	Targets for improved DO levels in phases: Phase1: freeze critical point and non-point nitrogen loading at 1990 levels; Phase 2: commit to low-cost reductions in annual nitrogen load below 1990 levels; and Phase 3: develop specific nitrogen targets for 11 geographic management areas.
Morro Bay	Require replacement of septic systems in Los Osos with a sewer system, improve riparian buffer areas in watershed, and implement BMPs.
New York-New Jersey Harbor Estuary	Upgrade municipal discharges to secondary treatment, establish DO targets and nitrogen controls based on eutrophication model, and control rainfall-induced discharges of organic materials.
San Francisco Bay	Encourage voluntary dairy waste discharge requirements. (Successes, though, are limited to cases where formal enforcement actions were taken.)
Sarasota Bay	Eliminate wastewater discharges to the bay and upgrade STP to advanced treatment standards (reducing nitrogen load by 25 percent).
Tampa Bay	Require all STPs discharging to the bay to be upgraded to advanced treatment standards, obtain voluntary agreement to maintain nitrogen loads at 1992-1994 levels, and develop a TMDL (EPA-approved) for nitrogen loading.
Western Peconic Estuary	Implement a nitrogen "freeze" on point sources and develop a TMDL to reduce nitrogen inputs.

TABLE C-1 National Estuary Programs and their regulatory and voluntary approaches.

effective strategies to address eutrophication and other effects of nutrient over-enrichment, and how the tools could be improved. In addition to the local managers and scientists, similar input was requested from state managers from several states. Managers from the following coastal areas responded to the questionnaire:

Casco Bay, Maine
Boston Harbor, Massachusetts
Long Island Sound, New York and Connecticut
Chesapeake Bay Mainstem, Maryland, Delaware, Virginia
Delaware Inland Bays, Delaware
Albemarle-Pamlico Sound, Neuse River, North Carolina
St. Johns River, Florida
Florida Bay, Florida
Charlotte Harbor, Florida
Sarasota Bay, Florida
Tampa Bay, Florida
Apalachicola, Florida
Upper Laguna Madre, Texas
Galveston Bay, Texas
Corpus Christi Bay, Texas
Newport Bay, California
San Francisco Bay, California
Puget Sound, Washington

Program Coordination

The local and state managers interviewed by the committee were asked which federal programs have been useful in developing their management strategy, and why. Those managers associated with a NEP generally identified EPA's NEP program as a critical federal player; one characterized NEP as "the glue that keeps the management program going." Those outside the national EPA's NEP or NOAA's NERR systems identified a variety of federal programs associated with monitoring as important, particularly USGS (primarily for stream flow) and NOAA. The U.S. Army Corps of Engineers is participating in large modeling efforts in several estuaries, and USDA and their National Resources Conservation Service were identified as important contributors to agricultural strategies. However, several managers indicated that federal involvement in the development of their management plans was minimal.

A consistent theme identified by managers was the need for better coordination between federal agencies, and between federal and state/local efforts. One program noted an "appalling lack of understanding at the federal level of what happens at the local level, and what local and regional managers need."

To address these issues, some of the larger local programs have initiated coordination/oversight bodies to their processes. The Florida Bay program noted that an oversight panel (including independent scientific review) is an important element of their multi-agency initiative, and is being used to integrate and coordinate research efforts among the agencies (federal, state and local). This program is finding that a periodic Florida Bay Science Conference for all researchers is proving very important for sharing of findings between agencies and initiatives.

Monitoring Effects and Sources

The larger estuaries (Long Island Sound, Chesapeake Bay, Florida Bay, St. Johns River, San Francisco) report that federal programs are the primary source of in-bay data, with little or no local support. Conversely, the smaller estuaries (including Casco Bay, Delaware Inland Bays, Albemarle-Pamlico, Charlotte Harbor, Sarasota Bay, Tampa Bay, the Texas estuaries, Newport Bay) report little or no federal assistance with in-bay monitoring; these programs rely on a mixture of local/state government or water quality authority, and volunteer monitoring. Several programs noted that the scale and objectives of large national monitoring programs (such as EMAP) are too broad to be useful at the local or regional level.

Almost all (15 local programs) report that federal USGS stream flow data have been a critical element to their loading estimates. Point source permit monitoring data are noted as important in eight of the 17 local programs (Boston Harbor, Long Island Sound, Chesapeake Bay, Charlotte Harbor, Tampa Bay, Newport Bay, San Francisco Bay, and Puget Sound).

Eight programs noted that atmospheric deposition data are important, but the majority of these programs (six of the eight) consider these data sources inadequate at this time. Relative contribution to nitrogen input budgets, transport through the watershed, sources, and impacts on coastal conditions from atmospheric deposition are all cited by local programs as largely unknown.

Access to monitoring data from different federal, state and local governments was noted as a barrier to development of management strategies in six of the 18 local programs included in the inquiry.

Models and Assessment Techniques

Results from the Managers Questionnaire indicate that local programs are using a wide range of modeling tools and assessment techniques, ranging from very complex linked watershed:hydrodynamic:water quality models to simplistic conceptual models. More than half of the programs indicated that modeling had not yet been completed. However, most of the managers who are at a point they can judge the effectiveness of their modeling tools indicate that the tools have been adequate to help develop management strategies, although all programs have noted limitations and additional needs.

Several of the larger programs that are developing complex modeling strategies (Chesapeake Bay, Long Island Sound, St. Johns River, Florida Bay) have not yet used them to identify management options (although several have used earlier versions to help with setting targets). These programs expect that the fully developed models, when functional, will be critical for their strategies.

To develop nutrient loading estimates, planners working on Long Island Sound, Chesapeake Bay, St. John's River, Charlotte Harbor, Sarasota Bay, Tampa Bay, Galveston, San Francisco Bay, Newport Bay, and Puget Sound all use land use-based spreadsheet models, with empirical loadings where available. Runoff coefficients used in the land use models were considered the "best available," primarily from literature sources. Hydrologic Simulation Program-FORTRAN-based models and/or Groundwater Loading Effects of Agricultural Management Systems are also used in Chesapeake Bay, St. John's River, and Newport Bay, and will be used in Long Island Sound. Florida Bay uses a suite of surface water flow models.

In-bay models range from 3-D hydrodynamic models linked to water quality models (Long Island Sound, Chesapeake Bay, St. John's River, Florida Bay) to an empirical regression-based model approach (Tampa Bay), to conceptual models (Newport Bay) to a technology-based approach (Sarasota Bay). Although the complexity ranges widely between programs, almost all programs believe that their modeling process is (or is expected to) providing adequate information for addressing the impacts of nutrient over-enrichment in their systems, with caveats noted by all programs. Charlotte Harbor and Sarasota Bay indicated that an in-bay model has not proven useful.

Effectiveness of the models used, as perceived by managers queried for this report, varied. Long Island Sound noted that "we'd be nowhere without the models" in helping to provide the scientific basis for multi-jurisdictional actions; Chesapeake Bay noted that the models provided a shared view of goals critical for political buy-in. Several programs noted that the complex estuarine circulation dynamics required that relatively complex modeling efforts be developed to help assess nutrient over-enrichment impacts and potential management options.

However, not all programs agreed that the limited funds available for assessment were best spent on complex modeling, particularly in estuaries with limited stratification. Tampa Bay has developed simple regression-based approaches relating nitrogen loading estimates to ambient water quality parameters to help develop nutrient loading goals. Sarasota Bay is using a "technology-based" approach in which potential reductions in the watershed are ranked according to cost-effectiveness, and implementation of actions with the largest potential for nutrient reduction are encouraged.

Management Strategies

When asked about management strategies during the committee's interviews with the managers of 18 estuaries where nutrient enrichment is a major management issue, the managers offered the following insights:

- Ten of the eighteen local programs interviewed reported that a management program to address the effects of nutrient over-enrichment is underway; five more noted that one was in development. Two programs reported that the method and means of implementation were "unclear". The Florida state-wide programs estimated that 50 percent of the coastal waters in Florida have some level of management at local levels ongoing.
- Strategies range from entirely educational and non-regulatory (Delaware Inland Bays, Casco Bay) to primarily regulatory (Long Island Sound, Newport Bay, Boston Harbor); the other programs reported mixed (regulatory and nonregulatory) approaches.
- Seven of the seventeen local programs (Long Island Sound, Delaware Inland Bays, St. John's River, Charlotte Harbor, Laguna Madre, Newport Bay, and Puget Sound) report that a TMDL is, or is expected to be, a driving factor for the management strategy. Two programs (Chesapeake Bay and Tampa Bay) expressed concern that a TMDL, if approved, could curtail ongoing voluntary nutrient management activities.
- Eight local programs report that numeric targets are in place or are being developed. Targets include nutrient reduction or loading targets (six programs) and in-bay indicator targets for four programs (Chesapeake Bay, Florida Bay, Delaware Inland Bays, and Tampa Bay).
- Two local programs (Tampa and Sarasota Bays) report that the strategy seems to be effective for reducing eutrophic conditions in their waterbody, resulting in increased water clarity and seagrass in both bays. These two programs are also the only two that report

that point source regulatory management efforts were in place prior the existing management program.

- Only one program, San Francisco, answered that their strategy did not appear to be working. However, three programs (Long Island Sound, Chesapeake Bay, and Albemarle-Pamlico) reported mixed results, with some areas improving and some not. Twelve of the 17 report that it is too early to tell.
- For those coastal areas which are far enough along in implementation of their strategy to judge effectiveness, a regulatory approach has been considered most effective for Long Island Sound, Albemarle-Pamlico Sounds, Sarasota Bay, and Newport Bay, while non-regulatory approaches (voluntary reductions, education) are considered by their managers to be most effective for Casco Bay, Chesapeake Bay, Delaware Inland Bays, and Tampa Bay.
- Newport Bay notes that the regulatory component is essential in that bay because it is "too late for a voluntary approach." In contrast, the state Coastal Management Program in Florida notes that, for nonpoint sources, probably a non-regulatory approach will be more effective due to the weakness of agricultural regulations.
- The Albemarle-Pamlico Sounds and the Neuse River programs report that they are moving toward a more regulatory process. The primarily voluntary efforts initiated in the early 1990s (including trading in the watershed) has not been successful with the huge increase in concentrated animal feeding operations. Nutrient trading was initially intriguing, but has not actually been effectively implemented, due to the details of implementing a formal trading program. However, the new regulatory rules have more flexibility for some areas in meeting requirements than previous point source discharge limits.
- Two programs (Delaware Inland Bays and San Francisco Bay) cited public policy involvement as essential elements for reducing the effects of nutrient over-enrichment, while Florida Bay note that a multi-agency approach has been critical.
- Setting quantitative reduction goals has been critical for the Chesapeake Bay and Tampa Bay.

In those same interviews, the most frequently cited barriers to development of management strategies noted by the managers included:

- Lack of participation in the management process by major sources and stakeholders. Several local programs reported lack of participation by agricultural interests, electric utilities, and jurisdictions upstream in watersheds as major barriers.

- Lack of regulatory authority (including numerical nutrient standards) to require nutrient reduction.
- Lack of coordination between local, state and federal programs, including implementation of regulatory programs such as TMDLs.
- Lack of credible source loading information, especially for atmospheric deposition, nonpoint sources, and sediment flux.

When asked which elements of their management strategy were most difficult to develop and implement, the managers offered a wide variety of responses, including:

- building trust among the management participants;
- source identification, particularly atmospheric deposition;
- nonpoint source controls, primarily due to lack of regulatory control on these sources;
- agricultural community difficult to engage (several programs mentioned this element);
- data on sediment flux and recycling;
- interagency cooperation;
- scientific basis a time-consuming process; and
- changing public attitudes about residential lawn maintenance.

When asked what could make their management strategy more effective, five of the 11 programs answering this question identified the need for additional or better information, and four identified additional funding. Several programs mentioned better cooperation between agencies.

Appendix D

Model Reviews

WATERSHED MODELS

Although some purely hydrologic and hydraulic models (e.g., those by the U.S. Army Corps of Engineers, Hydrologic Engineering Center) might contribute to analysis of nonpoint source water quality in watersheds, only models with an explicit water quality simulation capability are described herein, and more specifically, the models must be capable of simulating nutrient loadings. Some reviews of models are available, although their usefulness diminishes with time since publication due to the dynamic nature of model changes. Reviews of process models for simulation of nonpoint source water quality include those by Donigian and Huber (1991), DeVries and Hromadka (1993), Novotny and Olem (1994), Donigian et al. (1995), Singh (1995), the Environmental Protection Agency (EPA 1997), and Deliman et al. (1999). The references given in the discussion that follows are unlikely to be the best source of information about the latest model capabilities. The best approach is usually to find the webpage for the agency that distributes the model and obtain the most current information in that manner. URLs as of the time of publication of this document are provided in Appendix E. Discussion groups exist on the Internet for some of the models as well.

PROCESS MODELS

Non-Urban Watersheds

With few exceptions, simulation models sponsored by federal government organizations dominate the non-urban watershed environment.

AGNPS–Agricultural Nonpoint Source Pollution Model (Young et al. 1994)

The model was developed by the U.S. Department of Agriculture's (USDA) Agricultural Research Service. The primary emphasis of the model is on nutrients, soil erosion, and sediment yield for comparing the effects of various best management practices on agricultural pollutant loadings. The AGNPS model can simulate sediment and nutrient loads from agricultural watersheds for a single storm event or for a continuous simulation. The watershed must be divided into a uniform grid (square cells). The cells are grouped by dividing the basin into subwatersheds. However, water flow and pollutant routing is accomplished by a function of the unit hydrograph type, which is a lumped parameter approach. The model does not simulate pesticides.

AGNPS is also capable of simulating point inputs such as feedlots, wastewater discharges, and stream bank and gully erosion. In the model, pollutants are routed from the top of the watershed to the watershed outlet in a series of steps. The modified universal soil erosion equation is used for predicting soil loss in five different particle sizes (clay, silt, sand, small aggregates, and large aggregates). The pollutant transport portion is subdivided into one part handling soluble pollutants and another part handling sediment absorbed pollutants. The input data requirements are extensive, but most of the data can be retrieved from topographic and soil maps, local meteorological information, field observations, and various publications, tables, and graphs provided in the user manual or references.

ANSWERS–Areal, Nonpoint Source Watershed Environment Response Simulation (Beasley and Huggins 1981)

The model was developed by the Agricultural Engineering Department of Purdue University. It is a distributed parameter model designed to simulate rainfall-runoff events. Currently the model is maintained and distributed by the Agricultural Engineering Department, University of Georgia, Tifton, Georgia. To use the ANSWERS model, the watershed is divided into a uniform grid (square elements). The element may range from one to four hectares. Within each element the model simulates the processes of interception, infiltration, surface storage, surface flow, subsurface drainage, sediment detachment, and movement across the element. The output from one element then becomes a source of input to an adjacent element. Nutrients (nitrogen and phosphorus) are simulated using correlation relationships between chemical concentrations, sediment yield, and runoff volume. Snowmelt or pesticides movement cannot be simulated. A single storm rainfall hyetograph drives the model.

BASINS–Better Assessment Science Integrating Point and Nonpoint Sources (Lahlou et al. 1998)

This model is a multipurpose environmental analysis system for use by regional, state, and local agencies in performing watershed and water quality based studies. It was developed by EPA to address three objectives:

1. To facilitate examination of environmental information
2. To support analysis of environmental systems
3. To provide a framework for examining management alternatives

A geographic information system (GIS) based on ArcView provides the integrating framework for BASINS. GIS organizes spatial information so it can be displayed as maps, tables, or graphics. Through the use of GIS, BASINS has the ability to display and integrate a wide range of information (e.g., land use, point source discharges, water supply withdrawals) at a scale selected by the user. For example, some users may need to examine data at a state scale to determine problem areas, compare watersheds, or investigate gaps in data. Others may want to work at a much smaller scale, such as investigating a particular river segment. These features makes BASINS a unique environmental analysis tool. The analytical tools in BASINS are organized into two modules. The assessment and planning module, working within GIS, allow users to quickly evaluate selected areas, organize information, and display results. The modeling module allows users to examine the impacts of pollutant loadings from point and nonpoint sources. The modeling module includes the following: QUAL2E, Version 3.2, a water quality and eutrophication model; TOXIROUTE, a model for routing pollutants through a stream system; NPSM-HSPF, version 10, a nonpoint source model for estimating loadings. The latest versions of both QUAL2E and Hydrologic Simulation Program-FORTRAN (HSPF) are included in the BASINS package.

CREAMS–Chemicals, Runoff, and Erosion from Agricultural Management Systems (Knisel 1980) and GLEAMS–Groundwater Loading Effects of Agricultural Management Systems (Knisel 1993)

CREAMS is a field-scale model for evaluation of agricultural best management practices (BMPs) for pollution control. Daily erosion, sediment yield, and associated nutrient and pollutant loads are estimated at the boundary of the agricultural area. Runoff estimates are based on the Soil Conservation Service (SCS) method. The vertical flux of nutrients and pesticides in the root zone may be simulated using GLEAMS in order to provide a groundwater component to off-site loadings. Both CREAMS and GLEAMS are maintained by USDA Agricultural Research Service.

HSPF–Hydrologic Simulation Program-FORTRAN (Bicknell et al. 1993)

HSPF is a simulation model developed under EPA sponsorship to simulate hydrologic and water quality processes in natural and man-made water systems. It is an analytical tool that has application in planning, design, and operation of water resources systems. The model enables the use of probabilistic analysis in the fields of hydrology and water quality management through its continuous simulation capability. It uses such information as time history of rainfall, temperature, evaporation, and parameters related to land use patterns, soil characteristics, and agricultural practices to simulate the processes that occur in a watershed. The initial result of an HSPF simulation is a time history of the quantity of water transported over the land surface and through various soil zones down to the groundwater aquifer. Runoff flow rate, sediment loads, nutrients, pesticides, toxic chemicals, and other water quality constituent concentrations can be predicted. The model can simulate continuous, dynamic, or steady state behavior of both hydrologic/hydraulic and water quality processes in a watershed. HSPF also may be applied to urban watersheds through its impervious land module.

SWRRB–Simulator for Water Resources in Rural Basins (Arnold and Williams 1994) and SWAT–Soil and Water Assessment Tool (Arnold et al. 1995)

The SWRRB model was developed for large, complex, rural basins through modifications to the CREAMS model for simulation of daily-time-step hydrology, nutrient and other loads. Similarly, SWAT is an extension of SWRRB from small watershed scale to basin scale. SCS hydrology techniques are modified to allow for agricultural return flow and flow through the root zone. Nitrogen and phosphorus computations are based on regression relationships between chemical concentration, sediment yield and runoff volume. Both nitrate and organic nitrogen may be simulated for different soil layers. Both models are maintained by USDA's Agricultural Research Service.

WARMF–Watershed Analysis Risk Management Framework (EPRI 1998)

The Electric Power Research Institute (EPRI) has sponsored model development specifically for the purpose of developing Total Maximum Daily Loads (TMDLs) in accordance with EPA water quality regulations. The models embedded in WARMF draw upon other simulation models described herein, including ANSWERS and SWMM. Nonpoint loads are generated on the basis of land use and other data in a GIS format, coupled with simplified hydrologic techniques, but with complete coverage of the hydrologic cycle. Erosion processes are included in land runoff modules, as are integrated water quality fate and transport processes in riverine segments. Because point source loads are also included, the model may be executed in an iterative fashion to determine TMDL allocations.

Urban Watersheds

Perhaps more so than for non-urban watersheds, process-type simulation models for urban watersheds include a number of non-proprietary, mostly governmentally-sponsored models, plus a few frequently used proprietary private models (Table D-1). However, the discussion that follows describes only models sponsored by a federal agency, since these represent the bulk of the models used in practice. Information on other models may be found from their websites (Appendix E).

DR3M-QUAL–Distributed Routing, Rainfall, Runoff Model-Quality (Alley and Smith 1982)

This model developed by the U.S. Geological Survey (USGS) includes a quality simulation routine coupled to its earlier DR3M model. Runoff generation and subsequent routing is based on the kinematic wave method. Quality generation is based on buildup and washoff functions. The model may be used for single-event or long-term (continuous) simulation of hydrographs and quality constituents.

STORM–Storage, Treatment, Overflow Runoff Model (HEC 1977)

This model was developed by the U.S. Army Corps of Engineers, Hydrologic Engineering Center (HEC) for simplified long-term analysis (continuous simulation) of runoff from urban areas. The model addresses

TABLE D-1

Model	Agency/Source	Primarily Hydrology/ Hydraulics	Continuous Simulation or Storm Event	Complete Dynamic Flow Routing?	Graphical User Interface
DR3M-QUAL	USGS	Hydrology	CS/SE	No	ANNIE
HSPF	EPA	Hydrology	CS/SE	No	ANNIE, 3rd party
Hydroworks	HR Wallingford in the United Kingdom, Montgom. Watson in the United States.	Hydrology/ Hydraulics	CS/SE	Yes	Yes
MOUSE	Danish Hydraulic Institute	Hydrology/ Hydraulics	CS/SE	Yes	Yes
P8	Wiiliam Walker, Jr.	Hydrology	CS/SE	No	Menu
STORM	HEC/Vendors	Hydrology	CS	No	3rd party
SWMM	EPA/OSU	Hydrology/ Hydraulics	CS/SE	Yes	3rd party

TABLE D-1 Nonpoint source water quality simulation models commonly applied to urban watersheds in the United States.

combined sewer overflows (CSOs) in particular, although it may also be used to simulate stormwater runoff quantity and quality. Hydrologic methods utilize a simple runoff coefficient and depression storage for hourly time steps. Water quality loads are estimated on the basis of buildup and washoff functions. The trade-off between treatment and storage options at the catchment outlet may be evaluated for economic optimization of control strategies. The model is no longer supported by HEC, but is available from private vendors with an enhanced graphic user interface.

SWMM–Storm Water Management Model (Huber and Dickinson 1988; Roesner et al. 1988)

The SWMM model was originally developed by EPA as a simulation model for the quantity and quality of CSOs. However, it has been widely used for hydrologic, hydraulic and water quality analysis for urban stormwater and some non-urban applications as well. Runoff is generated on a single-event or continuous basis using nonlinear reservoir methods and Horton or Green-Ampt loss functions. Flow routing options range from simple to complete solution of the Saint-Venant equations in the Extran Block. Simulation of nonpoint source runoff quality may be performed by several options, including constant concentrations, regression methods and buildup and washoff functions. Quality routing and treatment processes may also be performed.

SPREADSHEET MODELS

A "spreadsheet model" is basically a generic category in which water quality predictions are based on unit loads (e.g., kg ha day^{-1}) or event mean concentrations, as a function primarily of land use, although other factors such as delivery ratios and effects of best management practices may be incorporated. Numerous ad hoc models have been generated in this category for application to both urban and non-urban settings.

STATISTICAL APPROACHES

Non-Urban Watersheds

SPARROW–USGS's Spatially Referenced Regressions on Watershed Attributes Model (Smith et al. 1997; Preston et al. 1998)

A highly sophisticated regression procedure based on spatially-referenced land use and stream channel characteristics has been developed by the USGS for prediction of total nitrogen and total phosphorus loads (kg day^{-1}) at the outlet of major U.S. watersheds. Independent variables for nitrogen prediction include load-related parameters such as point source loads, fertilizer application, livestock waste production, and atmospheric deposition, plus factors such as temperature, soil permeability, and stream density. Coefficients of determination are 0.87 and 0.81 for TN and TP load prediction for the conterminous United States, respectively, based on data from 414 National Stream Quality Accounting Network stream monitoring stations. The regression is based on land use data from 78,000 soil geographic units. An application for the determination of total nitrogen in the Chesapeake Bay watershed (Preston et al. 1998) indicated areas that are most important to the delivery of nutrients to the bay, mainly those that drain directly to large streams or those that are near the bay.

Urban Watersheds

EPA's Statistical Method (EPA 1983; Driscoll et al. 1989)

This method was adopted for the EPA's Nationwide Urban Runoff Program (NURP) and later refined in an investigation of highway runoff quality for the Federal Highway Administration. A derived-distribution technique is used to derive the lognormal distribution of event mean concentrations from urban sites. This is a considerable improvement over the simpler, constant concentration approach using a constant event mean concentration (EMC) because the frequency distribution is provided, which in turn may be used to simulate receiving water loadings.

USGS's Regression Models (Driver and Tasker 1990)

All of the EPA's NURP data were combined with some other urban stormwater quality data by the USGS to develop a set of regression equations for estimation of EMC and loads. The equations were developed for three different hydrologic zones of the United States and utilize commonly available watershed characteristics as independent variables. If a single EMC or loading value is needed, this work remains the only comprehensive synthesis of thousands of samples of stormwater runoff from urban watersheds. In the more than 16 years since NURP studies ended, thousands of urban runoff sites have been sampled as part of EPA's NPDES requirements. The USGS study should be updated to reflect these abundant newer and more spatially diverse data.

Once again, printed summaries of receiving water models tend to rapidly become outdated. However, useful background information may be obtained from: Ambrose et al. (1988, 1996), Bouchard et al. (1994), Barnwell et al. (1995), EPA (1997), Rauch et al. (1998), Shanahan et al. (1998), and Somlyody et al. (1998). Much more current information may often be gleaned from Oregon State University (1999) and USGS (1999b).

ESTUARINE AND COASTAL EUTROPHICATION MODELS

As with the preceding watershed models, eutrophication models for receiving waters, both estuaries and the coastal ocean, come in a variety of forms. These include desktop screening models, steady state and tidally averaged models, and dynamic simulation models. The degree of complexity and the computer and personnel resources necessary to use each effectively are highly variable. Much of the following information follows from the earlier work of the EPA (1990).

Desktop Screening Models

Desktop screening methodologies may be utilized with a hand-held calculator or computer spreadsheet and are based on steady state conditions, first order decay coefficients, simplified estimates of flushing time, and seasonal pollutant concentrations. The Water Quality Assessment Methodology (WQAM) provides a series of such analyses.

WQAM–Water Quality Assessment Methodology (Mills et al. 1985)

WQAM is a set of steady state desktop models that includes both one-dimensional and two-dimensional box model calculations. Specific techniques contained in WQAM are the Fraction of Freshwater Method, the Modified Tidal Prism Method, Advection-Dispersion equations, and Pritchard's two-dimensional box model.

Steady State and Tidally Averaged

Steady state and tidally averaged simulation models generally use a box or compartment-type network and are difficult to calibrate in situations where hydrodynamics and pollutant releases are rapidly varying. Consequently, they are less appropriate when waste load, river inflow, or tidal range vary appreciably with a period close to the flushing time of the waterbody. These are the simplest models available that are capable of describing the relationship between nutrient loads and some of the endpoints of concern of the eutrophication process (i.e., chlorophyll, minimum dissolved oxygen).

QUAL2E (EPA 1995)

QUAL2E is a steady state, one-dimensional model designed for simulating conventional pollutants in streams and well-mixed lakes. It has been applied to tidal rivers with minor adaptations to the hydraulic geometry and dispersion functions. Water quality variables simulated include conservative substances, temperature, biochemical oxygen demand (BOD), dissolved oxygen (DO), ammonia, nitrite, nitrate, organic nitrogen, phosphate and organic phosphorus, and algae. It simulates the major reactions of nutrient cycles, algal production, benthic and carbonaceous demand, atmospheric reaeration and their effects on the DO balance. It is applicable to well mixed, dendritic streams. It also has the capability to compute required dilution flows for flow augmentation to meet any prescribed DO level. QUAL2E is widely used for stream waste load allocations and discharge permit determinations in the United States and other countries.

WASP5 (Ambrose et al. 1993)

WASP5 is a general, multi-dimensional model that utilizes box modeling techniques. The equations solved by WASP5 are based on the principle of conservation of mass. Operated in the quasi-dynamic mode, WASP5 requires the user to supply initial box volumes, network flow fields, and inflow time functions. The user also must calibrate dispersion coefficients between boxes. WASP5 has the capability of simulating nutrient-related water quality issues at a wide range of complexity.

EUTRO5 is the submodel in the WASP5 system that is designed to simulate conventional pollutants. It predicts DO, carbonaceous BOD, phytoplankton carbon and chlorophyll *a*, ammonia, nitrate, organic nitrogen, organic phosphorus, and orthophosphate in the water column and, if specified, the underlying bed.

Dynamic Simulation Models of One or More Dimensions

Numerical one-dimensional and two-dimensional models that simulate variations in tidal height and velocity throughout each tidal cycle enable the characterization of phenomena varying rapidly within each tidal cycle, such as pollutant spills, stormwater runoff, and batch discharges. They also are deemed appropriate for systems where the tidal boundary impact is important to the modeled system within a tidal period. The application of tidally varying (intratidal) models has found most use in the analysis of short-term events, in which the model simulates a period of time from one tidal cycle to a month. Some seasonal simulations have also been conducted.

WASP5 (Ambrose et al. 1993)

WASP5 may be operated in the tidal dynamic mode through linkage with the associated hydrodynamic model DYNHYD5, which is a link-node model that may be driven by either constantly repetitive or variable tides. Unsteady inflows may be specified, as well as wind that varies in speed and direction. DYNHYD5 is best suited for one-dimensional longitudinal simulations, but has also been applied in two-dimensional mode to evaluate lateral variations in estuarine water quality. DYNHYD5 is not suited for systems with significant vertical stratification. There is, though, no reason to prevent the WASP5 eutrophication sub-model, EUTRO5, from being linked to a more complex hydrodynamic model than is supplied with WASP5 (see below).

In numerical two-dimensional and three-dimensional dynamic simulation models, dispersive mixing and seaward boundary exchanges are treated more realistically than in one-dimensional models. While not routinely used in nutrient analyses, they are now finding use by experts in special studies.

CE-QUAL-W2 (Environmental and Hydraulics Laboratories 1986)

CE-QUAL-W2 is a dynamic two-dimensional (x-z) model developed for stratified waterbodies. This is a Corps of Engineers modification of the Laterally Averaged Reservoir Model (Edinger and Buchak 1983; Buchak and Edinger 1984a, b). CE-QUAL-W2 consists of directly coupled hydrodynamic and water quality transport models. Hydrodynamic computations are influenced by variable water density caused by temperature, salinity, and dissolved and suspended solids. CE-QUAL-W2 simulates as many as 20 other water quality variables. Primary physical processes included are surface heat transfer, shortwave and longwave radiation and penetration, convective mixing, wind and flow induced mixing, entrainment of ambient water by pumped-storage inflows, inflow density current placement, selective withdrawal, and density stratification as impacted by temperature and dissolved and suspended solids. Major chemical and biological processes in CE-QUAL-W2 include: effects on DO of atmospheric exchange, photosynthesis, respiration, organic matter decomposition, nitrification, and chemical oxidation of reduced substances; uptake, excretion, and regeneration of phosphorus and nitrogen and nitrification-denitrification under aerobic and anaerobic conditions; carbon cycling and alkalinity-pH-CO_2 interactions; trophic relationships for total phytoplankton; accumulation and decomposition of detritus and organic sediment; and coliform bacteria mortality.

MIKE3

MIKE3 is a three-dimensional, time-dependent, free surface model with wetting and drying of shoals. It is maintained and marketed by the Danish Hydraulic Institute. It offers two different hydrodynamic engines. The first of these assumes a hydrostatic pressure distribution and solves the equations of motion on a sigma-coordinate grid. The second involves a non-hydrostatic formulation and solution on a z-level coordinate grid. Nested grids are allowed. Numerous turbulence closure schemes may be selected. The hydrodynamic model can be coupled to a water quality module, which focuses on dissolved oxygen, organic matter, ammonia, nitrate, phosphorus, bacteria, and chlorophyll *a*. It can also be coupled to a eutrophication model that includes carbon and nutrient cycling, phytoplankton and zooplankton growth, oxygen balance, and benthic vegetation.

The Danish Hydraulic Institute also markets one and two-dimensional coupled hydrodynamic and water quality models, MIKE11 and MIKE21 (Warren and Bach 1992).

ECOM/*EM

Hydroqual, Inc. has produced a series of similar models based on ECOMsi, a three-dimensional, free surface, finite difference, hydrodynamic code based on the community Princeton Ocean Model (Blumberg and Mellor 1987), and eutrophication code based on the EUTRO code contained within WASP5 (Hydroqual, Inc. 1998). These models include, among others, Bays Eutrophication Model (applied to Massachusetts and Cape Cod Bays), and System-Wide Eutrophication Model (applied to the New York Apex and adjacent estuaries). The models allow for wetting and drying during a tidal cycle and contain various options for the turbulence closure in both the vertical and horizontal dimensions.

The eutrophication code involves 25 state variables. These include different classes of phytoplankton, as well as both refractive and labile particulate and dissolved organic matter.

CH3D-ICM

CH3D-ICM is a linkage of CH3D, a hydrodynamic model, and CE-QUAL-ICM, a water quality model. CH3D is a three-dimensional, finite difference hydrodynamic model developed by Peter Sheng, recently modified for the Chesapeake Bay Program (Johnson et al. 1993), and maintained by the U.S. Army Corps of Engineers' Waterways Experiment Station in Vicksburg, Mississippi. The model can be used to predict system response to water levels, flow velocities, salinities, temperatures, and the three-dimensional velocity field. CH3D makes hydrodynamic computations on a curvilinear or boundary-fitted planform grid. Deep navigation channels and irregular shorelines can be modeled because of the boundary-fitted coordinates feature. Vertical turbulence is predicted by the model, and is crucial to a successful simulation of stratification, destratification, and anoxia. A second-order model based upon the assumption of local equilibrium of turbulence is employed.

ICM is an unstructured finite volume water quality model that may be applied to most waterbodies in one, two, or three dimensions (Cerco and Cole 1995) by readily linking it to any type of hydrodynamic model. The model predicts time-varying concentrations of water quality constituents and includes advective and dispersive transport. The model contains detailed eutrophication kinetics, modeling the carbon, nitrogen, phosphorus, silica, and dissolved oxygen cycles. The model also considers sediment diagenesis and benthic exchange interactions among state variables are described in 80 partial-differential equations that employ over 140 parameters.

EFDC–Environmental Fluid Dynamics Code (Hamrick 1996)

EFDC is a linked three-dimensional, finite difference hydrodynamic and water quality model developed at the Virginia Institute of Marine Sciences. EFDC contains extensive water quality capabilities, including a eutrophication framework based upon the ICM model. EFDC is a general-purpose hydrodynamic and transport model that simulates tidal, density, and wind-driven flow; salinity; temperature; and sediment transport. Two built-in, full-coupled water quality/eutrophication sub-models are included in the code.

EFDC solves the vertically hydrostatic, free-surface, variable-density, turbulent-averaged equations of motion and transport; and transport equations for turbulence intensity and length scale, salinity, and temperature in a stretched, vertical coordinate system; and in horizontal coordinate systems that may be Cartesian or curvilinear-orthogonal. Equations describing the transport of suspended sediment, toxic contaminants, and water quality state variables are also solved.

Further information is available for some models at websites (Appendix E). The important features of these models are summarized in Tables D-2 and D-3. The information provided in these tables is primarily qualitative and sufficient to determine whether a model may be suitable for a particular application. For complete information, the potential user must consult the appropriate user's manuals, the supporting agency, and other experienced users.

TABLE D-2

Model	Time Scales	Spatial Dimensions	Hydro-dynamics	Data Expertise Requirements	Supporting Agency	Scale of Effort
Fraction of Freshwater	SS	1D	0	Minimal	EPA	Days
Modified Tidal Prism	SS	1D	0	Minimal	EPA	Days
Advection-Dispersion Equations	SS	1D	0	Minimal	EPA	Days
Pritchard's 2-D Box Model	SS	2D (xz)	0	Minimal	EPA	Days
QUAL2E	SS	1D	I	Moderate	EPA	Few months
WASP5	Q/D	1D, 2D (xy), or 3D	I, S	Moderate to substantial	EPA	Few months
CE-QUAL-W2	D	2D (xz)	S	Substantial	U.S. Army Corps of Engineers	Several months
MIKE 3	D	3D	S	Extreme	Danish Hydraulic Institute	Several months
ECOMsi/*EM	D	3D	S	Extreme	Hydoqual, Inc.	Several months
CH3D-ICM	D	3D	S	Extreme	U.S. Army Corps of Engineers	Several months
EFDC	D	3D	S	Extreme	None	Several months

D = dynamic
Q = quasi-dynamic (tidal-averaged)
SS = steady state
xy = two-dimensional, longitudinal-lateral
xz = two-dimensional, longitudinal-vertical
0 = no hydraulics specified, inferred from salinity data
I = hydrodynamics input
S = hydrodynamics simulated

TABLE D-2 Summary of selected estuarine water quality model characteristics.

TABLE D-3

Model	Key Features	Advantages	Disadvantages/ Limitations
WQAM	Simplified equations to simulate dilution, advection, dispersion, first-order decay, empirical relationships between nutrient loading, and total nutrient concentration.	Few data requirements; can be easily applied with a hand calculator or computer spreadsheet.	Limited to screening- and mid-level applications.
QUAL2E	Steady-state model provides detailed simulation of water quality processes, including dissolved oxygen, biological oxygen demand, and algal growth cycles.	User-friendly Windows interface, which is widely used and accepted. Able to simulate all of the conventional pollutants of concern.	Limited to simulation of time periods during which stream flow and input loads are essentially constant.
WASP5	Based on flexible compartment modeling approach; can be applied in one, two, or three dimensions.	Has been widely applied to estuarine situations. Considers comprehensive dissolved oxygen and algal processes. Can be used in three-dimensional simulations by linking with hydrodynamic models.	Coupling with multi-dimensional hydrodynamic models requires extensive site-specific linkage efforts.
CE-QUAL-W2	Uses an implicit approach to solve equations of continuity and momentum. Simulates variations in water quality in the longitudinal and lateral directions.	Able to simulate the onset and breakdown of vertical stratification. Most appropriate model for cases where vertical variations are an important water quality consideration.	Application requires extensive modeling experience.

MIKE 3	Finite difference model for use in three dimensions. Predicts time-varying concentrations of constituents, including advective and dispersive transport.	Pre- and post-processing software that is user-friendly. Multiple turbulence closure schemes.	Computationally intensive. Requires extensive data for calibration and verification. Restricted set of state variables in water quality code. No access to source code.
ECOMsi/*EM	Finite difference model for use in three dimensions. Predicts time-varying concentrations of constituents, including advective and dispersive transport.	State-of-the-science eutrophication kinetics. Multiple turbulence closure schemes.	Computationally intensive. Requires extensive data for calibration and verification. Requires a high level of technical expertise to apply effectively. Limited access to source code.
CH3D-ICM	Finite difference model can be applied to most water bodies in one to three dimensions. Predicts time-varying concentrations of constituents, including advective and dispersive transport.	State-of-the-science eutrophication kinetics.	Computationally intensive. Requires extensive data for calibration and verification and a high level of technical expertise to apply effectively.
EFDC	Linked three-dimensional, finite difference hydro-dynamic, and water quality model contains extensive water quality capabilities. Water quality concentrations can be predicted in a variety of formats suitable for analysis and plotting.	Able to provide three-dimensional description of water quality parameters of concern. The entire range of hydrodynamic, sediment, eutrophication, and toxic chemical constituents can be considered.	Computationally intensive. Requires extensive data for calibration and verification and a high level of technical expertise to apply effectively.

TABLE D-3 Key features of selected estuarine water quality models.

Appendix E

Related Websites

Model Name	URL
AGNPS	http://www.infolink.morris.mn.us/~lwink/products/agnps.htm
ANNIE	http://water.usgs.gov/software/surface_water.html
	http://www.epa.gov/epa_ceam/wwwhtml/ceamhome.htm
ANSWERS	Not available
BASINS	http://www.epa.gov/ostwater/BASINS/
BEM	http://www.dhi.dk/general/dhisoft.htm
CE-QUAL-W2	http://www.wes.army.mil/el/elmodels/w2info.html
CH3D-ICM	http://hlnet.wes.army.mil/software/ch3d
CREAMS/GLEAMS	http://arsserv0.tamu.edu/nrsu/glmsfact.htm
DR3M-QUAL	http://water.usgs.gov/software/surface_water.html
	http://www.epa.gov/ednnrmrl/tools/model/dr3m/htm
EFDC	http://www.tetratech-ffx.com
HSPF	http://www.epa.gov/owowwtr1/watershed/Proceed/donigia2.html
Hydroworks	http://www.wallingfordsoftware.com/products_frame.htm

MIKE 3	http://www.dhi.dk/general/dhisoft.htm
MOUSE	http://www.dhi.dk/general/dhisoft.htm
P8	http://www2.shore.net/~wwwalker/#Software
QUAL2E	http://www.surfacewater.com/qual2eu_overview.html
STORM	http://www.epa.gov/owowwtr1/watershed/Proceed/donigia2.html
SWMM	http://www.epa.gov/owowwtr1/watershed/Proceed/donigia2.html http://www.dhi.dk/general/dhisoft.htm http://www.ccee.orst.edu/swmm/
SWRRB/SWAT	http://arsserv0.tamu.edu/nrsu/swrbfact.htm http://arsserv0.tamu.edu/nrsu/swatfact.htm
WASP5	http://www.cee.odu.edu/cee/model/wsp_desc.html
WARMF	http://systechengineering.com/warmf.htm
WQAM	Not Available

Index

A

B

C

D

E

H

I

K

L

M

N

Y